STUDIES IN ECONOMICS AND POLITICAL SCIENCE

Edited by

THE DIRECTOR OF THE LONDON SCHOOL OF ECONOMICS
AND POLITICAL SCIENCE

No 61 in the series of Monographs by writers connected with
the London School of Economics and Political Science.

THE INDUSTRIAL AND COMMERCIAL REVOLUTIONS IN GREAT BRITAIN DURING THE NINETEENTH CENTURY

The Industrial and Commercial Revolutions in Great Britain during the Nineteenth Century

BY

L. C. A. KNOWLES

M.A., LL.M. (Cantab), LITT.D. (*Trin. Coll., Dublin*),
*Late Professor of Economic History in The University of
London; Sometime Dean of the Faculty of Economics
in the University of London and Lecturer at the London
School of Economics*

LONDON

ROUTLEDGE & KEGAN PAUL LTD.

BROADWAY HOUSE : 68-74 CARTER LANE, E.C.4.

First published 1921
Second Edition (Revised) 1922
Third Edition (with a few corrections) 1924
Fourth Edition (Revised) 1926
Reprinted 1927, 1930, 1933, 1937,
1941, 1944, 1946, 1947, 1950, 1953, *and* 1961

**Printed in Great Britain by
Latimer, Trend & Co., Ltd. Plymouth**

To the Memory of

Dr. WILLIAM CUNNINGHAM
Archdeacon of Ely

A Great Teacher and Master
of English Economic History

This book is humbly dedicated by one
who is thankful to have been his pupil.

PREFACE

IN this book I have tried to bring out the causes which led to the coming of machinery and which made Great Britain the workshop of the world for a large part of the nineteenth century.

I have wished especially to lay stress on the world position of the United Kingdom during the past century owing to the developments of mechanical transport which were the inevitable outcome of the mass production by machines. English experiments in the control of labour conditions by legislation and the development of the labour movement along the lines of trade unionism and co-operation have attracted much attention both at home and abroad, but the changes wrought in the world's trade and in the economic relations of States to one another by the coming of the railway and steamship have hitherto not received the attention so stupendous a revolution deserved. I have endeavoured to emphasize this and to show how the developments in transport produced the new British Empire, the new constructive imperialism and the new agriculture of the last quarter of the nineteenth century. I have also tried to show that the necessity for controlling the new methods of transport combined with the new national rivalries they created by the penetration of continents and the diminution of distance, were important factors in the growing State control, so characteristic of the late nineteenth and early twentieth centuries.

I am moreover of opinion that it is easy to exaggerate the social evils of the industrial transformation known as the Industrial Revolution. There were many compensations and the progress of the change before 1830 was so gradual that it allowed considerable time for readjustment. It was the newness of the cotton and coal industries which attracted attention and brought old standing industrial evils to light. When the factory system spread more rapidly after 1830, with the introduction of machine tools to make machinery, the social safeguards had been devised in the shape of Truck Acts, Factory Acts and Trade Unions, which afforded an ever increasing measure of protection to the workers.

I have further tried to account for the great change in public opinion after 1870, which led to the growth of State control, not merely in industry, but in commerce, agriculture, transport and imperial relations.

I regard the nineteenth century as the product of French ideas of personal freedom combined with English technique. This book is an attempt to describe the development and effects of the new technique in the country of its origin, and it will shortly be followed by another describing the evolution of France, Germany, Russia and the United States under the influence of the ideas of liberty, equality and fraternity combined with machines, railways, telegraphs and steamships.

It has always been a tradition of the Staff of the London School of Economics to give unwearied and ungrudging assistance to colleagues, as I have found to my great advantage. I owe special thanks, however, to Mr W. T. Stephenson, who read through the whole of Parts IV and V on Mechanical Transport and saved me from many errors, though he is in no way responsible for any opinions I may have advanced. Professor Bowley has also been ever ready to place his unrivalled knowledge at my disposal and has kindly allowed me to use the tables on *p.* 168 and *p.* 217 from his works. Dr T. Gregory has generously permitted me to utilize his " Tariffs. A Study in Method," for the information on *p.* 353 as to tariffs and preferences in the Crown Colonies. Mrs George and Miss Buer have read Parts II and III in manuscript and have given me valuable suggestions. Mr Headicar has been most helpful, as always, about books. Only those who know my handwriting will realize the debt I owe to my typist, Miss Blackburn.

L. KNOWLES.

Killagorden, Truro.

THIRD EDITION

Since writing this book I have been engaged in expanding Part VI into two volumes on *The Economic Development of the Overseas Empire.* I have, however, endeavoured to bring this section up to date. I have also added an Appendix on recent railway developments in England, and have made some other alterations in the text.

FOURTH EDITION

The publication of Ashton's " Iron and Steel in the Industrial Revolution " has made it necessary for me to re-write those parts dealing with this subject. I have made several other changes with a view to bringing the fourth edition up to date.

L.C.A.K.

CONTENTS

Page

PART I: INTRODUCTION -   -   ❡   -   I

    The Principal Features of Nineteenth Century
    Economic Development.

PART II: THE INDUSTRIAL REVOLUTION CAUSED BY
        MACHINERY   -   -   -   15

    (i) Features of the Industrial Revolution.
    (ii) Why the Industrial Revolution came first in
        Great Britain.   26
    (iii) The Inventions in the Textiles.   47
        (a) The Spinning of Cotton and Wool.
        (b) Weaving.
        (c) Linen, Lace and Hosiery.
        (d) Industrial Chemistry.
    (iv) Slow Progress of the Factory System and the
        Development of Ironworks, Engineering and
        Coal Mining.   61
        (a) The Reluctance to abandon Family Work.
        (b) Reluctance of the Employer to embark
            on Factory Production.
        (c) Growth of Population relieved the
            Scarcity of Hands.
        (d) Development of Iron Manufacture,
            Engineering and Coal Mining.
    (v) Economic and Social Effects of the Change   79

PART III: INDUSTRIAL AND COMMERCIAL POLICY IN
        GREAT BRITAIN DURING THE NINE-
        TEENTH CENTURY -   -   -   110

    (i) Laissez-faire and the Reaction.   112
        (a) 1793-1815.   The period of the French Wars.
        (b) 1815-1830.   Reaction after the Wars
        (c) 1830-1850.   Period of Reforms.
        (d) 1850-1873.   The Good Years.
        (e) 1873-1886.   The Great Depression.
        (f) 1886-1914.   State Control.

Page

PART III—*contd.*

    (*ii*) Causes of the Supremacy of Great Britain during the Nineteenth Century.   162

    (*iii*) Growth in the Welfare of the Working Classes.   167

    (*iv*) The Contrast between the Individualism of Great Britain and the Paternalism of France and Germany.   171

    (*v*) The Economic Position of Great Britain in 1815 and 1914.   177

PART IV : THE COMMERCIAL REVOLUTION CAUSED BY MECHANICAL TRANSPORT -   -   180

    (*i*) The Revolution in the Importance of Continental Areas.

    (*ii*) The Revolution in Commercial Staples and Commercial and Industrial Organization.   198

    (*iii*) The Creation of a New Financial Era.   212

    (*iv*) Social Effects of the Commercial Revolution.   215

PART V : THE DEVELOPMENT OF MECHANICAL TRANSPORT IN GREAT BRITAIN AND THE PROBLEM OF STATE CONTROL OF TRANSPORT -   -   233

    (*i*) Roads.   236

    (*ii*) Canals.   240

    (*iii*) Railways.   253

        (*a*) 1821-1844. The Period of Experiment.

        (*b*) 1844-1873. The Consolidation of the Lines.

        (*c*) 1873-1893. The Development of State Control.

        (*d*) 1893-1914. Amalgamations and the Question of Nationalization.

    (*iv*) The Steamship and Shipping Problems.   291

        (*a*) Free Trade in Shipping.

        (*b*) The Coming of the Steamship and the Continuous Change in Technique.

        (*c*) The Supremacy of the United Kingdom in Shipbuilding and in the Carrying Trades.

        (*d*) The Growth of Foreign Shipping.

        (*e*) Combination in the Shipping World.

        (*f*) The Government and Shipping.

Page

PART VI : THE INDUSTRIAL AND COMMERCIAL REVOLU-
              TIONS AND THE NEW CONSTRUCTIVE
              IMPERIALISM   -   -   -   314

    (*i*) Periods of Colonial History.   316
          (*a*) 1603-1776.   The First Empire and its
                Disruption.
          (*b*) 1783-1870.   The Period of Drift.
          (*c*) 1870-1895.   The Creation of New Col-
                onial Values by Mechanical Transport.
          (*d*) 1895-1920.   Reaction from World Eco-
                nomics to Imperial Economics.
    (*ii*) The Empire in Alliance.   330
    (*iii*) The Empire in Trust.   341

PART VII : THE EFFECT OF THE DEVELOPMENT OF
              MECHANICAL TRANSPORT ON BRITISH
              AND IRISH AGRICULTURE   -   -   360

    (*i*) The Effect of the Development of Mechanical
          Transport on English Agriculture.   361
          (*a*) 1793-1850.   The Victory of the Large
                Farm.   363
          (*b*) 1850-1873.   The National Market
                created by the Railway and the Good
                Years.   370
          (*c*) 1873-1894.   The World Market and
                American Competition.   371
          (*d*) 1894-1914.   Agricultural Reconstruc-
                tion and Social Experimentation.   374
    (*ii*) The Effect of Mechanical Transport on Irish
          Agriculture and the Relations between Great
          Britain and Ireland.   381
          (*a*) The attempted Anglicisation of Ireland.
          (*b*) The Suppression of Competition in
                Ireland.   1660-1783.
          (*c*) The Equal Treatment of Great Britain
                and Ireland.   1801-1870.
          (*d*) Constructive Policy for Ireland.
                (1) Improvement of Tenures.
                (2) Improvement of Methods.

CONCLUSION   -   -   -   -   - 391

BIBLIOGRAPHY   -   -   -   -   - 393

APPENDIX.   THE RAILWAY ACT OF 1921   -   - 404

INDEX   -   -   -   -   - 408

Industrial and Commercial Revolutions in Great Britain during the Nineteenth Century

PART I

INTRODUCTION

THE PRINCIPAL FEATURES OF NINETEENTH CENTURY ECONOMIC DEVELOPMENT

SYNOPSIS

THE nineteenth century is an outcome of the French achievement and advertisement of personal freedom combined with the new mechanical inventions which emanated from England. The result was the simultaneous removal of legal and physical disabilities.

The five characteristics of the economic development of the century are (1) freedom of movement and the consequent agricultural revolution ; (2) the coming of machinery, creating a new industrial class and a new labour movement ; (3) mechanical transport, producing a revolution in the relative importance of countries, in commerce and social life ; (4) the development of new national economic policies, leading to an increasing State control of industry and commerce and of (5) a new effort of race expansion which inaugurated a new colonial era, making for world inter-dependence and world rivalry.

THE period which falls between the French Revolution of 1789 and the outbreak of the European War in 1914 may be styled the nineteenth century. It witnessed the general application of mechanical power to manufacture, transport and mining and was therefore a period of momentous economic change. The new inventions not merely altered all the old methods of production and distribution but the human factor in that production and distribution, man, was powerfully affected by machinery which enlarged his capacities and potentialities and by railways and steamships which increased his mobility. A revolution in ideas inevitably accompanied such far-reaching changes in the physical world.

A new conception of personal liberty emerged and the mass of the population of Europe became free in the nineteenth century as it had never been free before. Governments had to face new classes and new problems and a new conception of national policy also emerged. The new methods of manufacture and transport created new demands for raw materials and food, new areas were opened up, new wants created and new markets developed, so that by the end of the period the whole globe was knit up in a world economy of world interdependence and exchange and world rivalry.

The only century that can compare with the nineteenth for the rapidity and fundamental nature of its changes is the sixteenth. In this latter century the enormous importance of the discovery of the sea route to India and the two Americas became evident in new trade routes, new commercial and colonial rivalries, new struggles among nations, a new merchant class and a considerable acceleration of the growth of capital with all that it implied in the reorganisation of industrial and agricultural life. The linking up of Europe with the Indies and the New World was followed by the revolution in economic thought brought about by the Reformation and the substitution of the royal and secular governments for the Church as the directing power in economic life.

Of the five countries that may be styled " Great Powers " in the nineteenth century, viz., England, France, Germany, Russia and the United States,* only the first two counted as important economic entities in the sixteenth century.

The two " Great Powers " of that period were those that had made the discoveries referred to, Portugal with her Eastern Empire and the rich spice trade, Spain with the New

*The dominions of the Hapsburgs, called Austria-Hungary after 1867 cannot be reckoned as a " Great Power " in the economic sense in the nineteenth century. The country consisted of eleven main races, ten principal languages and twenty-three legislative bodies at the end of the nineteenth century (Seton Watson, " German, Slav and Magyar," p. 10). The governments of both Austria and Hungary were almost wholly occupied with the task of keeping their subject nationalities in due obedience and avoiding liberal constitutions while putting them on paper. Economic development played but little part in two countries so torn by political dissensions. Neither Austria nor Hungary became one of the dominating factors in the world's agriculture, industry or commerce as did the five other powers mentioned above.

World and its silver mines. When in 1580 Spain absorbed Portugal and controlled both bullion and spices she seemed to be a colossus astride the narrow world. The economic dominance of this great Catholic power was challenged by Protestant Holland and England. Between them they shattered the sea-power of Spain and nothing then stood in the way of the expansion of the English race in North America. The foundations of the United States were accordingly laid by English merchants combining in chartered companies partly to break the power of Spain, partly to build up a self-sufficing Empire on the basis of tobacco and sugar for which the English had hitherto depended on " the courtesy of strangers." Just as the sixteenth century belonged to Spain and Portugal, the seventeenth belonged to Holland. She became the great sea-power of that century with a world-wide trade and Amsterdam was the exchange place of Europe. The Dutch maritime superiority then excited as much jealousy in the English mind as had the overwhelming economic resources of Spain in the previous century. The result was Dutch wars and Navigation Acts and a conscious imitation of Holland by England ending up with a Dutch king, William III.

France was the great industrial country of the seventeenth and eighteenth centuries with a population estimated at twenty millions in 1700, and Paris with its 600,000–720,000 persons was the most populous city of Europe.* No other European town, except London, which numbered about 600,000 persons, had over 200,000 inhabitants.† At that time the Dutch numbered about three millions, the English five-and-a-half and the Spaniards about seven millions. France was, moreover, expanding rapidly at the end of the century in the two richest regions of the world, namely, India and the West Indies. She had also settlements on the continent of North America, which stretched down from Canada to Louisiana, hemming in the English. Indeed were she to obtain the crown

---

*Levasseur " Population Française," I. *pp.* 204-206.

†It must be remembered that all population estimates are very unsatisfactory and differ greatly. There was no census in Great Britain or France till 1801 ; nor in Prussia till 1810. The first census for the United Kingdom was 1821. The figures, however, give some idea of the relative man power of the countries.

of Spain, as seemed likely at the end of the seventeenth century, she would also become the heir of the Spanish dominions in Central and South America. She, in her turn, would then be as great an economic menace to the world as Spain had been earlier.

England ranked in the seventeenth century after both France and Holland in economic importance. She was in 1700 a prosperous agricultural island with a considerable woollen industry but no other manufactures of any importance. She had some settlements fringing the Atlantic in North America, she held some islands in the West Indies and some trading posts in India and Africa. She had been driven out of the spice islands by the Dutch and was as inferior to them in shipping and wealth as she was to the French in industry.

In the struggle for colonies and trade which took place during the seventeenth and eighteenth centuries Germany was not a competitor. The effect of the discoveries was to make the Atlantic the great highway for commerce and to bring into prominence the countries bordering that ocean. The inland seas such as the Baltic and the Mediterranean ceased to be the main arteries of trade and the important commercial towns bordering those seas also declined. The Hanse League —the great federation of towns in North Germany—suffered enormously from the diversion of routes. The once flourishing cities of South Germany which had been the intermediaries of the traffic between the two seas were also affected by the decline in importance of Venice and Genoa. Spices which used to be distributed by Augsburg were distributed *via* Lisbon and Antwerp after the Portuguese made their footing good in India.

Finally, the destruction of German economic life was completed by the devastation of the Thirty Years' War (1618-1648) which paralysed German economic development. For the next two centuries and a half Germany remained both politically and economically a mediæval state. At the beginning of the nineteenth century she was an agglomeration of over three hundred states separated from one another by tolls and tariffs, with many different coinages, weights and measures and laws, while communications were hampered by almost impassable roads. She was still a country of serfs and mediæval gilds in 1800.

Russia had been submerged by the Tartar invasions for two hundred years. Only at the end of the fifteenth century was that alien domination thrown off. In the sixteenth century she was only connected with Western Europe by the English merchants of the Muscovy Company who, greatly daring, explored the country *via* Archangel and connected barbaric Russia with Western civilisation. Russia was, therefore, in no position to compete for the Indies or the New World. Not till the reign of Peter the Great (1682-1725) did she really become part of Europe. She was even more mediæval and primitive than Germany at the beginning of the nineteenth century.

By the eighteenth century the Dutch Republic had begun to decline, while Spain and Portugal had ceased to be great powers. France was by far the most important economic power of the eighteenth century, England ranked second. The challenge of the growing economic dominance of France was taken up by England, now re-enforced by Scotland. She fought the War of the Spanish succession to prevent France joining Central and South America to her extensive possessions in India, the West Indies and North America. The two great powers of the eighteenth century thus joined issue and the great land power, France, with her teeming population fought the great sea power, Britain, inferior by far in numbers but better organized as regards finance, with the result that Britain increased her colonies and dependencies at the expense of France. From these two rivals were to emanate the new inventions which revolutionized physical conditions and the new ideas which revolutionized the position of man as a human being.

Great Britain was responsible for the successful development of steam power during the eighteenth century, while from France were to spread those ideas of personal liberty which, differently applied in different countries were, in combination with the steam engine and machinery, to transform Europe and by way of Europe the economy of the rest of the world. The nineteenth century is the outcome of French ideas and English technique.

The reason for the revolutionary effect of the steam engine is to be found in the fact that it provides a power independent of climate or geography which can be applied to an infinite number of different purposes. It can be used to drain mines, drive machinery in factories, work flour mills, bore tunnels, build houses, construct dams, empty ships, haul masses of goods from place to place, or cross oceans, deserts, or mountains. It is a power that can be applied in any country that can supply it with the necessary fuel. Feed it with coal and water, and it will economize labour and work night and day in cold or heat. Invented by Newcomen in 1710 to pump water out of coal mines, it could be worked away from the pit's mouth owing to the economies in fuel made possible by Watt in 1776, and came gradually into use in England from 1782 onwards, when it was applied to work machines and create the blast for the new iron furnaces. From England it spread to the continent with greater or less rapidity, according to the country, after 1815, and gained fresh conquests by proving itself the effective motive power of rapid transport during that century. Prior to the introduction of steam, man had been almost helpless before the forces of Nature, such as floods, storms and droughts. He was hemmed in by mountain barriers and deserts and limited by climatic conditions and sheer distance. The steam engine enabled him to surmount these phenomena and became the great instrument of man's control over Nature.

The French Revolution had so far-reaching an effect because it introduced suddenly into France a degree of personal freedom never before experienced in Europe, except in England. The ideas of the French Revolution were comprised in the words, " Liberty, Equality, Fraternity." This meant in the economic sphere the abolition of the rights of one man over another, the equalization of taxation, the right to move from one place to another, the abolition of internal hindrances to the movement of people and goods, free choice of an occupation and equality before the law. After 1789 the individual Frenchman was legally free to change his abode or manner of living, he could choose his occupation in life without let or hindrance from feudal superior or from gilds, he could cultivate his land in the manner that suited him best, he could buy and sell in th_ same manner as everyone else, while he paid no more than his fair share of taxation. All these things were new in France for the bulk of the population. They had previously been enjoyed only by a limited and privileged class. The Frenchman became independent, self-dependent and possessed of the rights of equal citizenship. In their characteristic method of effecting change by means of revolution rather than by gradual reform, the French swept away in one night, August 4th, 1789, many of the feudal limitations which had been undergoing the slow dissolution of the centuries in other countries. France advertised by one tremendous event that personal liberty which even then existed in Great Britain but which had barely been appreciated by other countries because it had come so slowly.

The French Revolution seemed to crystallize into actual fact and make possible all the ideas of those who in other countries had maintained that serfdom and slavery were anomalies but who felt unable to handle the enormous problems that a change to freedom would involve. Where the French armies went they spread this new gospel of the economic liberty of the individual and the abolition of restrictions and privilege. The result was that during the nineteenth century the countries of Central and Eastern Europe freed their serfs and reconstructed their agricultural methods, legal systems and administration.

The combination of steam and liberty of movement was momentous in its results in spite of the reaction that followed the excesses of the Revolution. When economic freedom had been accomplished the bulk of Europe became legally free to move, free to grow rich, free to starve. Then came the railway and the steamship, making possible a degree of mobility hitherto undreamt of. Legal and physical disabilities were removed almost simultaneously. At the very time when men found themselves free to choose their means of livelihood, new instruments of production in machines lay to their hands and new occupations opened out on every side. The result was new peoples, new classes, new policies, new problems, new Empires.

The three other " Great Powers " of the nineteenth century —Germany, Russia and the United States—were the outcome of the new inventions and new ideas. The application of steam to land transport produced the railway which opened up the interiors of these three Continental countries, hampered hitherto by the difficulties of land transport. They developed into economic powers of the first rank because the railways facilitated their agricultural exports, brought their iron and coal together and distributed their products at cheap rates over large land areas. Ease of distribution had hitherto been a monopoly of the sea powers bordering the Atlantic.

While they were indebted for the new technique to Great Britain, these new powers drew their inspiration as to the proper relations of human beings from France. The influence of French revolutionary ideas is to be seen in the fact that both Russia and Germany freed their millions of unfree cultivators and the United States emancipated her slaves.

By the beginning of the nineteenth century only Great Britain and France could be reckoned as great economic powers : by the end of the century the mediæval countries of Germany and Russia had become modern states developing their resources with free men. The thirteen revolting colonies of England had survived a civil war lasting four years and had expanded over a continent and had also joined the ranks of the great powers.

It is thus easy to see why the nineteenth century begins in 1789. It was the starting point of the new ideas of personal freedom in continental Europe. It was also in that decade that the steam engine, the new motive power that was to revolutionize human capacity and mobility, came into use for other purposes than pumping water out of mines. Watt invented in 1782 the rotary movement of the steam engine which made it possible to utilize steam to drive machinery. He had already in 1776 made steam a cheap power by his modifications of the old " fire engine " which had been very extravagant in the use of coal and this enabled steam to be widely used.

The new mechanism and the new liberty thus arrive within six years of each other. Within the same period, 1782-1789, i.e., in 1783 the independence of the United States was recognized and they started their national career apart from Great Britain. The economic ruin of France during the ten years after 1789 produced Napoleon I., who not merely reorganized economic life along modern lines in his own country but was the creator of modern Germany.

It is also clear that the nineteenth century ends in 1914 with the economic eclipse, temporary or permanent, of the

German Empire and Russia, two of the great economic forces of the past century. No great war leaves the economic condition of a country as it found it. The historian of the future will be able to judge whether 1914 be the opening of a new period of the economic federation of Europe, of new economic powers delegated to the State, whether it has inaugurated a new era with regard to transport in the air which will minimize the importance of access to the coast and further facilitate the opening of interiors and whether it has proved to be the beginning of a new epoch with regard to labour questions.

While nothing is ever really absolutely new in economic evolution there is often such an accentuated pace of evolution that the whole conditions of life are radically changed. This is the reason that the nineteenth century is so clearly marked off from the eighteenth. The accumulation of changes, shadowed before but quite definite after 1789, make the following century the age of mechanism and personal mobility.

The main outstanding features of the economic development of the great European powers and the United States as a result of this new mechanism and the new idea of personal liberty, are five in number.

There is first the abolition of restrictions on personal freedom comprised in the sweeping away of serfdom and all the mediæval and feudal limitations on free movement. The whole economy of the great agricultural estates of Central and Eastern Europe, based as it was on keeping a supply of labour fixed to a definite spot, had to be readjusted, and a new agriculture, working with free labour and carrying out individual instead of communal agriculture and intensive instead of extensive cultivation, had to be initiated. Per-' sonal liberty meant for Europe an agricultural revolution as did also the freeing of the slaves for the Southern States of North America.

The second great economic change was connected with the physical effects of machinery driven by steam. England/ and France, already considerable industrial nations in the eighteenth century, had both adopted machinery by the end of that century, and became the two leading industrial powers during the first half of the nineteenth century. Similar industrial transformations took place in Germany, Russia and the United States in the last half of the nineteenth century. The growth of industrial states would, however, have been impossible without the freedom of movement introduced by the abolition of serfdom. A population which could not leave the land could not have provided a labour force for factories, mines, railways, engineering works and blast furnaces. Nor would it have been physically possible to handle the mass of raw material, coal and finished goods, without the railways and steamships.

The nineteenth century is still further distinguished by the fact that mechanical transport and machinery caused the concentration of people on the coal and iron areas and created new towns and a new industrial class. The labour questions of the nineteenth century became radically different from those of preceding centuries and the treatment of the workers by employers and the State, the satisfaction of the demands of the artisans, the limit to be placed on the power of their organisations and the co-operation of labour and capital are still questions pressing for solution.

The third characteristic economic feature of the nineteenth century is the application of steam to sea and land transport and the coming of the railway and the steamship. The new methods of transport were capable of bridging great distances very rapidly, they could also carry heavy loads at cheap rates and were independent of heat or cold, frost or snow, summer or winter. The result was that countries hampered hitherto by distance from a coast or by climate almost suddenly became great economic possibilities. Russia and the United States emerge as two of the determining and disturbing economic factors of the world after 1880, when their great grain yields became available for export for the first time in large quantities.

Iron and coal were brought together in places where no iron industry had hitherto been possible owing to the expense of moving quantities of mineral long distances to the coal fields. The United States, Germany and Russia became great iron and steel producing nations and thereby reached the first stage of their industrial revolution. With rapid transport new articles became available for exchange and a commercial revolution followed the industrial revolution. Food stuffs, especially cereals, were transferred as they were never transferred before, perishable goods like meat and fruit, bulky goods such as machinery, masses of raw materials of all kinds all came to be the principal objects of commerce. Many of these things were quite new ; in other cases the scale on which they were transferred was new. Hence commercial methods were revolutionized. As communications became more rapid the growth of huge business concerns with world-wide interests emerged and equally large trade union or labour combinations became possible. All countries were knit together into closer economic relationship. A social revolution followed. People massed in towns to an increasing extent or migrated in millions to the New World and opened up new countries as markets and as sources of raw material. The whole world became interconnected as it had never been before.

The fourth characteristic of the nineteenth century is the emergence of new national economic policies. People were striving all through that century to form new political units which should represent what they thought were their common affinities in matters of race, history or religion. United Germany, Italy, Belgium, Greece, Roumania, Hungary, Norway, Bulgaria and Serbia all bear witness to the power of national aspirations to create new forms of government —nations. It was inevitable that these new nations should evolve new methods of dealing with economic problems and that the older nations should be affected thereby. It was also probable that they would try to expand their own territorial limits when new instruments such as the railway and the steamship facilitated the expansion of nations to new regions.

The nations of the nineteenth century had to face entirely new conditions of industry, commerce, transport, agriculture and colonization, and the question was what ought to be the national attitude. Should the State conduct industry itself or leave it to individuals ? If the latter, should it regulate and direct those individuals or leave them entirely free to make their own contracts and bargains ? In commerce, ought the nation to adopt the policy of free imports or that of protecting home industry by a tariff ? In agriculture, should it intervene to save the peasant or leave the growth of the large farm to proceed unchecked ? In transport, ought the railways to be State railways or State subsidized or should the building and control of them be left to private companies ? In colonization, how far should the State finance or assist the development of the new areas and to what extent should it leave the work to individuals and chartered companies ? These questions had to be faced, not by the old governing aristocracy but by the newly enfranchised masses.

The industrial and commercial revolutions had created new social classes ; a new trading class, a new industrial class, and a new moneyed power arose, and the old landed interest declined correspondingly in importance. These new classes constituted the new democracy of the nineteenth century and it was this new democracy that had to evolve the new policy. Its political aim was to obtain the extension of the franchise and a liberal constitution for its own particular state. It did not, however, believe that a government could carry out any economic function efficiently and it felt strongly that the less government intervention there was in any sphere the better. Hence the new democracies were all on the side of leaving everything to the individual, who was to be as far as possible unhampered by government regulation. In industry they believed in laissez-faire, in commerce in free trade. The result was that they attacked and swept away the old protectionist and development policy of the autocratic kings which had stood for regulation and which was known as mercantilism. To this succeeded, after 1848, an era of liberalism and cosmopolitanism, when the removal of commercial restrictions and the freedom of individual initiative and enterprise was the goal. This is reflected in the commercial treaties of the period, all of which were negotiated on the basis of a low tariff, and in the small amount of legislation which was enacted. A reaction followed, and a third change of national policy becomes obvious after 1870, when there was a return to protection and State regulation on every side increased. There was an abandonment of laissez-faire in commerce, industry, transport and agriculture. This was due to intensified national feelings which rejected the cosmopolitanism of the previous twenty years and strove to make the new unit of the nation more self sufficing by developing its own resources inside a barrier of tariffs. The admission of the working classes to greater political power also worked in the same direction. With the constant spread of the industrial revolution and the consequent change in working conditions, the artisans demanded and obtained in every great country an increasingly ela' orate code of labour legislation for their protection. The railways began to amalgamate and form great transport monopolies. It became necessary to control them in both England and the United States ; in Germany and Russia most of them were transferred from private to State ownership. The trusts and combines increased, and the question of their control became urgent with a corresponding increase of State activity. In agriculture, the imports of wheat and meat from North America and Australasia produced an acute crisis in Europe with further State intervention. On every side the power of the Government has been extended ; even in the United States, the most individualist of all the Great Powers during the century. The functions exercised by the State in the last quarter of the nineteenth century differed, however, from the old mercantilism and paternalism which prevailed from the sixteenth to the beginning of the nineteenth century as fundamentally as did the liberal era of the fifties and sixties.

Fifthly, the new nations were anxious to extend their power and influence over-seas, which gave rise to fresh State activity in the colonial sphere. Raw materials and markets became vital questions for the great industrial powers, the railways enabled continents to be opened up, the steamships brought the produce to Europe and the whole world was brought under the economic influence of the new Europe by a new effort of national expansion and colonisation. Distance was largely abolished as a barrier by the new methods of transport, capital was increasingly invested in the undeveloped lands, a new colonial era was inaugurated and the whole world became economically linked up in spite of the striving after self-sufficiency which was characteristic of the new reaction to protection after 1870.

The national idea still persisted, however, in the desire to include the mother country and the colonial areas in bigger units which should favour each other or penalize other countries by some form of discrimination in tariffs or shipping. The ideal was the creation if possible of self-sufficing economic Empires—a tendency constantly counteracted by the developments of transport which make for a world economy, *i.e.*, for world production and world distribution irrespective of national boundaries.

The nineteenth century witnessed the rise of a new British Empire, a new Russia in Asia, a new France in Africa, a new Germany in Europe, Africa and Polynesia, to say nothing of the United States which has transformed itself from thirteen scantily populated States along the Atlantic into a great federation reaching from the Atlantic to the Pacific and extending to the Philippines.

These great Empires could not have arisen had there not been a vast increase and movement of population to fill the new countries. The newly gained freedom of movement was not exhausted with the migration to towns and factories within the same country. Following on the abolition of the legal barrier to movement and the facilities provided by railways and steamships there was yearly a large European exodus to the new lands beyond the seas, amounting in 1913 alone to nearly two million people.* In Russia where no barrier is interposed by sea, the Russian population, helped again by the railways, spread eastwards. In North and South America the railways colonized two continents and initiated the penetration of Africa. There is an unparalleled expansion of Europe beyond the seas and European history becomes world history in consequence, just as trade becomes essentially world trade.

Personal freedom and the consequent revolution in agricultural methods and tenures, the industrial and commercial revolutions brought about by steam power, the labour movement, the new national policies and the new colonization and migration are the outstanding economic features of the nineteenth century.

*1,964.000 of which 1,198,000 went to the U.S.A. *Cd.* 9092 (1918) p. 6

PART II

THE INDUSTRIAL REVOLUTION CAUSED
BY MACHINERY

SYNOPSIS

I.—FEATURES OF THE INDUSTRIAL REVOLUTION

The industrial revolution hinged on the development of steam power, iron, coal, machinery and chemical factories. The new industrial population concentrated on the coal areas.

New methods of transport became necessary to cope with the masses of raw material and finished goods.

There were two phases of the industrial revolution : the first was limited by the amount that could be transferred by road, river, canal and sailing ship ; the second witnessed a vast extension of machinery and production owing to the railways and the steamships. The labour movement and the forms of business organization corresponded to the developments of transport, being first local, then national, then international.

II.—WHY THE INDUSTRIAL REVOLUTION FIRST CAME IN GREAT BRITAIN

The reasons for the development of the " industrial revolution " in Great Britain were that she had a ready command of capital, a scarcity of hands, large and expanding markets, a free population, political security, a training in large scale business for over-seas markets, ease of access to those markets through her geographical position and her shipping, while her iron and coal fields provided her with the most valuable raw material and motive power for machinery and for iron smelting. The early machines were made of wood and were worked by water. When machinery was worked by steam and made of iron the possession of cheap coal and iron was a further asset.

The immediate impulse in the development of the iron trade was the lack of timber for household fuel and iron smelting and the need to use coal. This stimulated the adoption of the steam engine in coal mining and the making of canals to move the coal while the spread of the steam engine and mining gear created a fresh demand for iron. The impulse in the case of the textiles was the scarcity of hands to supply the increasing demand caused by the prohibition of the import of cotton piece goods from India for wear in England, which induced Englishmen to undertake the manufacture of cotton at home. In a new trade hands were lacking. There was a famine in yarns. A machine called the " spinning jenny " which could be used in the home was supplemented by a cotton spinning machine worked by water.

The use of power meant concentration in factories. The steam engine was adopted later to form a stronger and more reliable power than water and there was a fresh demand for iron for machines and steam engines and an increased demand for coal to create steam power. Thus there was continuous action and interaction between the engineering, mining and textile developments.

III.—THE INVENTIONS IN THE TEXTILES

(a) Spinning and Weaving were the crucial processes.

There were usually three stages (1) a hand machine used in the home ; (2) hand machines massed in one building for better supervision ; (3) the application of power to those machines.

In cotton spinning Hargreave's jenny (1764) for wefts was developed alongside of Arkwright's water frame for warps (1768). Both were gradually superseded by Crompton's mule (1775).

In the spinning of worsted (long staple wool) and wool (short staple) the jenny and the mule were adopted later but hand spinning had disappeared by 1820 in the woollen as in the cotton industry.

(b.) In cotton weaving there was a great scarcity of weavers owing to the abundance of yarns. This stimulated the invention of a machine. Kay's flying shuttle, a hand machine, was gradually superseded by the inventions of Cartwright, Johnson and Horrocks, who produced a practicable power loom by the end of the French Wars. Hand-loom weaving still continued and formed an important part of the cotton industry in 1840.

The weaving of worsted by power dates from the decade 1820-1830 but woollen weaving did not become a factory industry till the decade 1850-1860.

(c) The spinning of flax by machinery was a practical success owing to the introduction of wet spinning during the decade 1820-1830 and then rapidly became a factory industry. The power loom was not applied to linen weaving to any appreciable extent till after the fifties. Lace and Hosiery became factory trades between 1840 and 1880.

(d) Industrial chemistry was developed to deal with the new masses of bleaching and dye-stuffs required.

IV.—SLOW PROGRESS OF THE FACTORY SYSTEM AND THE DEVELOPMENT OF ENGINEERING AND COAL MINING

The transformation was gradual especially in the older textile trade of wool as also in iron making, engineering and coal mining, partly because one process or trade hinged on another and all were not revolutionized at the same time, partly because the workmen were unskilled workers who had to be trained. In this respect the introduction of machine tools was the turning point of the industrial revolution as they enabled machines to be made rapidly and accurately. Transport was also defective and limited markets. Other reasons for the slow development of machine production are to be found in the reluctance of the worker to abandon home work, the reluctance of the employer to engage in factory production, and the growth of population which provided plenty of " hands " and delayed the necessity for labour saving machinery.

V.—ECONOMIC AND SOCIAL EFFECTS OF THE CHANGE

The term " industrial revolution " does not mean rapid change but it does ultimately mean a fundamental change in the character of a country. The social system that gradually gave way before the factory system was one of family work and family earnings, most workers having an agricultural bye-employment.

So long as water was the principal motive power industry retained much of its rural character. The steam engine meant the growth of towns.

The effect was obvious in England between 1830 and 1840. The principal changes were : the rise of the North, South Wales, the West of Scotland and the Midlands as great mining and manufacturing areas ; the development of a new commerce ; the rise of a new system of production ; the increase of urban areas ; the development of new relations between capital and labour.

The advantages of the change were : the regulation of children's work and education ; the separation of the home and the work-place ; better sanitary conditions ; more regular hours ; greater efficiency in production ; higher wages ; the power of combination and class expression ; more openings for employment.

The disadvantages were : the loss of independence ; the subjection to the foreman's orders ; the monotony of the work ; the dependence of the married woman on the man's earnings.

Great Britain saved herself and Europe from Napoleon by means of the development of her resources and became the workshop of the world.

I.—FEATURES OF THE INDUSTRIAL REVOLUTION

THE industrialization of the Great Powers one after another has been one of the striking features of the nineteenth century and it is in this direction that the influence of Great Britain has been all important. Her inventions have helped to change agricultural into industrial States and have been instrumental in bringing the whole world into a common system of economic relationships.

Both the industrial and the commercial revolutions hinged on coal and iron and the power to transport them. As soon as Great Britain, after experimenting with water as power in the eighteenth century, began to organise her industry along the lines of steam in the nineteenth, new possibilities arose. Steam as a motive power never dries up like water, is never in flood, is never frozen ; it only requires a small amount of coal and water and it can be used as a tireless force economizing labour and supplementing man's puny hauling and lifting powers, and those of his tamed animals. It is also transferable to a far greater extent than water power.

Compare the tonnage that can be moved by a railway goods train in the twentieth century with the few hundredweights that could be moved by the eighteenth century train of pack animals, and it is easy to see the tremendous material advance that steam caused in the transfer of goods. Compare the feeble scratching of the surface, which was called mining, in the seventeenth century, and which could not go very deep because of flooding by water, with the vast amounts of coal and ore that can be extracted from great depths by modern methods of pumping and hauling by steam. Compare the output of a modern blast furnace with the charcoal oven, or that of a power loom with that of the hand loom weaver, and the enormous importance of modern mass production and modern driving power—mainly up to the present, steam power*—becomes obvious.

Steam, however, requires coal, and there arose a great demand for coal in every country. In the nineteenth century coal was indispensable both for power and household fuel and those countries that did not possess it have had to import it. In the eighteenth century coal was only used to a limited extent in England, about four-and-a-half to five million tons being raised in 1750. It was scarcely used at all outside England. The ordinary household fuel was turf, wood or charcoal. It was impossible to break up the great wastes which were so characteristic of manorial cultivation until an alternative fuel could be obtained. The commons and private plantations simply had to be kept intact in each area for wood, turf or peat. As soon as coal was available for burning, enormous tracts could be taken into cultivation. Therefore the agricultural revolution was bound up with coal development. Coal was also bound up with the industrial changes since it was required for driving power ; it was required for smelting iron, it was required as

*Countries like France, Italy, Norway, Sweden and Switzerland have been of recent years using their water to produce electricity, and this will considerably alter their industrial position. Electricity as a power is being made increasing use of in England, Germany and the United States. See Hobart " James Forrest " Lecture, " Minutes of Proceedings Institute of Civil Engineers," Vol. CCI., p. 132. Norway is even exporting current by cable to Denmark. (Address by Sir J. Aspinall to Institute of Civil Engineers, November, 1918.) Iron is now being smelted by electricity in the North of England.

the basis of the chemical industry, and in the nineteenth century it became indispensable for the new transport by railway and steamship.

There was in consequence of the supreme importance of coal a great concentration of population on or near coal fields ; part of the population being occupied in extracting the ore ; part being occupied in utilizing it either for iron smelting, in engineering works, or as motive power for working machinery.

Steam in its turn created a new demand for iron. Wood was no longer strong enough to stand the strain of the new driving force, and machines had to be made of iron, hence an increased demand for iron. To make these machines, new tools such as steam hammers and lathes were needed, and large and entirely new branches of industry developed, viz., engineering shops to make machines. But masses of iron ore could never have reached coal in sufficient quantities had not new methods of transport been utilized, and these new methods of transport—the railway and the steamship—began in their turn to make fresh demands on iron and coal ; iron for locomotives, rails, and parts of coaches and wagons, iron for marine engines, and iron for the ship itself ; coal for the locomotives and engines of both. In addition to this there were great demands for iron for renewals of machinery, railway rolling stock and new ships.

The population, therefore, continued to mass in the region of the coal and iron areas, extending outwards as railways extended the distribution of pig-iron and coal.

Alongside of steam, iron, coal and machinery came the chemical factories, and these too required coal, partly as a basis of the chemical products and partly for power. The result has been a phenomenal development of mining, and the amounts of coal and iron produced, or imported, became the tests of a country's development. Those countries, therefore, came to the front in the nineteenth century that possessed, and used, their coal and iron resources, viz., Great Britain, Germany and the United States. France, with her comparatively poor coal and iron production, fell relatively speaking behind. Having, however, an enormous asset in her artistic taste she was enabled to produce the higher-priced artistic articles, which yield a large profit. As these *Articles de Paris* depend on their individuality for their sale they are not suited to the mass production of " the great industry." Hence, although France was industrialized, she was not industrialized to the degree reached by the other three Great Powers before 1914. Russia had hardly begun to tap her mineral resources, but even in European Russia where 86.8 per cent. of the people still lived outside towns* there had been a rapid development between 1890 and 1914 of the textile industry and of iron.

Apart from special and exceptional circumstances, industry in Europe and the United States tends to grow up within easy railway access to the great coal areas and on these areas the population is massed in towns.

In the case of Europe there is a broad population belt coinciding with the coal and iron fields, which commences in Scotland and stretches right across the centre of the continent with an inclination upwards to the North at one end and downwards to the South on the other. Starting round Glasgow in the North it comes down through England, continues to Belgium and Northern France, runs on into the Rhine land and the Saar valley, through Westphalia, Saxony and Silesia and dips down to the Donetz basin in Russia, and on this line the bulk of the manufactures are located.

The so-called " industrial revolution " comprised six great changes or developments all of which were interdependent. It involved in the first place the development of engineering. Engineers were required to make and repair the steam engines, to make machinery for the textiles, to make machinery for lifting coal out of the pit, to make machine tools and locomotives. The only engineers before the middle of the eighteenth century were men who repaired the mechanism of the flour mills—the millwrights ; the iron workers were blacksmiths. Skilled engineers had to train themselves from the beginning by learning as they went on. Engineering, however, depended on the iron-founders. Unless the iron was cast in quantities and of sufficiently good quality the engineers could not get material on which to work, so that a revolution in iron-making was the second development which almost necessarily preceded machinery. The iron works in both England and France before 1750 were scattered all over the country, near

* " Russian Year Book, 1916," p. 59. An urban area is generally considered to be an area containing 2,000 persons and over.

woods to get charcoal for smelting, and near water for power and the transport of bulky awkward articles like iron goods since roads were only earthen tracks. But the iron-founders would not have been able to concentrate and develop their works on a large scale had not a large demand existed for iron for the wars and had not the steam engine enabled them to free themselves from the limitations of water power.

The third change came when mechanical devices moved by water or steam power were applied in the textiles. They began in the simple operation of spinning. This created a surplus of yarn and a weaving machine was gradually adopted to use up the yarn. The inventions started in cotton, spread to wool, then to flax and to silk.

A fourth development then became necessary. The bleaching, dyeing, finishing or printing processes had all to be accelerated or transformed to keep pace with the output of the piece goods and this meant the creation of the great chemical industries. They in their turn required engineering plant with a consequent reaction on the metallurgical industries which were already experiencing a fresh demand owing to the adoption of iron machinery in the textiles. Indeed the tendency of the textiles to develop in the neighbourhood of the iron works was very marked because then they could get their machinery repaired.

Engineering, iron-founding, textile machinery and industrial chemistry all hinged ultimately on coal. The great development of coal mining is the fifth great change comprised in the term " industrial revolution." Coal in the form of coke was needed by the blast furnaces to smelt the iron ore so that it should take on the cohesive form known as pig-iron ; coal was needed to refine the pig-iron or cast it into the form in which it was required by the engineers ; it was needed for the new motive power—steam.

Coal could not, however, have been obtained from the pits in sufficient quantities had not the engineers devised and made a steam engine which pumped the water out of the mines.

Each of these series of inventions depended in turn on the others and the reason for their spread in the nineteenth century lies in the fact that they all reached a point in the eighteenth where they could be utilized together so that they reacted

C

on and stimulated each other. It is no accident, however, that they should have developed in Great Britain soon after the founding of the Bank of England in 1694. Capital had to be accumulated in sufficient quantities to allow of the expensive preliminary works to be undertaken and the experiments applied on a large scale.

Finally the mass production by machinery in factories, of iron in blast furnaces, the development of great engineering and chemical works and the growth of coal mining could not have attained their present overwhelming importance had there not been a corresponding development in the means of transport which facilitated the movement of food to feed the population gathered on the coal and iron areas, which enabled the transference of the vast quantity of ores, fuel and raw materials, cotton, wool, oils, fibres, timber, and chemicals required to feed the factories, and which was instrumental in distributing the vast bulk of the manufactured articles.

The industrial revolution, as a matter of fact, falls into two epochs corresponding to the means of transport at each period. The first phase coincides with an improvement in roads and inland water-ways and is mainly concerned with the early development of coal and iron mines, engineering works and textile factories. It was confined in the first half of the nineteenth century to England and France and was limited by the amount that could be hauled in waggons or small barges. It was carried through by individuals commanding only a small amount of capital. The employer had often risen from the ranks and the typical business of that period was the one man or family firm. Labour unions were prohibited in England till 1825, in France till 1884, and in Germany till 1892, but in any case could only be local in action owing to the difficulties of communication.

With the railways and steamships the transformation proceeded at an enormously acclerated pace and a second phase of the industrial revolution began. The inventions spread to other countries notably Germany, the United States, Russia, Austria, Switzerland, Italy, Japan and India.

Mechanism began to transform trades other than the two textiles, cotton and wool—the boot and shoe industry, the loading and unloading of goods, carpentry, building, furniture, ready-made clothing, hosiery, lace, silk, linen, flour milling, food preserving, printing, fishing, and laundry work were among the trades radically affected. In addition, new trades were rapidly developed connected with Bessemer steel, electricity and electrical appliances, jute, rubber, petroleum, electro-plating and linoleum, to say nothing of the vast extension of mining and engineering activities that mechanical transport occasioned by its demand for locomotives, rails, steamers and marine engines. There continued to be an unprecedented demand for iron for railways alone, not merely for their erection and extension but for renewal and repairs.* When once larger markets were available through the opening up of new areas by transport facilities it was worth while to manufacture in larger quantities and hence businesses increased in size, the whole bulk of production mounted up and the stimulus reacted on every raw material and food producing country in the world.

The vast scale on which business was done in the railway era needed increased command of capital. There was a rapid growth of banking and of the form of business organization known as the joint stock company, which enabled a much more speedy and successful mobilisation of capital to carry out operations on a scale hitherto undreamed of. Very few private individuals could undertake the " great industry " on the scale it has now reached. The result is that the old individual employer or group of partners has now largely become an impersonal corporation. The typical employer at the end of the nineteenth century is the shareholder who subscribes capital, puts in a manager and wants high dividends but is not personally responsible for the business. Instead of receiving the wages of management the new employer— the joint stock company—pays them. With the creation of companies competition became fiercer,† with their shareholders behind them a company could make calls or new issues and had a greater staying power for competitive

---

* " The annual value of material and stores of all descriptions bought by British railways on revenue account may be put down roughly at £26 million." Report of the Board of Trade Railway Conference, 1908. Cd. 4677, p. 22. The whole railway equipment is renewed about every twenty-five years.

†Royal Commission on Depression of Trade, 1886. Final Report, p xviii

purposes than the typical one-man firm or family business, but after a period of violent and cut-throat competition between companies they settled down in the nineties to mutual arrangements to limit competition. Trusts, pools, cartells, syndicates, banking amalgamations, shipping rings and railway conferences are all names for this phase of combination to avoid competition which is noticeable in every industrial country and indicates the rapid impetus observable everywhere towards larger aggregate productive units. This movement was equally noticeable before 1914 in free trade England or protectionist United States, in industrial Germany or in agricultural Russia, in Austria-Hungary and in France. These amalgamations were not merely confined to operating in one country but have become international in scope, in some cases, as in the case of cotton thread, tobacco and steel rails, parcelling out the world between them by treaties in which they agree not to entrench on each other's territories. This tendency to organized monopolistic production on a large scale gave in its turn a great stimulus to organized labour. The trade union and socialist movements grew in strength and showed signs of becoming in their turn international in scope.

It was not until communications were developed that trade unions could be organized into national groups of one trade instead of local lodges and this made for far more effective action. Not until the railways were built on the Continent was it possible to hold international meetings or communications which could in any way correspond to Marx's appeal : " Proletarians of all lands unite ! " The railways and steamships facilitated imports from all parts of the world and goods made with sweated labour in one country might be sold at cheap rates in another to the injury of those who enjoyed a high standard of wages or leisure. Hence the desire of labour leaders to try to level up labour conditions in all industrial countries, which afforded a further stimulus to international action on the part of labour.

In the second period, the railway period, combination became physically possible for both masters and men as it never had been before. To both it seemed desirable. Both wished to limit competition, the former the competition of rival businesses, the latter that of underpaid workers, while

the men also wished to be able more effectually to enforce
their demands against the growing power of the rings and
trusts. Organized capital is therefore increasingly confronted
with organized labour with world-wide connections on both
sides.

As far as British economic development is concerned the
two periods may be expressed as follows :—

	Trades affected.	Organisation of Capital.	Labour.
METALLED ROAD AND CANAL PERIOD, 1770–1840.	Textiles Cotton. Wool. Engineering and Metallurgical Industries. Mining. Chemicals.	One man business or family firm. Partners.	Local Unions. Friendly Societies. Revolutionary Outbreaks.
RAILWAY PERIOD, 1840–1914.	Widespread application of machinery to other trades. Rise of new trades :— steamships, railways, acid and basic steel. electrical appliances.	Joint stock Companies and Joint-stock Banks. Amalgamations and combines :— a. national in scope. b. international. horizontal i.e., all businesses of same type. vertical, including all or most processes from raw material to finished article. Banking amalgamations.	National Unions of one Trade. National federations of various trades. International action.

II.—Why the Industrial Revolution came first in Great Britain.

It is at first sight remarkable that the industrial revolution should have started here in a country only containing about nine million people between 1780 and 1790, when France had twenty-six million in 1789* and ought to have afforded a better market for manufactures produced on a large scale. The French, too, possessed capital. Their exports and imports were larger than those of Great Britain,† they, too, had a vast colonial trade and were great re-exporters of colonial goods in Europe. They also had a large and increasing export of manufactures, which proves that they could increase their production and command markets abroad as well as at home. There was also a steady purchasing of land by the French peasants, which shows that there must have been money in the country. Possibly the explanation lies in the fact that the English population was so small that, to deal with the growing export trade, machinery was essential, as there were not people enough to satisfy by hand-work the increased demand. On the other hand, France with her twenty-six million had plenty of available labour that could be and was occupied in domestic industrial production. In other words, to cater for an export and import trade of £40 million, France had twenty-six million people, while Great Britain only had nine million to deal with a foreign trade of £32 million. After the abolition of the restrictions on country industry in 1762, there was a rapid development of industrial production on a domestic basis in France,‡ but England had to supplement her population with machines.

It is easy to see why Great Britain became a great iron-making country. She had the coal—the basic material of cheap power for smelting—her coal and iron also lay together

*Levasseur " Population française," I., *p.* 288. The population of England is given as 6,736,000 in 1760 ; 7,428,000 in 1770 and 7,953,000 in 1780. The population of Scotland was about a million. Cunningham " Growth," III. *p.* 935.

† Levasseur " Hist. du Commerce de la France," II., *p.* 517, gives the figures as 1,018 million livres just before the Revolution (approximately £40 million), while the English exports are given as £16,845,000 and the imports as £15,416,000, *i.e.*, £32,261,000 as against £40,000,000. Leone Levi, " History of British Commerce," *p.* 64.

‡ Tarlé, L'industrie dans les campagnes à la fin de l'ancien régime.

and the mines were close to the coast in Wales, Northumberland and Scotland, which minimized the difficulty of transporting the finished goods, but it is curious that she should have become supreme in cotton which she neither grows nor uses, when made, to any large extent.* Her whole previous development had been bound up with wool and cloth, in which she had a large overseas trade in the eighteenth century and for which she grew the bulk of the raw material herself.

" The wool, as I have said, is an exclusive grant from Heaven to Great Britain, 'tis peculiar to this Country and no other nation has it or anything equal to it in the world. While England has the wool her trade is invulnerable, at least no mortal, final, destructive blow can be given to it. . . . The woollen manufacture is singular to our Nation, no People in the world can come up to us in the Workmanship or have the materials . . . 'tis evident other Nations would go a great way if they had the Wool the main Principle of the Manufacture to work upon ; but it cannot be, they have it not nor can have it, the whole world cannot supply it."

This was the verdict of so great an authority as Defoe in 1730.† By 1830 the pre-eminent textile industry of this country was cotton—a new trade, drawing its supply of raw material from abroad and relying on foreign markets for sale, while the iron and engineering industries, almost unknown in the days of Defoe, had made Great Britain the forge of the world.

The engineering development in this country in the latter half of the eighteenth century is due to the combination of several causes. There was a timber famine in the eighteenth century. There was accordingly a demand for coal for domestic fuel. Sea-coal, as it was called, had been unpopular ; it was thought to be unhealthy, but people were driven to use it and hence a great stimulus was given to coal mining. As it was difficult to get at the coal owing to the flooding of the mines with water, the steam engine was invented to pump

*Great Britain exports about nine-tenths of her cotton output if bulk is considered and eight-tenths in value, and the amount she retains for home consumption is worth approximately £30 million (" Census of Production, 1907 ") quoted by Todd in " Staple Trades of the Empire," ed. Newton, p. 84.

† " Plan of the English Commerce," Second Edition, p. 173.

the mines dry : more coal became available and a new motive power was placed at the service of man. The iron trade had fallen into decay because there was a lack of timber for charcoal used for smelting, and it was not possible to smelt the ore with coal as the sulphur of the coal mingled with the iron and made it brittle. This shortage of iron was got over by a family of iron masters, the Darbys, at Coalbrookdale, who devised the method of coking the coal first and made it possible to use the English iron ores in spite of the shortage of timber. This device got out into the trade by 1760 and created a demand for more coal. By 1784 the iron industry was still further revolutionized by Cort's adaptation of processes which rendered it possible to utilize coal and mechanical appliances in the final stages of the manufacture of iron goods, not merely its smelting. A steel industry (crucible steel) had also been developed in Sheffield, by Huntsman. The canals were then constructed to transport the growing quantities of coal required for household purposes and iron smelting, the first being built in 1760 by the Duke of Bridgwater to join his colliery at Worsley to Manchester. Manchester could then get cheap power. Cheap transport to and from the coast was secured when the Duke built another canal, at his own expense, from Manchester to Liverpool. After that canals were rapidly developed by companies all over the country ; natural inland water-ways were also improved and a system of inland navigation, excellent for its time, was developed by the end of the eighteenth century. The main roads were also metalled and reconstructed during the century by the various turnpike trusts.*

All these developments might have taken place without affecting the textiles, and as a matter of fact the early machines in the textiles were made of wood and worked by water and were independent of the coal and iron developments. The coming of machinery in the textiles was due to a growing demand at home, large markets abroad, a scarcity of hands which made it necessary to employ mechanical devices if those markets were to be filled, large accumulations of capital which enabled men to try experiments, the knowledge of how to cater for markets in every part of the globe, freedom to take advantage of those markets and political security to

*For the history of the development of transport, *see p. 234f.*

enjoy the fruits of enterprise. Great Britain was the only country in the eighteenth century that combined all these factors.

Watts' steam engine with the rotary movement provided after 1782 a power that was more certain than water; the wooden machinery was not strong enough to stand the strain of steam, and iron machinery or iron parts were substituted. Iron machinery was superior to wooden machinery for other reasons. It took up less space; it could be made larger and more powerful and being more durable was less extravagant. The oil with which the wooden machinery was saturated spoiled much of the woven material and iron was preferable because it was easier to clean and did not absorb the oil. The motion was far more regular and uniform with an iron machine and this made for a more equal product.

When iron was available for machinery and when engineers existed to make and fit it, it was regarded as the only possible material. Iron machinery and parts were employed first of all in the forges and foundries for wheels, hammers and other parts.* A French traveller in 1786 records the wide use of iron machinery in British cotton mills at that date and one may take the last decade of the eighteenth century as the period when the new textile industry and the new engineering trades were married, so to say.† For long after, however, machines were made of wood with iron parts.

With the use of iron to make machines the two lines of invention reacted on one another, the demands of the textiles for coal and iron in the shape of machines and steam engines giving an added impetus to the metallurgical and mining industries and the two branches became so closely connected by the end of the eighteenth century that the newer cotton factories tended to grow up in the neighbourhood of the iron-works in order to get the machinery repaired.

*The cylinders of the early steam engines before Watt were made of brass. Fairbairn, " Life," p. 33.

† " I admired here (at Paisley) as in all the great factories which I had occasion to visit in England their cleverness in working iron and the extreme value of the result, the length of life of the machine and its accuracy. All the cog-wheels and generally the whole is carried out in cast iron but of a quality so hard that it polishes like steel with friction and never stops the general movement." Rochefoucault-Liancourt. Voyage aux Montagnes, May 9th, 1786, quoted Mantoux, " Révolution Industrielle," p. 315.

As far as textiles were concerned, machinery was introduced to cope with the rapid expansion of British trade during the eighteenth century.

At home the new roads and the canals provided a better home market and a richer England could afford to buy more goods ; abroad there was a growing trade, especially with semi-tropical countries where cotton goods would be specially in request.

For textiles England had large and constantly growing markets for her wares in all parts of the world, therefore if she made goods amounting to millions of yards by machinery she might expect to sell millions of yards. Her trade had quadrupled in value and her shipping had quintupled in tonnage between 1702 and 1792. The annual average value of the exports had doubled between 1700-1702, when they were worth £6,045,452, and 1749-1751 when they were worth £12,599,112.* The colonial expansion also stimulated the coming of machinery. The colonies afforded increasingly good markets for British products though it is worth noticing that the French colonial trade with America before the Revolution was larger than the English.‡

BRITISH TRADE.

	†Total Exports.	Exports to Plantations in America.	Africa.	India.
1712–13	£7,352,655	£1,053,739	£111,805	£94,179
1750–51	£13,967,811	£1,911,700	£214,640	£798,077
1770–71	£17,161,146	£5,742,532	£712,538	£1,184,824
	Total Imports.	Imports from Plantations.	Africa.	India
1712–13	£5,811,077	£1,104,563	£11,515	£933,013
1750–51	£7,943,436	£2,293,576	£56,292	£1,096,837
1770–71	£12,821,995	£4,225,476	£97,486	£1,882,139

This growing trade was a direct incentive to adopt machinery to cope with the markets when the population was so small and workers difficult to obtain because they were partly agricultural, partly industrial, and very independent. There was a standing difficulty in getting the yarn

*Chalmers, " An Historical View of the Domestic Economy of Great Britain," *p.* 325.

†Whitworth, Sir C. : " State of the Trade of Great Britain in its Imports and Exports." ‡*See p.* 35.

spun. Spinners were scarce at any time, but in spring and summer, when the women and children were employed about the hay and corn harvests or other pressing rural work, the weaving trade and the manufacture of cloth were almost at a standstill.*

The connection between the growing markets, the scarcity of hands and the coming of machinery is clearly brought out in the following passage from a contemporary :

"About 1760 the Manchester Merchants began also to export Fustians in considerable quantities to Italy, Germany and the North American colonies and the cotton manufacture continued to increase until the spinners were unable to supply the weavers with weft. It was no uncommon thing for a weaver to walk three or four miles in the morning and call on five or six spinners before he could collect weft to serve him for the remainder of the day and when he wished to weave a piece in a shorter time than usual, a new ribbon or gown was necessary to quicken the exertions of the spinner. It is evident that an important crisis for the Cotton Manufacture of Lancashire was now arrived. . . . The spinners could not supply weft enough for the weavers. The first consequence of this would be to raise the price of spinning . . . this would have rendered the price of manufactured cloth too great to have been purchased for home or foreign consumption for which its cheapness must of course have been the principal inducement."† The result was the invention of mechanical appliances for spinning.

Other writers bear witness to the same effect. "Nottingham, Leicester, Birmingham, Sheffield, etc., must long ago have given up all hopes of foreign commerce if they had not been constantly counteracting the advancing price of manual labour by adopting every ingenious improvement the human mind could invent, by which means their foreign demands have continued," was the verdict of a pamphleteer in 1780.‡ "No exertion of the manufacturers or workmen could have answered the demands of trade without the introduction of spinning jennies," was the opinion of another in 1783.§

*James, "History of the Worsted Manufacture," p. 312.
†Guest, "Compendious History," p. 12.
‡T. Letters on the liberty and policy of employing machines.
§ "Historical Description of Manchester by a Native of the Town" (James Ogden), p. 87.

In consequence of a Petition against machines a Committee of the House of Commons was appointed to go into the question. Their finding was " that the Cotton Manufactory has been supported by the use of the machines without which that Manufacture must have sunk as well as the Linen and that the Demand for Cotton goods could not have been supplied if the Machines had not been made use of . . . That if the Spinning Machines were prevented from working it would not be possible to supply the Weaver with Warp equal to the present demand."*

A witness before the Committee in 1802-1803 which enquired into the Woollen Clothiers' Petition stated that he was " clearly of opinion that if Machinery was laid aside that there would not have been found Hands enough to manufacture anything like the Quantity of Cloth that has been wanted for Home and Foreign Consumption."† Other witnesses stated that labour was so difficult to obtain that in the West of England women were increasingly employed in weaving and that machinery was being introduced there for the finishing of cloth. Weavers were so scarce in Lancashire that a meeting of the merchants was held in 1800 to try and devise improvements in the power loom‡ so that cloth could be made in England instead of exporting the yarn for foreigners to work up, which was creating a dangerous foreign rivalry.

As early as 1728 Defoe had chronicled an enormous rise in the rates paid for spinning and the consequent scarcity of servants,§ and the tale of woe is continued all through the eighteenth century. The scarcity of labour arose not merely from the smallness of the population, but from the large demand for labour for other purposes than textiles. Coal mining was expanding, so was iron-working, the road system of England was being re-made by the turnpike trusts, the canals were being rapidly developed after 1760 and men were required not merely to excavate them but to work the barges and load and unload the growing mass of goods. Wharves and harbours were being constructed, towns built,

* " Commons Journals," *p.* 926, 1780, Vol. 38.
† Maitland, Merchant and Warehouseman, Importer of Spanish Wool, in Reports, 1802-1803, V., *p.* 265.
‡ *p.* 55.
§ " Behaviour of Servants," *p.* 84-85.

and everywhere land was being enclosed and hedges made. Some pamphleteers ascribe the dearth of labour to the idleness of those engaged in manufacture. They would not, so it was said, work more than three or four days a week because they were so well off.* Their conduct was contrasted most unfavourably with that of the French workers who were said to work " cheaper by a third " than the English.

It is obvious that this scarcity combined with the growing

* The " Essay on the Causes of the Decline of Foreign Trade, 1744," explains that the reason why the wages of our Servants and Labourers are so " excessive high " is that " provisions are so cheap " that they " will not work above half the week." " Our Labour is grown so excessive Dear that we lose all Trades where Foreigners come in Competition with us." (p. 44.)

The same complaint is made by another writer a quarter of a century later, " Thoughts on Trade and Commerce, 1770." " Another cause of idleness in this kingdom is the want of a sufficient number of labouring hands. One would naturally suppose that where hands are scarce they should be all fully employed but this is far from being the case as is well known to the master manufacturers in this kingdom. Whenever from an extraordinary demand for manufactures labour grows scarce the labourers feel their own consequence and will make their masters feel it likewise : it is amazing but so depraved are the dispositions of these people that in such cases a set of workmen have combined to distress their employer by idling a whole day together." (p.27.)

" My next business is to enquire what it is that enables the French to undersell us in foreign markets. And we find almost all writers agree on this point, viz., that the principal reason why the French are able to undersell us in foreign markets is that labour is much cheaper in France than in England. Indeed when we consider how much labour enters into the value of a commodity that it frequently advances it from five to fifty times the first cost of the raw materials we must readily own that a small advance in the price of labour is of great consequence in the trade of a State . . . the high price of labour in England has been the principal cause of the decline of our trade to Turkey, Spain and Italy in which States we have been undersold by the French." (p. 66.)

" The principal evil is allowed on all hands to be the high price of labour in our manufactories. The principal cause of the high price of labour I have all along supposed to be the disposition of our manufacturing population for idleness and debauchery. (p. 299). . . All we want is that the manufacturing people should labour cheaper or which would be better for them and for the State, that they should labour six days for the same money they now earn in four and I am confident they could do this and yet live much better than a Frenchman or a Dutchman. This alone would recover the trades we have lost and greatly extend those which remain." (p. 301.)

foreign demand for the goods was one of the great impulses
to the adoption of machinery.

However urgently Great Britain may have needed
machinery she would not have been able to work out the
experiments and instal the plant without a considerable
expenditure of capital, and capital she fortunately possessed,
as the accumulation of capital had gone on rapidly during
the eighteenth century. In this direction the importance of
the Colonial and Indian trade is especially marked. England
had made large profits out of the tobacco, sugar, spices and
other products brought home from her colonies and India.
She rivalled France in being the great distributor of these
goods in Europe. She had accumulated capital through
this sale and resale and could afford to sink money in coal
mines and iron works and wait for the returns.* Her banking
was organized so as to make this capital easily obtainable
and private persons with money were willing to enter into
partnership with inventors. France with her larger export
and import trade might have had more capital but there
was no banking system which made credit readily available.
The banking system of France did not really take shape till
the nineteenth century, so great a shock had it suffered
through the failure in 1720 of the speculative schemes started
by Law, and her moneyed people were afraid to take the
risks of new ventures and preferred land as an investment.
Moreover, the political security of Great Britain was so good
that people did not hesitate to sink their money in the fixed
form necessary for large scale enterprise. They understood
capital ; they understood large scale production ; and they
knew they would reap where they had sowed. It is only
when one contrasts the utter destruction of industrial and
commercial life for ten years after the French Revolution
and grasps the fact that it took France till 1830 to get back
to the same pitch of commercial prosperity that she enjoyed
before the Revolution† that one realizes how destructive
to economic progress political insecurity may become.
Indeed, one is tempted to think that but for the French

*The number of inventors who came from Scotland to England
because they could obtain capital is very striking, *e.g.*, Watt, Nasmyth,
Fairbairn.

†Levasseur, " Classes ouvrières après 1789," I., *pp*. 623-4.

Revolution, France and not England might have been the pioneer country of the industrial revolution in the textiles. Coal, such an enormous advantage to England and the lack of which was later such a serious handicap to France, might have been imported as in Ulster, and in any case the early machines were worked by water and here France was as well equipped as Great Britain. The French had a great industrial tradition, an industrious and numerous population, large markets at home and abroad and a great reputation for their products. They also had considerable inventive genius, as witness the Jacquard loom and the great discoveries in industrial chemistry of Leblanc and others. It is true France had lost important colonies to England, but in 1787 the French colonial trade was larger than that of England in 1771-72 when the latter still possessed all her North American colonies and the French had lost Canada.* While England's foreign trade had quadrupled French foreign trade had quintupled between 1715 and 1787. Had it not been for the Revolution the trade with the United States would have developed enormously, as the Americans were friendly to France and hostile to Great Britain. As a result of the French Revolution, however, war broke out between Great Britain and France in 1793 ; the English cut off the French overseas trade, and this readjustment of commercial relations could not take place.

The French roads were improving after 1750, and all sorts of hindrances to trade were being swept away under physiocratic influences before 1789 and the King and his ministers were doing their best to introduce machinery. There were already

---

*The import and export trade between Great Britain and her colonies in Africa and America in 1771-1772 was £9,566,418 (Whitworth, " State of the Trade of Great Britain in its Imports and Exports, 1697-1774 [1776] ") that of France amounted to over £11 million. The East Indian trade of Great Britain according to Whitworth was £3,414,553 in 1771-1772 (£2,473,192 import and £941,361 export), while Arnould gives the figures for France as £2,086,200, reckoning the livre as equivalent to a franc and twenty-five francs to the pound sterling. France had a re-export trade of colonial products of 152 million livres in 1789, i.e., approximately £6 million. The total exports of France had risen from 118 million livres in 1715 (i.e., £4,700,000) to 517 million livres (£21,600,000) in 1787 and the " products of French industry " exported rose from 45 million livres in 1716 to 133 in 1789 (i.e., from £1,800,000 to £5,320,000).

in France before the Revolution the beginnings of a cotton industry worked by machinery and a small modern iron industry at Creusot using coke. But the economic disturbances caused by the Revolution put France back for forty years and by 1830, when she had recovered, Great Britain was the work-shop of the world.

The way in which German industrial development was held up through lack of capital, bad roads, lack of enterprise owing to the parochialism of her people, the internal tariff divisions of the country and want of internal and external markets also points to the remarkable advantage England had in her widespread trade, organized credit system, good roads and canals and widely distributed colonies in two hemispheres.

English shipping was so efficient for its time that English manufactures could be conveyed to their destination overseas with the utmost facility, and in this respect England's power of distribution far surpassed that of other nations with the possible exception of the Dutch.

It is interesting to speculate why Holland did not become the pioneer of the new factory system. She had capital, good shipping facilities and a small population which might have stimulated the adoption of machinery ; she was a linen making country and had large markets in the Indies. Holland was, however, a country whose trade was declining ; it was not worth while to cater for dwindling markets by producing increased quantities. Her political system was cumbrous in the extreme and intensely local,* her gilds were monopolistic and it would have been difficult to get hands to work the machines in view of the gild opposition. Large scale industry was not familiar to the Dutch, who had no great industrial tradition as had the English and the French, and industrial freedom was lacking. But above all the attainable raw material of the world was very limited, and France and England were successful in obtaining it to the detriment of the Netherlands. Thus, even if they put in machinery, they would have lacked the material to feed the machines.†

---

*Loon, " The Fall of the Dutch Republic."

†Pringsheim : Beiträge zur wirthschaftlichen Entwicklungs geschichte der Vereinigten Niederlande.

There were also other reasons that favoured the development of the factory system in this country.

The geographical position of Great Britain on the outskirts of Europe at the head of the Atlantic and commanding the approach to Northern Europe gave her unrivalled opportunities for selling in any market. With the development of inland water-ways and roads after 1760 it was easy to get raw materials from overseas to the coal fields and easy to ship the manufactured stuff down to the coast. The English internal market was enlarged at the same time owing to the increased transport facilities provided by roads and canals. England had a further advantage in her freedom from any system of inland tariff barriers. When one realizes the thousands of internal tariffs that obstructed traffic in Germany up to 1834 and the innumerable tolls and charges that hindered trade in France before 1789, to say nothing of the three great tariff districts into which that country was divided over and above the local tolls, it is clear that the political and economic freedom in England was one of the contributing causes of her industrial expansion. Factories could not have got hands with a population fixed to the soil as serfs or by the necessity of making payments in kind, and this alone would have hindered factory development in most European countries in the eighteenth century. It would not have been difficult for continental countries to have obtained " hands " in the towns but the gilds were strongly entrenched and would have prevented the town artisan from taking up factory work Moreover, as the early machines were worked by water they were necessarily set up in country places where the prevalence of serfdom would prevent the population from moving into a factory.*

Serfdom had disappeared in England, Scotland, and Ireland by the end of the sixteenth century, so that by the middle of the eighteenth century the inhabitants of Great Britain were free to move as perhaps no other people were at that time. The gilds had come under royal control in

*When the Creusot ironworks were started in France it was necessary to scour the country to get labour although the works were only on a small scale. Ballot : " La révolution technique et les débuts de la grande exploitation dans la metallurgie française. Revue d'histoire des doctrines économiques," 1912.

the sixteenth century and could place no obstacles in the way of the new machines nor could they prevent men from taking up any occupation they chose. The companies for foreign trade had ceased to be monopolies in the eighteenth century and imposed no limit on the amounts that might be bought and sold.

To the Englishman who had catered for a century for large foreign markets of the most diverse character, large scale production was perfectly familiar,—he had trading connections all over the world, with hot climates and with cold ones. He could and did sell his stuff from the Arctic to Mexico.

" But take our English Woollen Manufacture and go where you will find it ; 'tis in every Country, in every Market, in every Trading Place ; and 'tis receiv'd, valued and made use of, nay call'd for and wanted everywhere. In a Word, all the World wears it, all the World desires it and all the World almost envies us the Glory and Advantage of it. Nor is it the Dress of the Mean and the Poor in the several Countries where it spreads but of the Best and Richest. The Princes, nay at the Time I may say, the Kings of the Earth are cloth'd with it. The King of Spain vouchsafes, even on his Days of Ceremony to appear in a Bays Cloak ; the Grand Seignior Lord of the whole Turkish Empire has his robe of English cloth, and the Sophy of Persia amidst all his Persian and Indian silks wears his long Gown of Crimson Broad Cloth and esteems it as it really is, the noblest Dress in the World. . .

" Not a capital City in the Empire, but you may find the Shops of the Tradesmen stor'd with English Cloth, as far as the Navigation of the Elb, the Oder or the Weissel can convey them ; the Rhine, the Maes, the Moselle, the Saar, the Main, the Neckar, the Danube, they all assist to hand it on, not at Prague only, not at Vienna, not at Munich, but even at Buda and Belgrade, it is to be sold and the best Gentlemen in the Country buy it, if they do not, 'tis for want of Money and not Want of Will. . . . You see the Italians generally clothed in English Cloth or thin Stuffs ; the Clergy in black Bays, the Nuns are vailed with fine Says and Long Ells and even the noble Venetians wear our fine Cloth for their best Dress. What one Manufacture like this can boast of so general a Reception or of being the Favorite Dress of the whole Christian World ? If we should go over to America whether to the Brazils, the flourishing Colony of the Portuguese how many Hundred Thousand Moy d'ors a year do we receive from thence, for the English Manufactures worn and consumed there, notwithstanding the intense Heat of the Place. In Value 'tis the same ; not too cheap for the Nobility, no not for the Kings and Emperors of the World ; not too dear for the Burghers and Tradesmen, no not for the Boors and the Peasants, not too gay for the Men, not too grave for the Ladies."*

This wide-spread trade naturally necessitated an elaborate organization. Merchants imported the merino wool from Spain which was invaluable for the manufacture of fine cloth ; they also imported the Irish clip, and their great endeavour was to prevent the smuggling of English or Scotch wool from England into France. There was a shortage of wool in the world in the eighteenth century. The Government of Great Britain had absolutely prohibited the export of wool so as to give the British the whole of the raw material, but they were unable to prevent large quantities going to France. †

The merchants sorted out the various qualities of wool and distributed it all over the country to be spun. There were very few cottages where the woman and children did not spin or card wool, even the very young children of four or five years could in this way contribute to their maintenance. The merchants who reassembled the yarn would give it out to the weavers and see to the dyeing and finishing of the cloth themselves. The cloth trade was a highly organized capitalistic industry in the eighteenth century. There were, however, persons who would buy or grow the wool themselves, weave it into cloth, dye it and sell the finished piece either to a known customer or at a local cloth hall. This was typical of the Yorkshire trade. The specialization of the

---

*Defoe, " Plan of the English Commerce," *p.* 180-186.

†The prohibition of export was renewed as late as 1788. By 28 George III. c. 38, any person exporting wool was liable to a fine of £3 per lb. or £50 on the whole with three months' imprisonment for the first offence and six months' for the second. French ships used to frequent the Irish channel and pick up the wool ships going from Ireland to England. There was also a large smuggling trade from Ireland as well as from England.

raw material and the catering for so many different markets was, however, bringing quite a different type of man to the fore, namely the large scale organizer and this meant that Great Britain was being trained in large scale production for the most diverse markets.

After wool, silk was the next most important textile industry in the early eighteenth century* thanks to the stimulus given to it by the Huguenots after 1685. Linen came third in importance. The Cotton industry, as can be seen from the fact that only 2,173,287 lbs. of cotton-wool were imported on an average in the years 1716-1720, was insignificant. The English East India Company fetched the cotton piece goods required from India and re-exported them. The cotton industry was unpopular with those in authority because the supply of raw material was so uncertain. They held it better for the nation to concentrate on wool, where the bulk of the raw material was available at home. Raw cotton came from the Levant and was peculiarly liable to be intercepted by the French, while the second great source of supply, the West Indian Islands, was equally uncertain owing to the powerful position of the French as colonists and traders in those regions.

The great cloth-making regions were East Anglia, especially the counties of Norfolk, Suffolk and Essex, the West of England where the manufacture was spread over the counties of Wiltshire, Devonshire and Gloucestershire and in the North, Yorkshire, was the important clothing county. Silk was manufactured at Spitalfields, in Essex, in Macclesfield and in Manchester; linen in Scotland and the North of Ireland.

It must always be borne in mind that there was a great scarcity of raw material in the eighteenth century and that both France and Britain, the two great industrial nations, were struggling to obtain wool, silk, cotton and flax for their respective countries and this lay at the back of much of the colonial rivalry in the West Indies and India. Walpole reorganized the tariff at various times from 1721 onwards so as to give Great Britain the chance to obtain raw material at cheaper rates of duty,† and after 1766 raw cotton might be imported duty free into England if imported in British ships.

---

* Hertz, " The English Silk Industry in the Eighteenth Century," *English Historical Review*, 1909, *p.* 721.

†Brisco, " The Economic Policy of Robert Walpole," *p.* 131-139.

As far as wool was concerned the English had the bulk of the raw material in the home clip but they required the long staple Spanish merino wool for some of the fine cloths, as did also the French. The English seem, however, to have been more successful in obtaining it. As large branches of the French cloth trade depended on English wool, and, as its export was prohibited, there was a constant struggle between the English manufacturers to prevent its being smuggled out and the French to obtain it by any means.* Both countries struggled for raw silk in Italy, though here the French had a great advantage in that silk was produced in the Rhone valley. The English were however able to obtain some of the raw material from the East Indies. The East India Company did their best to stimulate its further production in India, while in the American colonies the English Government tried to obtain a regular supply by bounties.

In cotton the French seem to have been successful in getting the bulk of the raw material, which probably accounts for the small development of cotton in Great Britain until she held the seas and cut France off from access to the raw material in the Levant and the West Indies.

Before the machine era English cotton goods were made on a linen warp, as cotton could not be spun strong enough for the warps, and in 1751 it was obvious that the English were finding increasing difficulty not merely in getting cotton but in getting linen yarn. The Irish were using up the linen yarns themselves in making linen piece goods for export ; England produced very little flax and there was a struggle between the French and the English to get the German flax. The cotton industry asked for a remission of the duty on linen yarns to help their trade and in the enquiry the great rivalry for raw cotton becomes evident.†

*See p. 39. There was also a struggle between the various districts in England to obtain raw material especially between Yorkshire and East Anglia.
" The competition for wool is now so vigorous that at Norwich they are working up at home the wool that should make narrow cloths in Yorkshire. The North probably is, too, keeping at home for baize manufacture the wool wanting at a cheap rate at Bocking." 'Annals of Agriculture, IX.," p. 312 (1787).
†Report on Chequed and Striped Linens. Committees of House of Commons, Reprints II., p. 291, etc. :—
One witness giving evidence as to the state of things in Smyrna

To relieve the scarcity of flax bounties were given on its production in the colonies. The complaints about the shortage of cotton seem to cease after the Seven Years' War (1756-1763) and it seems not improbable that the English superiority at sea was successful in diverting the supplies of raw material. One of the reasons why many English wished to retain Guadeloupe rather than Canada at the end of the Seven Years' War was the fact that Guadeloupe was an important cotton producing island. All through the eighteenth century the colonies enjoyed bounties and preferences on such articles as indigo, tobacco, flax, hemp, raw silk, vegetable oils, pearl-ash, potash, cochineal, logwood and naval stores of all kinds.* In 1780 a preference was given on colonial cotton, *i.e.*, a low duty was imposed on foreign grown cotton and the proceeds were devoted to the encouragement of cotton in the Leeward Islands.†

In this way the growing colonial power of Great Britain and her supremacy at sea stimulated the development of the factory system, since it would have been no use to put in machinery unless large supplies of raw material were available and obtainable. The development of the West Indies and Virginia was specially valuable to the cotton trade, because the slaves employed on the plantations were clothed in English-made cottons and therefore afforded a safe market

between 1728 and 1750 spoke of the rise in price of raw cotton to almost treble and ascribed it to the large demand made for it by the French and the Dutch, " that the French used to export 5,000 sacks, last year 8,000 from Salonica besides 5,000 from Smyrna. The Dutch during the time the witness resided in Turkey used to fix the price of cotton but since the French introduced themselves into the trade they have fixed the price and if the present high price of cotton continues he does not believe it will be worth the while of the British merchants to follow that trade. Being asked how the French could afford to give such prices : That he knows the French have all possible encouragement from the Crown for the importation of that commodity, that they not only manufacture all the cotton they import but have bought up cotton here to be manufactured in France."

Other witnesses gave evidence that cotton eight years before had been 11*d*. to 1*s*. a lb. in the West Indies and had risen to 2*s*. 1*d*. This was attributed " to the number of French, Dutch and German vessels employed in buying it up."

*Hertz, " The Old Colonial System," *p*. 39-40.

†20 George III., c. 45. As to the efforts of the English to develop cotton in the West Indies, *see* Report on the African Slave Trade, 1789, passim.

for a manufactured article which was not much worn in England itself.

The rivalry of the French and the English over raw material continued all through the eighteenth century. Nothing is more interesting than to watch Napoleon's struggle to get cotton for the French cotton industry. He was successful in getting some of the raw material brought overland from the Levant *via* Vienna and Strassburg or *via* Dalmatia and Italy, when the English were masters of the Mediterranean. His domination of Northern Spain was probably not unconnected with the desire to obtain merino wool, the production of which he did so much to stimulate in France by his encouragement of sheep breeding of the merino type. The English and French were not merely struggling for markets in the eighteenth century in their great duel with each other, they were also struggling for raw materials, the failure of which meant wide-spread unemployment in both countries.

As far as manufactures were concerned French goods were either prohibited in Great Britain or excluded by an almost prohibitive tariff. A lady was fined £200 at the Guildhall in 1766 for having in her possession a handkerchief of French cambric.

It seems remarkable at first sight that wool should have been ousted from the first place by a new and exotic trade like cotton.

The reasons for the coming of machinery in cotton were many but the immediate impulse probably came from the shutting out of the Indian imports.†

*" Annals of Agriculture, IX.," *p.* 312 (1787).

†It is possible that the starting of the cotton industry in this country on a new and enlarged scale was also connected with the collapse of the Mogul Empire and the uncertainty of the Indian imports. The English merchant had been in the habit of importing cotton piece goods from India for the markets of West Africa, the West Indies and South America. Some of this latter trade was done through Spain and Portugal as they did not allow other nations to trade with their colonies, but much of it was done in defiance of the prohibition by a smuggling trade from Jamaica.

Then the Mogul Empire went to pieces in the middle of the eighteenth century and it became a struggle as to whether the English could keep their footing in India or whether the French would drive them out. The result was that a regular supply of cotton piece goods became more difficult to obtain and English merchants were forced to rely on the home manufacture for the West African trade. " Ogden, Historical Description of Manchester " (1783), *p.* 70.

The English had taken to wearing the East India cottons and this was much resented by the woollen and silk trades. Defoe stated in 1708 that :

" The general fansie of the people runs upon East India goods to that degree that the chints and painted calicoes which before were only made use of for carpets, quilts, etc., and to clothe children and ordinary people now became the dress of our ladies and such is the power of a mode as we saw our persons of quality dressed in Indian carpets which but a few years before their chambermaids would have thought too ordinary for them . . and even the queen herself was pleased to appear in China silks and calico. Nor was this all, but it crept into our houses, our closets and bedchambers ; curtains, cushions, chairs and at last beds themselves were nothing but callicoes or Indian stuffs."*

The government, afraid of anything which would injure the woollen trade, had prohibited in 1700 the import of printed cotton goods from India, China and Persia except for re-export.† Cotton goods were then imported in the white and were printed here and as they continued to be worn, bitter were the complaints of the silk and woollen trades. " That double the quantity of printed Callicoes and linnen have been worn these last twelve months past, than in the year 1717 is the universal opinion of all observing Men. . . . Again our Women kind us'd to line their English and Dutch Callicoes with slight Silks called Persians and Sarsnets ; which Silks employed many Hundreds of Looms : whereas at present there are not half of them employ'd because of late our Women line their Callicoes with some of the same kind."‡

There was obviously a greatly increased use of cotton fabrics and the woollen and silk weavers were successful in obtaining an Act in 1720 which prohibited the use or wearing of printed calicoes whether printed in England or elsewhere. It was still possible for the East India Company to import white calico or muslin for use in this country and they continued to do this.

To satisfy the demand which had been created the cotton industry began to turn out a material which was not cotton

---

*Defoe, 1708, quoted Baines, *p.* 79.
†11 William III., c. 10.
‡ " The Weaver's True Case, 1719," B.M. 1029, **e.** 17, *p.* 8-9.

but was half linen and half cotton. For the purposes of this manufacture Manchester was admirably situated for getting linen yarn from Ireland for the warps. In 1736 some doubt seems to have occurred as to whether this was legal and an Act (9 G. II. c. 4) was passed expressly legalizing the trade in mixed stuffs. The prohibition of the import of printed cotton goods from abroad continued even after 1774 when the Act of 1720 was repealed so far as to allow the English to wear pure cotton goods made and printed at home. Thus the English cotton industry had reserved for' it the whole of the national market. Bounties of $\frac{1}{2}d.$ to $1\frac{1}{2}d.$ a yard were given on the export of cotton piece-goods in 1781 and 1783 and the excise duty was given back (21 G. III. c. 40 and 23 G. III. c. 21). Although plain muslins and white calicoes might still be imported, there was a considerable duty on these goods, which was increased as the wars made revenue more urgent.* It is therefore untrue to say that the cotton industry differed from the woollen in getting no assistance from the government.

The explanation of the almost sudden starting of cotton making by machinery is probably to be found in the fact that English women who had become used to cotton, when deprived of the Indian goods supplied their wants from the product half linen and half cotton now offered them.†

*In 1787 the duty on white calicoes was £16 10s. per cent. *ad valorem*, in 1814 £67 10s., per cent. *ad valorem*. On Muslins and Nankeens it was £18 per cent. *ad valorem* rising to £37 10s. per cent. in 1814. Baines, *p.* 325.

†The import of cotton wool does not, however, show any very rapid expansion of the trade :—

			IMPORTS.	
1720	-	-	1,972,805	lbs.
1730	-	-	1,545,472	,,
1741	-	-	1,645,031	,,
1751	-	-	2,976,610	,,
1764	-	-	3,870,392	,,

Of this, some was always re-exported. The raw cotton was, however, supplemented by linen yarn of which 13,734 cwts. were imported from Ireland in 1731 and 18,519 cwts. in 1740 and 22,231 cwts. in 1750. This was in addition to the Hamburg yarn. Baines, *p.* 108-109.

When the cotton industry had to cater for a new demand it became especially difficult for those engaged in it to obtain yarn. There was already a scarcity of spinners and the women fully occupied as they were with wool would not readily take to cotton which was paid at a cheaper rate.* It was a machine or nothing if the trade were to expand and the yarn were to be furnished in quantities. The absence of any vested interests in the shape of trained workmen who might object to displacement by machinery made its adoption fairly easy.† The fact that the raw material had to be imported from abroad and the linen yarn brought from Ireland and Hamburg made the trade a capitalistic one from the first. The supply of wool was very limited but raw cotton could be obtained from the Levant and from the West Indies in large quantities, and as time went on the United States furnished an ever growing supply. The possibility of an indefinite increase in the raw material supply made cotton especially suitable for machinery which needed large quantities of raw material so that the machines should be constantly employed. In 1787 cotton seed was sent from the Bahamas by loyalist refugees to Georgia and was successfully developed there. The invention of the cotton gin in 1793 by Whitney combed out the cotton seeds and made abundant supplies of cotton-wool available from the United States so that from that date there was no limit to the expansion of the cotton industry owing to lack of raw material if the English could only ship it to England. As they began to dominate the seas to an ever increasing extent the French were never able to interrupt the regular supplies and there was no fear that the machines would stand idle for shortage of raw material.‡

Moreover, the English had such large trade connections that they could expect to sell easily any quantity of cotton goods that were made and this was a further inducement to manufacture on a large scale. The introduction of

* Baines said that higher prices for the yarn might have attracted hands but that it would have made cotton goods too costly ; they had to be cheap to out-rival wool. " History of the Cotton Industry," p. 116.

†There were riots against the jennies by the weavers who feared they might be forced to weave closer with the finer yarns. " An Historical Description " (Ogden), p. 88.

machinery in cotton was further facilitated by the fact that the technical difficulties in spinning raw cotton by machinery were easily overcome. Thus if machines were put in, the markets were there, the raw material was available and lent itself to machine production and the investment promised to pay.

III.—The Inventions in the Textiles.

(a.) *The Spinning of Cotton and Wool.*

Before the cotton or wool could be spun it had to be opened out, all the lumps or knots got out of the raw material and the fibres made into an even fleecy roll. This preliminary process was called carding and was usually done in the home. In the case of wool required for spinning into worsted as distinct from woollen yarn a long staple wool was required for the worsted thread and this was obtained by a process of combing the wool. But in cotton and wool the raw material was carded, *i.e.*, after being scratched open the fibres were interlaced for wool and straightened for cotton by wires fixed on to cards. When the wool required for worsted yarn was combed, the short fibres were used for woollen cloth and the long for worsted stuffs. A machine had been devised for carding as early as 1748, but was not utilized until the inventions of mechanical spinning created an enormous demand for carded wool to work upon. It also required further improvements which were supplied by Arkwright who took out his patent in 1774.* The standing difficulty, however, was to get the yarn spun as it took from six to eight spinners to keep one weaver going, to say nothing of the annual summer shortage when the women were engaged in field work. Many attempts were made to evolve a spinning machine to relieve the scarcity but the first practical success

‡Average Annual Import of Cotton Wool :—

1701-1705	-	-	1,170,881 lbs.
1716-1720	-	-	2,173,287 ,,
1781-1785	-	-	10,941,934 ,,
1786-1790	-	-	25,443,270 ,,
1800	-	-	56,010,732 ,,

Guest, " Compendious History of the Cotton Manufacture," 1823 *p.* 51.

*Baines, *op. cit. p.* 177.

was due to Hargreaves, of Blackburn, who invented an improved hand machine in 1767 which he called the " Jenny " after his wife. This spun eleven threads instead of one and was soon improved upon so that it spun as many as a hundred threads at once, but it was almost immediately followed by a machine worked by water. The water frame, as it was called, was invented by Arkwright in 1768 and rapidly got out into the trade after 1785, when he failed to maintain his patent on the ground that he had been anticipated by two other men, Highs and Paul. The social importance of the water frame was that people had to be concentrated in one building for the sake of the water power ; the jenny could be worked in the home. Arkwright's machine, therefore, meant the coming of the factory system. He was partly financed by the stocking manufacturer, Strutt, who wanted a stronger yarn for his stockings than the jenny could turn out and Arkwright accordingly set up a factory at Cromford, near Derby, in 1771, to be near the stocking manufacture.* In some cases the weavers kept a jenny or two in their homes to do their spinning, and the supply of yarn was re-enforced from two quarters, from the jennies of the home workers and from the factories worked by power. Technically the water frame was superior to the jenny. The jenny could only spin a thread strong enough for the weft, the warp had as before to be made of linen yarn brought from Ireland or Hamburg. The water frame gave the thread such a firm twist that it was suitable for warps and pure cotton goods could be made for the first time in England. The result was that the Act prohibiting the wearing of cotton goods was repealed in 1774, if the goods were made in England.

So slowly did the water frame spread owing to lack of trained hands, unwillingness to go into the factories and difficulties with the machinery itself, that the number of water frames in 1780 was said to have been only 20 but they had increased to 150 by 1790.† The jennies, therefore,

*Arkwright had adopted the idea of water power from a silk mill which had been set up by Sir Thomas Lombe in Derby in 1719, who had brought the idea from Italy.

†Guest, op. cit., p. 31. " The difficulties which Arkwright encountered in organizing his factory system were much greater than is commonly imagined. In the first place he had to train his work-people to a precision in assiduity altogether unknown before against which

continued to make the weft as the water frames were so few that they were fully occupied with the warps. Spinning continued to be carried on in the homes on the jennies from twenty to thirty years after their invention (1767-1790) and only slowly did the factory system evolve.

When the jennies came in many men began to devote themselves wholly to spinning and gave up their weaving or their farming. Many small yeomen farmers who felt the effects of the agricultural revolution and who had hitherto not engaged in industry also began to buy a jenny or two and spin yarn. Then the jennies began to get more and more elaborate and expensive. The newer jennies ousted the earlier and simpler jennies but they became so costly that only rich men could afford them and so the capitalist began to set them up in workshops and the workers had either to use them as wage earners in a place provided by the employer or, if he could not get hands, he ran his machines with water power. The home-worker found that a machine bought this year was obsolete the next and ceased to invest in jennies. So great was the difficulty of getting machines at all that many cotton spinners became makers of machinery as well. The result was a strong impulse to the growth of mills on the one hand and a greatly improved agriculture on the other. As men became wholly dependent on their farms for their living, better methods were adopted.*

In 1775 the cotton trade was still further stimulated by the invention of the mule by Crompton who by combining the jenny and the water frame evolved a cotton spinning machine that spun a yarn so fine that it was possible to develop the manufacture of muslins in this country. The mule was, like the jenny, a hand machine that could be worked in the home. The mules were "erected in garrets or lofts and many a dilapidated barn and cow shed was patched up in the walls, repaired in the roof and provided with windows to serve as lodging room for the new muslin their listless and restive habits rose in continual rebellion ; in the second place he had to form a body of accurate mechanics very different from the rude hands which then satisfied the manufacturer ; in the third he had to seek a market for his yarns. . . . So late as 1779, ten years after the date of his first patent, his enterprise was regarded by many as a doubtful novelty." Ure, " History of the Cotton Manufacture, 1836."

* Gaskell, " Artisans and Machinery, 1836," pp. 25, 30.

wheels."* "Many industrious men commenced business with a single mule worked by their own hands, who as their means increased added to their machinery and progressively extended their business until they rose to honourable eminence as the most useful and extensive manufacturers in the kingdom."† The mule was adapted to water power by Kelly, of Glasgow, in 1792, and Dale, of New Lanark, was the first to use power for working it. Crompton himself worked his mules by power in 1803. By 1812 it had superseded Arkwright's water frame for all fine yarns.‡

In using both the jenny and the mule a great deal of skill was required on the part of the spinner. It was not a question of pulling a handle and letting the machines do the work. The early water frames were low and simple and on them children could be employed, and this was the case with the mule or the jenny at the very first but they rapidly became so elaborate as to be unsuitable for children and were worked by women or men. The children continued to be employed as piecers.

Sometimes the jennies and mules were collected into workshops, and as they became heavier and more elaborate horses were employed as power, then water, and finally steam was adopted. The self-acting mule which made spinning an automatic affair was not invented till 1825 and then only turned out coarse yarns. These machines were expensive and their adoption was slow. They seem to have become fairly common, however, between 1850 and 1860 for medium and coarse yarns.§

The woollen industry as a whole was revolutionized later than cotton, partly because it was an older trade and very prosperous and the *entrepreneurs* were therefore less likely to make a change. When people are doing well in one direction it is difficult to persuade them to alter for the unknown. Their work-people were very hostile to machinery and the early machines were clumsy things always breaking down. Moreover water is an uncertain power. There is sometimes too much of it when the river is in spate, in summer there

*French, " Life of Crompton," p. 89.
†*Op. cit.*, p. 131.
‡Article on the " Rise, progress, present state, and prospects of Cotton Manufacture," Edinburgh Review, 1827.
§Chapman, " Lancashire Cotton Industry," p. 70.

is often too little ; in winter it occasionally freezes. The
uncertainty of water as a motive force made manufacturers
slow to adopt water power to work the early machinery,
and steam in its turn presented the difficulty of constant
breakdowns, the problem of repairs and the scarcity of coal
due partly to transport and partly to a shortage of miners.
The general tendency was to gather the workers into workshops where they could be supervised and to put them to work
on raw materials and looms provided by the employer, which
stage lasted longer in the woollen industry than in cotton.

There were technical obstacles, too, which made it more
difficult for mechanism to be introduced in connection with
wool than cotton, the softer thread of the former being
more apt to snap.

As in cotton, machinery came in wool first in the preliminary
processes of carding and spinning and then spread to weaving,
but it came more slowly in the spinning of wool and worsted
for the reasons indicated above and also because there was
not the same scarcity of hands. As, however, the spinning
of wool was carried on in most cottage homes all over England
machinery, when it did come, was bound to have a great
social reaction on women's earnings and on family earnings.

The first stage of the change was the adoption of the jenny
for spinning. This seems to have taken place in the North
from 1785 onwards, *i.e.*, about eighteen years after the jenny
had been applied to cotton. Although riots against the
jenny took place in Somerset in 1776, it is spoken of as a new
introduction there in 1794 by the writers on the agriculture
of Wiltshire and Somerset. As the jenny was merely an
improved hand implement and did not require water or
steam power, it was used in the cottages in place of the
spinning-wheel. The adoption of the jenny did not break
up the system of family work but it penalized the poorer
persons who could not afford jennies. As, however, there was
so much employment in weaving owing to the increased
abundance of yarn, the amount of unemployment created was
slight. Wool spinning by power was another matter
altogether, but the transition to factory spinning was by no
means rapid. In worsted, where a long staple wool is required,
spinning factories seem to have come at the end of the
eighteenth century. The worsted industry had migrated to

the North from the Eastern counties. It was, therefore, a new industry in the North and the adoption of machinery was hastened by a scarcity of hands.*

It was, however, easier to introduce machinery for spinning worsted than wool owing to the fact that the long staple wool used for worsted presented fewer technical difficulties for machinery. As many as ten mills for spinning worsted existed in Bradford in 1800.† A false reed or slay was invented in 1809 which guided the shuttle and made it much easier for the weaver to use mill-spun yarn which was rougher than the hand-spun yarn, and the worsted spinning factories seem to have increased rapidly after that date.

In wool the transition to spinning mills was slower. That it did come was due to the fact that the raw material became scarce and the employers became unwilling to give out the wool to be spun or woven in the home as so much seems to have been embezzled. Moreover, it was troublesome to send about the country to give out the wool and collect the jenny-spun yarns from the cottages. They therefore preferred to set up jennies in a workshop or spinning shop and they hired persons who worked under supervision. From this it is but a short step to harness the machines to water or steam with the workmen as overlookers. There was here also a transition stage of workshop production between that of domestic production and factory production.

Power spinning in wool as distinct from worsted was introduced into Leeds, probably by Gott, early in the nineteenth century.‡ The rates for spinning in cottages had risen to a considerable height owing to the general demand for labour, and this stimulated the use of machines.§ The adoption of power for spinning was by no means general even in the North and in 1828 mule spinning had only just been introduced in the West of England though the jennies appear to have superseded hand labour by 1803. By 1830, however, we may say that the spinning of both wool and worsted had become definitely a factory industry. This meant a great loss in the country districts of the South of England and the

* " Annals of Agriculture, XVI.," *p.* 423.
† James, " History of the Worsted Manufacture," *p.* 355.
‡ Bischoff, " Comprehensive History of the Woollen and Worsted Manufacture, I.," *p.* 315.
§ " Annals of Agriculture, X.," *p.* 281.

poor law had to step in and supplement wages with a dole which proved to te extremely demoralizing in practice.

(b.) *Weaving.*

In weaving the change to the factory system was much slower than in spinning. It must be remembered that there were two kinds of weavers in the eighteenth century. One class carried on weaving as a bye-employment and did a little farming, the other depended wholly on weaving. The latter lived in the country or in a town* and was in a very depressed condition when yarns became so difficult to obtain. He could not, like the first class of weaver, fall back on his agriculture as he possessed no land, and as small farms became more and more difficult to get during the eighteenth century, this class of weaver tended to increase. The abundance of yarns turned out by the jennies, water frames and mules created a veritable famine in weavers. The class that had hitherto combined a little agriculture and a little weaving gave up the former and took wholly to weaving, so large were the profits. Their places on the small farms were taken by men who devoted themselves wholly to farming. The second class got from the machines all the yarn they required, their work did not have to stand still for lack of raw material and their earnings steadily increased. There then began a great competition for weavers. The cotton manufacturers tempted the woollen weavers away from wool to make cotton, and there was a scarcity of woollen weavers who were also in demand owing to the new output of woollen yarn from the jennies. The growth of the muslin industry after Crompton's mule introduced the fine yarns also increased the scarcity of weavers. Women and children ousted by the spinning and carding machinery took to weaving and found it very profitable.† In the year 1793 the weaving trade was said

*Compare Silas Marner.

†Guest, *op. cit.*, *p.* 31. A description of the prosperity of the weavers is given by Radcliffe, *p.* 67. " The operative weavers on machine yarns both as cottagers and small farmers even with three times their former rents might truly be said to be placed in a higher state of wealth, peace and godliness by the great demand for and high price of their labour than they had ever before experienced, their dwellings and small gardens clean and neat, all the family well clad, the men with each a watch in his pocket and the women dressed to their own

E

to be " that of a gentle man." " They (the weavers) brought home their work in top boots and ruffled shirts, carried a cane and in some instances took a coach. Many weavers at that time used to walk about the streets with a £5 Bank of England Note spread out under their hat-bands."[*]

A contemporary says : " It was at this time impossible to get more weavers than were then employed although many descriptions of goods . . . were in such demand that any quantity might be sold. . . . There was not a village within thirty miles of Manchester on the Cheshire and Derbyshire side in which some of us were not putting out cotton warps and taking in goods, employing all the weavers of woollen and linen goods . . . in short we employed every person in cotton weaving who could be induced to learn the trade, but want of population, want of hands and want of looms set us fast."[†]

The deficiency was partly supplied by the adoption of Kay's flying shuttle, sometimes called the spring shuttle. Patented by Kay, of Bury, in 1733, it had encountered so much opposition that the inventor was obliged to take refuge in France. It did not therefore make much headway, but towards the end of the eighteenth century it came into its own. Like the jenny, it was an improved hand machine. The scarcity of weavers was, however, so great that there was an excess of yarns which were being sent abroad and a cotton industry was being stimulated on the continent which the English manufacturers considered to be very dangerous. As a result they held a meeting in 1800 to try and devise a power machine for weaving cotton goods in order to use up the yarns at home. Cartwright had invented a power loom

fancy, the church crowded to excess every Sunday, every house well furnished with a clock in elegant mahogany or fancy case, handsome tea services in Staffordshire ware with silver or plated sugar tongs and spoons. Many cottage families had their cow, paying so much for the summer's grass and about a statute acre of land laid out for them in some croft or corner which they dressed up as a meadow for hay in the winter. The yarn . . . which was wanted by the weaver was received or delivered as the case might be, by agents, who travelled for the wholesale houses ; or depôts were established in particular neighbourhoods to which they could apply at weekly periods."

[*] French, " Life of Crompton."
[†] Radcliffe, op. cit., p. 11.

in 1784, but it was imperfect in many respects. As a result of this meeting a great stimulus was given to the invention of a practicable power loom. One of the great obstacles to the adoption of the power loom was the fact that the loom had to be stopped frequently to dress the warp. This difficulty was surmounted by Johnson in 1803. The successful development of the power loom is ascribed to Horrocks, of Stockport, in 1813. The power loom was widely adopted in cotton by 1835 but had only slightly affected the other textiles at that date. The figures are as follows :*

TOTAL NUMBER OF POWER LOOMS IN 1835.

		Cotton.	Woollen.	Silk.	Flax.	Mixed.	Total.
England	-	96,679	5,105	1,714	41	25	103,564
Scotland	-	17,531	22	—	168	—	17,721
Ireland	-	1,416	—	—	100	—	1,516

122,801

Therefore out of 116,801 power looms, only 7,175 were to be found outside the cotton industry. One would expect that in cotton at least the hand loom weaver would have disappeared by 1835. But this was not the case. The Commissioner in 1840 recorded the surprise he felt " at the discovery that notwithstanding the gigantic competition of the power loom the number of hand looms employed in this branch of the trade of weaving is not only very considerable but from almost universal testimony almost as great as at any former period. . . . It would seem that the power loom has created for itself a market almost sufficient to carry off its own productions leaving the demand for cotton cloth nearly as great as before."†

It is interesting to notice that although men displaced women as spinners when the jennies and mules became large and heavy yet women were the first power loom weavers‡ thus taking over a trade which had largely been carried on by the men hitherto although women had always done some weaving.

The five thousand power looms recorded as existing in

* " Report on Hand Loom Weavers, 1840, XXIV.," *p.* 591.

† " Report, 1840, XXIV.," *p.* 650.

‡ " Report, 1824, V.," *pp.* 302, 481.

the woollen industry in 1835 were chiefly used in worsted
weaving where they had been introduced by Horsfall, in
1824. Woollen as distinct from worsted weaving was said
to be " in its infancy " by the Commissioner appointed to
investigate the condition of the weavers in 1840.* It did
not become definitely a factory industry in Yorkshire till
the decade 1850-1860. The finishing processes of woollen
cloth such as fulling and shearing had, however, been subjected
to machinery early in the nineteenth century.† Although
Cartwright had invented a wool combing machine between
1790 and 1792, wool combing still continued to be a hand
trade and was not revolutionized till after 1840.‡

The standing difficulty about the introduction of machinery
was the scarcity of raw wool, so that machines, if adopted,
might stand idle a large part of the time. It was not until
large supplies of Australian wool came in after 1830 that
we get the rapid spread of the factory system in the woollen
industry. The other difficulty was the shortage of machines
owing to the lack of trained engineers to make them, and
that again was not remedied till the general adoption of
machine tools between 1820 and 1840. After 1815, when the
troops were demobilized and population increased rapidly
weavers became so plentiful, the trade being easily learnt,
that there was no lack of hands and machines would not be a
great economy. In a new trade like cotton it was largely
machinery or nothing but in wool there was a very old
industrial tradition and there was a choice between domestic
and factory production. Thus weaving came first in cotton,
then spread slowly to worsted and later to wool. It would
need weighty reasons to make the woollen manufacturers, as
long as they could get home workers, abandon the system of
organized domestic production for the expenses of a factory.
We find, however, between 1815 and 1835 a growing tendency
to group hand looms in workshops, i.e., there is in weaving as
in spinning the stages of domestic production, supervised
group production without power, and the factory system with
steam or water power and all three systems may be found at

* " Hickson's Report, 1840," p. 20.

† See Charlotte Bronte's " Shirley " Between 1806 and 1817 the
number of gig mills in Yorkshire increased from 5 to 72 and the
number of machine worked shears from 100 to 1462.

‡ Burnley, " Wool and Wool-Combing."

work in 1840. After 1835, however, the power loom spread rapidly; more machinery was available, more wool was forthcoming from Australia, and the condition of the hand loom weavers, whether wool, cotton or silk, became increasingly miserable.* But they disappeared slowly even then.

It thus took seventy years at least (1770-1840) before the two principal textile trades were radically transformed by machinery. Although the abundance of cotton yarn did much to extend the manufacture of cotton stockings, net and lace, yet hosiery, net work and lace and linen weaving did not become factory industries till after the forties.

(c.) *Linen, Lace and Hosiery.*

The spinning of linen yarn by machinery had been regularly carried on by Marshall, of Leeds, as early as 1788, but only coarse yarns were turned out. The first really successful machine was due to Girard, a Frenchman, who spun the flax wet. He came to England as he could not raise the capital in France nor could he obtain there the skill to make his machinery. A patent was taken out in 1814 for his machine under the name of Hall.† The great obstacle in flax spinning is the sticky nature of the fibres and the consequent difficulty of separating one from the other. This was got over by Kay, in 1826, who devised a preliminary process of preparing the flax with the result that flax spinning machinery spread so rapidly after that date that the Commissioner in 1840 spoke of home spun linen yarn as "superseded." Between Marshall's mill in 1788 and Kay's invention in 1826, thirty-eight years had elapsed. The reason for the slow progress of machinery in linen is given by a writer in 1819 as "the low price of labour." The Irish women who no longer spun linen yarns for the warps of cotton piece goods after Arkwright's water frame had produced a strong cotton warp, were only too anxious to take any price for spinning linen, with the result that hand spun linen yarns were cheaper than factory yarns and they were also much finer. These factors, joined to "the expense of machinery with its wear

---

* " Reports on Hand Loom Weavers, 1835, XIII. ; 1839, XLII. ; 1840, XXIII and XXIV."
† Horner, " The Linen Trade of Europe."

and **tear**," tended to keep the spinning and weaving of flax
in the stage of domestic industry. The power loom in linen
was of such recent introduction in 1840 that it was said to
exercise no influence on the trade.*

Linen weaving, therefore, continued to be an important
hand trade throughout the fifties.

The net trade and lace trade began to feel the effects of
machinery in the fifties. It was stated in 1841 that there
were twenty-nine or thirty power factories in that trade
and forty workshops with hand machines,† but the bulk
of the work was still done by hand in the workers' own homes
on looms or frames hired to them by the givers-out of work.
By 1861 the lace trade had so far become a factory industry
that it was included in the Factory Act of 1861. The
Inspector, however, still chronicles a large amount of home
work in this trade at that date.‡ Although stocking frames
worked by steam came in as early as 1846§ the great enquiry
into truck in 1870 reveals to what a large extent hosiery was
still a home work trade. The masters hired out the frames
and charged a rent which was often oppressive. The people
who did not hire frames did not get the work given out to
them and the masters made such a profit from letting out the
frames that they often placed out frames in excess of any
work that they could supply to fully employ the frames.
As a result of the enquiry frame rents were prohibited by
the Act of 1874. The profits out of these frame rents put
a premium on the maintenance of hosiery as a cottage
industry ; with their abolition the transition to factories was
rapid. The embroidering of socks and stockings by women
also became a machine industry after 1880.∥

* " Hickson's Report," *p.* 14. " Hand Loom Weavers, 1840."
† " Export of Machinery, 1841, VII." Felkin's evidence. It is
interesting to trace here also the stage of workshop production without
power before the introduction of power. " The small machine owner
or the single machine owner ceases to operate in the trade and it is
becoming more of an aggregate of machines under one roof." *p.* 138.
‡ " Tremenheere's Report, 1861, XXII."
§ Hutchins and Harrison, " History of Factory Legislation," *p.*
156. In 1862, of 120,000 persons employed in this trade only 4,063
came under the Factory Acts.
∥ The spread of machinery can best be traced in the annual reports
of the factory inspectors. A perusal of these reports will show how
difficult it is to generalize about the spread of machinery. Hand loom

It is thus easy to see that the transformation of the textiles was going on all through the first three-quarters of the nineteenth century.*

(d.) Industrial Chemistry.

The developments in the textiles stimulated the discoveries in industrial chemistry. The old method of bleaching linen or cotton had been to steep it in sour milk and expose it for some months to the air. Bleaching often took eight months. Roebuck evolved oil of vitriol and set up works at Manchester in 1746 and Prestonpans in 1749. This revolutionized the industry of bleaching. A French chemist, Berthollet, took the matter further and developed chlorine, the knowledge of which was brought back from France by Watt in 1786, when it was tried in Glasgow in the bleaching field of his father-in-law† with the result that bleaching was reduced to a few days. The process was made a commercial success by Tennant in 1799. It would have been impossible for the cotton industry to have increased rapidly if the cloth were hung up for months in the bleaching.

weavers still exist in Scotland, Wales and Ireland and the writer possessed in 1920 two costumes, the cloth for one of which was woven by a hand loom weaver in Scotland and for the other by a weaver in the South of Ireland.

* From the numbers of persons employed in textile factories in the United Kingdom (figures from " British Commerce and Industry," ed. Page, II., p. 230) we can see the position of cotton as the pre-eminent factory industry in 1835 and the spread of the factory system in the other textiles.

		Cotton.	Wool.	Flax, Jute, and Hemp.	Silk.
1835	-	219,286	55,461	33,212	30,745
1850	-	330,924	154,180	68,434	42,544
1861	-	451,569	173,046	94,003	52,429
1870	-	450,087	258,503	145,592	48,124
1880	-	528,795	301,556	162,965	41,277
1901	-	522,623	259,909	150,319	31,555

The figures are said to be taken from the Annual Reports of the Chief Inspectors of Factories and Workshops and from the Statistical Abstracts for the United Kingdom. The effect of the Lancashire cotton famine is traceable in the figures 1861-1870. Otherwise it is clear that the expansion in wool, flax, jute and hemp took place chiefly between 1860 and 1870.

† Baines, " History of Lancaster, 1788," p. 476.

In the *Commons Journals* there are several requests after 1779 by inventors for rewards for discovering dyes ; Turkey red, a scarlet dye, a green and yellow dye for cotton and linen are all mentioned, which shows the way the new cotton trade was reacting on industrial chemistry.* New chemical works were a necessary corollary of the textile factories, and the new chemical works required machinery and steam power and reacted in their turn on the metallurgical industries.

Finally the colours had to be got on to the calico by some quicker method than printing with hand blocks and cylinder printing developed after 1785.† This meant the development of great bleaching, dyeing and printing industries in close proximity to the textile factories and they in their turn clustered round the district where coal and machinery were to be obtained and repairs were easy to execute. The proximity of the cotton trade and the machine shops made for rapid improvement in the industry.‡

The general process of the change was that the early factories set up in country districts where water was available for power, and industry then retained much of its rural character. As, however, steam became increasingly employed, the newer factories set up on the coal areas near the iron foundries, partly because they could get coal easily and partly because they were in touch with the machine makers and the subsidiary industries.

The new factories were a great improvement on the old type. Many of these had been ordinary country houses converted into factories. These older places would not stand

* "Commons Journal," Vol. 34, *p.* 104 ; Vol. 37, *p.* 392 ; Vol. 41, *p.* 467.

† Baines, " History of Lancaster, 1788," *p.* 265. The Peels were the great calico printers.

‡ " The cotton manufacture is improving because the manufacturers of machinery are co-operating with the cotton manufacturers. If they do not co-operate they can never produce any very important improvements and it is from want of that co-operation and experience that France must continue behind us."

" A cotton manufacturer who left Manchester seven years ago would be driven out of the market by the men who are now living in it provided his knowledge had not kept pace with those who have been during that time constantly profitting by the progressive improvements that have taken place in that period : this progressive knowledge and experience is our great power and advantage." " Report on Artisans and Machinery, 1825, V.," *p.* 44.

the weight and vibration of the steam-driven machinery and so buildings specially built for the purpose were erected after 1815 with a great improvement in the conditions under which the hands worked. The work was also much more regular when steam was the motive power. Water was apt to dry up or freeze, and overtime was worked to catch up arrears. Moreover, the amount of water power was limited, and this limited the number of factories. Steam enabled an almost indefinite expansion to take place.*

It must be borne in mind that the change did not consist of a sudden jump from home work to factory work but that it consisted of many stages. There was first an improved home implement like the jenny for spinning or flying shuttle for weaving; secondly there was a grouping of persons in a workshop to work these improved hand implements under supervision, the reason usually being the desire to prevent the theft of raw material or to get work more punctually executed. The scarcity of hands produced a stage of small, ill-regulated factories worked by water in country districts and these were followed by greatly improved larger factories set up in towns and worked by steam. All these various phases of industrial organization went on side by side and only gradually did the factory with its steam-driven machinery become the prominent type.†

IV.—SLOW PROGRESS OF THE FACTORY SYSTEM AND THE DEVELOPMENT OF ENGINEERING AND COAL MINING.

(a.) *The Reluctance to abandon Family Work.*

The typical form of production eventually ousted by machinery was the same in England as in France.‡ A merchant employer gave out orders and raw material to the people who worked for him either in the towns or scattered

---

* For the effects of steam on children's work, *see p.* 92.

†The clothing trade in 1920 presents many of these features to-day. There is the little dressmaker with her sewing machine, the clothing contractor who cuts out the garments and gives them out to workers in their own homes. Sometimes he gathers them into workshops for supervision, and the sewing machines used in these big work-rooms are frequently worked by electricity. Then there are the clothing factories.

‡In Yorkshire the domestic worker was far more of an independent producer, growing his own wool, making his own cloth in all its processes and taking it to a cloth hall to be sold to the dealer or buyer.

over the country districts. While many of these latter carried on a little agriculture alongside of their industrial work, the effect of the jenny and the mule was to cause industry and agriculture to become specialized and for men to devote themselves to one or other exclusively. As long as work was carried on in the home the earnings were family earnings—children, mother and father all worked and the money was pooled. There was not much independence for the members of the group. " So long as families were thus bound together by the strong link of interest and affection, each member in its turn, as it attained an age fitted for the loom joined its labour to the general stock, its earnings forming part of a fund the whole of which was placed at the disposal of the father or mother as the case might be ; and each individual looked to him or to her for the adequate supply of its wants. No separate or distinct interests were ever acknowledged or dreamt of. If anyone by superior skill or industry earned more in proportion than another, no claim was made for such excess on the part of that individual : on the contrary it was looked upon equally as a part of the wages of the family."*

If a wage was paid it was never calculated to keep the family, the wage was intended to suffice for the maintenance of the man, and the wife and children practically kept themselves either out of the produce of the farm or their earnings at spinning, carding, weaving or mining.† Although

*Gaskell, " Artisans and Machinery."

†The following extracts from official sources show how prevalent this was :—
" A weaver would scorn to marry a servant girl but chooses a weaver who earns as much or half as much as himself." Report of the Commissioners on Hand Loom Weavers, X., 1841, *p.* 47 (ribbon weavers).
" In the first place the weaver's pecuniary position is improved by uniting himself with a woman capable of earning perhaps as much as himself and performing for him offices involving an actual pecuniary saving." Report, *op. cit., p.* 45. (Woollen trade.)
" I have seen cases when the wife was more expert at the loom than her husband and although this is not a general rule the wife and children of a weaver in most cases contribute very materially to their own support." Report of Commissioner on Hand Loom Weavers, 1840, *p.* 650.
" While the males were employed digging in the pits with pick-axes and shovels the women were engaged in carrying the coal on their backs from the extremity of the mines to the pit bottoms or in

conditions of work in the home were by no means ideal, the handling and the stability of a family wage were advantages which a man would not lightly forego. Under the factory system each unit became independent and received its own payment. The domestic manufacturer clung, therefore, to his old method of manufacturing to the last. The family earnings made it worth his while to do so. He might only earn 7s. himself in a week but the labour of the united family might bring the amount up to 21s., and so he held on. If he had a little patch of land or garden that provided a bye-employment he would also be very unwilling to give it up for regular factory work. Both the family wage and the bye-employment were a kind of insurance making for stability of conditions.

One reason for the slow transformation of the English industrial system from domestic to factory industry was the difficulty of getting hands to go into the factories. Had it not been for the large numbers of Irish who migrated owing to the over-population of Ireland it is difficult to see how the

dragging that mineral there by hutches or hurleys along the underground roads. Muscular strength in a female, not beauty, was the grand qualification by which she was estimated and a strong young woman was sure of finding a husband readily. There is an old Scotch saying, ' She is like the collier's daughter better than she is bonny,' proving the value put upon that description of female excellence." Letter addressed to the Duchess of Hamilton by the principal agent of the Hamilton Estates on the subject of the Education of Females in the Mining District of Lanarkshire, quoted in Report XXIII., 1851.

" The crying reproach, however, in regard to the great mass of the mining population of the district and those engaged in analogous occupations remains in full force—that of sacrificing the best interests of their children by sending them to work at the earliest possible age in order to profit by the small sums they can then earn. The excuse that such additions to the receipts of the family are necessary can seldom be valid in this neighbourhood where wages are so high." Report of Commissioner on Observation of the Mining Law, 1851, XXIII., *p.* 3.

That this prevalence of women and children's work was also the accepted rule in Agriculture can be seen from the following remark in the " Gentleman's Magazine," 1834, Vol. I., *p.* 531 :

" Formerly it was of no importance to the farmer whether he employed a single or a married labourer, inasmuch as the labourer's wife and family could provide for themselves. They are now dependent on the man's labour or nearly so."

See also *p.* 96 for an additional example of children's work and family earnings.

factory population could have been recruited. The peasant farmers and agricultural labourers dispossessed by the agricultural revolution also formed a large proportion of the new " hands." An official report on the state of commerce, (1810-11) states that if a manufacturer were obliged to stop work for a time " his workmen get dispersed throughout the country and he cannot collect them again but at considerable trouble and expense,"* which shows the difficulty of recruiting for the factories. It must also be remembered that the housing difficulty was acute. If a man set up a factory or iron works in the country he had to house his child apprentices or erect cottages for workmen and this took some time to do. In the towns the most appalling overcrowding took place and the difficulty of finding accommodation limited the numbers employed in factories.

(b.) Reluctance of the Employer to embark on Factory Production.

From the economic point of view the employer or giver-out of orders was in a very strong position. If he stopped giving out orders he incurred no loss in factories or machines standing idle with rates, taxes and interest at the Bank going on all the time. He could practically make the home-worker who had no agriculture to fall back on take what price he liked to offer—no combination was possible between people so dispersed. One cotton manufacturer, Radcliffe, states that his firm used to employ 1,000 weavers scattered over three counties at the beginning of the nineteenth century.† The home-workers used to underbid each other in their desire to get work allotted to them and this made the rate of remuneration low. In order to get work they were often forced to pay high rents for the frames or looms supplied to them by the employer and in addition they had to put up with payments in kind and deductions that often made home work open to the worst kind of sweating.

*Quoted A. Cunningham, " British Credit," p. 79.
†" Origin of the New System of Manufacture, 1828," p. 12.
In France we find the same system on a larger scale. We hear just before the Revolution of one merchant who had 2,100 country workers in his employ and another was said in 1770 to have no less than 6,000 scattered through Languedoc. (Tarlé, L'industrie dans les campagnes à la fin de l'ancien régime.)

An employer was not likely to change readily a system in which, economically speaking, he held the whip hand and incurred very little capital expense other than warehouse room, for one where he had to provide machines, factory, coal, light and cartage, borrow money from the Bank and take the risks of stoppages on himself. Hence the coming of machinery is nothing like the rapid cataclysm that it is apt to appear from the use of the word " revolution." It is easy to see that in what were practically new trades, such as cotton spinning or engineering, there would not be hands enough and therefore that there would be an inducement from the outset to employ mechanical appliances. In new districts with the difficulty of transport there would also be a scarcity of hands as in the worsted industry in Yorkshire, and this again would stimulate the adoption of labour-saving inventions. That machinery spread eventually to the older trades such as woollen weaving was apparently due to the delays and dishonesty involved in the system of home work. So long as there was no alternative the employer had to make the best of existing conditions ; when machines provided an alternative he gradually, but only gradually, reconstructed his method of production. In 1840 one of the Commissioners on Hand Loom Weavers cited the case of a man who had recently adopted machinery for the following reasons : " He could turn over his capital in much quicker time. He could command a much better price for his cloth by its being more regular in the weaving and more honestly manufactured. That he was neither robbed of the warp or weft and now he could take an order and know *when* he could execute it, which while he employed hand loom weavers he could never do."*

Another Commissioner speaking of the great irregularity in the weaver's work owing to the fact that he could " at any moment throw down his shuttle and convert the rest of the day into a holiday," went on to say that in Manchester in a workshop under the supervision of the master one hundred webs would be finished in the time " in which fifty would not be finished by an equal number of domestic weavers." †

* Mr. Muggeridge, " Report on Hand Loom Weavers," 1840, XXIV., *p.* 607.

† Hickson, " Report on Hand Loom Weavers," XXIV., 1840, *p.* 10.

In some cases the introduction of machinery was directly due to trouble with the workmen, ending in strikes which induced the masters to substitute mechanism.*

" The shearmen before the introduction of machines in that department were notorious for their drunken and careless habits. They would sometimes refuse to work when they knew that their employers were under contract and penalties as to time, unless he gave them drink ; and it was to clear themselves from these drunken dictatorial liabilities that the manufacturers eagerly adopted the use of machinery to rid themselves of the shearmen."†

In the same way in other branches of the textiles and in the engineering trades some of the most important inventions were developed in consequence of labour troubles. This was the opinion of one of the great machine makers as recorded by his biographer :

" Notwithstanding the losses and suffering occasioned by strikes, Mr. Nasmyth holds the opinion that they have on the whole produced much more good than evil. They have served to stimulate invention in an extraordinary degree. Some of the most important labour-saving processes now in common use are directly traceable to them. In the case of many of our most potent self-acting tools and machines manufacturers could not be induced to adopt them until compelled to do so by strikes. This was the case with the self-acting mule, the wool-combing machine, the planing machine, the slotting machine, Nasmyth's steam arm and many others."‡

(c.) Growth of Population relieved the Scarcity of Hands.

The coming of machinery was probably delayed owing to the rapid increase of population in the nineteenth century, which made it less urgent to introduce machinery in the older trades as more hands became available. The causes

* In France it was the turbulence of the workmen after the Revolution that led the manufacturers of Sedan to try to introduce machinery. Schmidt, C. : " Enquête sur la draperie à Sedan " in Revue des doctrines économiques, 1912, p. 94.

† Hand Loom Weavers' Report, 1840, XXIV., p. 373, Mile's Report.

‡Smiles, " Industrial Biography, Iron Workers and Tool Makers," p. 294.

of this growth of population are many. In the first place there had been throughout the eighteenth century in England a cessation of the plague, and although typhus and small-pox still took enormous toll, the devastating epidemics of the sixteenth and seventeenth centuries seem to have ceased. It has been suggested that burying the dead in coffins, which seems to have come into fashion at the end of the seventeenth century, may have had something to do with lessening the plague.*

The agricultural revolution of the eighteenth century had provided winter fodder for the cattle in the shape of good hay and turnips. This meant that fresh meat was available all the winter instead of the salt junk that had been the usual fare when the cattle had to be killed and salted in at Martinmas, because they could not be kept alive over the winter.

Winter fodder also meant winter milk, while a regular supply of fresh food and vegetables to towns, especially in winter, was facilitated by the new turnpike roads and thus decreased the high death rate from scurvy. Towards the end of the century inoculation seems to have lessened the mortality from small-pox. After 1750, owing to the knowledge gained in the new Lying-in Hospitals, there was an extraordinary fall in the number of deaths in child-birth and far more young infants were kept alive.

Apart from a cessation of epidemics and famines, several other factors contributed to the growth of population. One of the causes limiting marriage had been the difficulty of getting houses. With the coming of the factory system it was necessary to build new houses either in the towns or in the country districts for the new hands, and although there was a great difficulty in getting houses and much overcrowding it did become much easier to get married and set up a home. Indeed there are many complaints of the unruliness of young people who leave home and set up for themselves. The law had forbidden a man to marry while he was still an apprentice, and apprenticeship lasted till 21 or 24 according to the trade. Apprenticeship tended to decline after 1720 and compulsory apprenticeship was abolished in 1813. It therefore ceased to act as a limitation on marriage. Nor did a young couple need to hesitate before getting married for

*Creighton, " History of Epidemics."

prudential reasons. Children could always get employment in factories or in domestic work and from a very early age would become profitable to their parents.*

(d.) The Development of Iron Manufacture, Engineering and Coal Mining.

We have seen that it took seventy years to develop the factory system in the principal processes of the cotton, woollen and worsted trades, but in the iron, engineering and coal trades the transformation was even slower and occupied the whole of the eighteenth and the first three decades of the nineteenth centuries. The succession of wars in the eighteenth century created a constant demand for iron for ordnance and ships. England was, however, suffering from a timber famine and up to the end of the seventeenth century had not been able to use coal instead of charcoal as the sulphur in the coal spoilt the iron. The iron industry consisted of two main branches, smelting and forging.† The first was transformed between 1709 and 1782 by using coke but the second was not affected till after that date. Previously iron ore was smelted by charcoal in a blast furnace. Water power was employed to work the bellows. Wood and water determined the location of the industry which in the early eighteenth century was forsaking old centres like the Weald to find new woods near water.

The iron produced by blast furnaces was either ladled hot into casts or moulds, making articles like cannon, or pots and kettles, or was allowed to cool into pig iron when it had to be re-melted in a foundry and " cast." The iron required for special strains such as ship's pumps, bolts, nails, anchors and tools had to be refined. This second stage was done in a forge by frequently reheating the pig iron and hammering it to get out impurities. Water was essential for the hammers and for the transport of ore and finished goods. Charcoal was also needed for reheating. The bar iron from the forge was rolled into plates, or slit into rods for nail makers in rolling and slitting mills. There was not sufficient water power in any one spot to work all these operations, so each process was carried on in a different place in small works. At the beginning of the eighteenth century about two thirds of the iron needed was imported in the form

*Report on Hand Loom Weavers, 1840, *p.* 689.
†On the whole subject Ashton " Iron and Steel in the Industrial Revolution."

oi bars (refined) from Sweden and Russia and was worked up at Newcastle, Hull and London. Round Birmingham 45,000 domestic metal workers were partly dependent on this imported iron in the manufacture of which they were able to use coal.* Imported bars also went to Sheffield for conversion into steel swords and knives. A small amount of pig iron from British America came to London and Bristol and was cast there. In 1720 only 20,000 tons of bar iron were estimated to have been produced in England, an amount diminishing up to 1750.† This was serious both for the iron workers and the nation, as the Swedish supplies might be interrupted in time of war, to say nothing of a rise in price.

The change from wood to coal was inaugurated by Abraham Darby who cast pots and pans at Bristol. He migrated to Coalbrookdale in Shropshire in 1709 where he coked his coal and smelted iron with it successfully.‡ The practice did not, however, spread beyond the district till 1760. John Wilkinson, operating at Bersham, Bradley and Broseley between 1763 and 1770 was famous for his new method of boring cannon, invented in 1774, which, when applied to Watts' steam engine made it a success. The first iron bridge over the Severn was cast in 1779. Other new centres were inaugurated by the Walkers at Masborough in 1746 and by Roebuck at Carron in Stirlingshire in 1759-60. Meanwhile Huntsman had revolutionized the Sheffield steel industry by inventing cast or crucible steel between 1740-2.

The use of coal in the forging or refining of iron was made a commercial success in 1783-4 by Cort, who was working for the Navy at Fareham. By combining the reverberatory furnace in which coal was used, with the puddling process which eliminated the carbon and by putting the iron in a pasty form under rollers to get out other impurities, he not merely made it possible to use coal but turned out 15 tons of refined iron instead of one ton in twelve hours. His patents, owing to the misappropriation by his partner of government funds, were seized and thrown open and were energetically developed in the South Wales area. Thus the eighteenth century saw the rise of a new blast furnace industry using coke and situated on the coalfields. The steam engine made it unnecessary to consider water power, and being regular gave about twelve extra weeks' work in the year. The new canals and roads enabled the iron industry to select

*Ashton, *p.* 104.   †*Ib., p.* 13.   ‡*Ib., p.* 30.

P

their sites without considering river transport. The import ceased and an export of iron had begun by the end of the century. Iron became cheaper and was used for new purposes. New districts developed, and the Midlands, South Yorkshire, South Wales and the West of Scotland became centres of iron manufacture. These provided a new market for coal and for the steam engine. A new population was recruited for these areas and a new race of iron workers, engineers and wealthy iron masters emerged. The size of the business was no longer limited by the available water power. With the steam engine power was unlimited and all processes could be combined in one place. The typical large scale capitalist business at the end of the eighteenth century was the iron works, not the textile factory.* The iron trade suffered considerably after 1815 but the war demand was gradually replaced by the need of the new towns for pipes for sewerage, gas and water. Gas lamps for the street, cooking stoves, machines and steam engines all provided new markets. In 1828 Neilson invented the hot blast which reduced the amount of coal required for smelting by more than half. This cheapened iron. Then the growing demands of the railways after the thirties created another era of expansion in the iron industry.

The development of coal mining was also gradual. In 1712 the problem of pumping water out of the pits had been solved by Newcomen. Only at the end of the eighteenth century was it possible to apply steam to raise the coal from the pit by using Watt's double acting steam engine, and women were still employed in considerable numbers as late as 1842 in carrying the coal up the ladders out of the pits on their backs. Previously the steam engine had been used to pump water on to a water wheel which raised the coal. Then there was the further difficulty of getting the coal moved from the pits to the canals for general distribution. The roads, though improving, were by no means good, as Macadam had not yet begun to revolutionize the surface. Wooden railways had been laid down in the seventeenth century to convey the coal from the pits to the rivers. They were, if possible, arranged on an incline and the coal trucks went down by their own

*In 1812 it was said that it cost at least £50,000 to erect a set of iron works (Ashton, p. 63), and there were no less than ten works near Birmingham which had cost over this sum, each employing 300-500 men. Crawshay employed 2,000 men at Cyfarthfa.

impetus. The rails were, however, very unsatisfactory, the wood soon decayed or got broken with the heavy loads passing over ; there was great loss of time and heavy expenses in constant repairs. An enormous improvement occurred when iron rails were substituted after 1767, and these connected up the pits with the new canals. These rails gave a great impetus to the moving of coal. By 1812 a steam engine, invented by Blenkinsop, was successfully used to haul the coal along one of these waggon-ways, as they were termed, from the Middleton Colliery to Leeds. It was in connection with coal that those experiments in modern mechanical transport were worked out which have wrought a revolution in the transfer of bulky articles.*

The figures of the estimated amount of coal extracted show how small was the coal trade during the eighteenth century.

			Tons.	
1700	–	–	2,148,000	
1750	–	–	4,773,828	
1770	–	–	6,205,400	
1790	–	–	7,618,728	
†1795	–	–	10,080,300	Canal Period.
1854	–	–	64,700,000	Railway Period.
1913	–	–	287,411,869	

By 1816 the industry showed considerable expansion but the fifteen million tons which was the then estimated total shows how restricted must have been the amount of steam power in use when one considers the demand for household fuel. The shortage of miners was so great that women and children went down into the mines in increasing numbers but the lack of miners no doubt hindered the rapid development of the coal fields.‡

The invention by Andrew Smith, in 1839, of an iron wire rope for winding up the coal made it no longer necessary to employ women to carry it up the ladders. The hempen ropes were costly as they wore out so soon and women were cheaper.

*Galloway, " History of Coal Mining," Chapter XVII.
†Report of Coal Commission, 1871, XVIII., *p.* 852.
‡Women did not go down in the mines after 1780 in Northumberland and Durham. In that region the coal-owners utilized the steam engine to lift the coal out of the pits and they had tramways in the mines along which the coal was dragged. Children were, however, extensively employed in this region to open and shut the ventilating doors. Galloway, " Annals," The large demand for coal in this region probably led to the introduction of machinery instead of women.

In the same way the exhaust fan patented by Fourness, of Leeds, in 1837, inaugurated the modern system of ventilation and it was possible to dispense with young children.

The steam engine is another instance of the slow march of the " revolution." Newcomen's engine was very extravagant in the use of coal and for over sixty years steam was only used for pumping water out of mines. Watt made the steam engine much more economical in the use of coal, reducing the amount required to a quarter of what had previously been used, with the result that after 1776 his engine superseded Newcomen's in the Cornish tin mines. One has only to read Watt's life to see the endless difficulties he and his partner Boulton encountered owing to the lack of skilled workmen. All their men had to be taught engineering from the beginning. The engines when put up were often breaking down and no one could mend them. Watt himself had to live in Cornwall, a place he detested, simply because no one but he could rectify the engine troubles that were frequently occurring. Until Wilkinson invented his method of boring the cylinders it was almost impossible to get the steam engine parts made true to specification. In 1782 Watt invented the rotary or planetary movement and this enabled steam to be applied to drive machinery of all kinds. Pumping ceased to be its only use. By 1786 the steam engine had been adopted as the motive power in a corn mill, a paper mill, a cotton spinning factory, a brewery, in iron works and the Wedgwood potteries.* The steam engine could, however, only spread slowly. The skilled engineers were lacking. The firm of Boulton & Watt had a monopoly owing to their patents and the output of the one firm would be quite inadequate to transform the methods of manufacture of Great Britain with any rapidity. The engines themselves were so defective that their use spread slowly† and there was a standing difficulty

*When used in cotton factories the steam engine was first applied to pump the water that worked the mill. The steam engine to work the machinery itself was installed by Peel in 1787, Drinkwater in 1789 and Arkwright in 1790. When it was proposed to introduce it into Bradford in 1793 the would-be introducer was threatened with prosecution by his neighbours. Wedgwood introduced one into his pottery manufacture in 1782. Lord. Capital and Steam Power, *p.* 179, note 1.

†" Even after Watt had removed to Birmingham and he had the assistance of Boulton's best workmen, Smeaton expressed the opinion when he saw the engine at work that notwithstanding the excellence of the invention it could never be brought into general use because of the difficulty of getting its various parts manufactured with precision.

in getting men to work them to say nothing of the repairs. By 1800, a quarter of a century after the Watt's steam engine and eighteen years after the invention of the rotary movement, there were only sixty-three in three of the principal manufacturing centres, viz., eleven in Birmingham, twenty in Leeds and thirty-two in Manchester. In all England there were but 289, with 4,543 horse power. Of these eighty-four were in cotton mills.*

After 1815 steam as a motive power began to be more widely used, but water wheels did not disappear rapidly even after 1815. Water mills were frequently encountered by the Commissioners who were inquiring into the condition of women and children in factories in 1833. Indeed, one gains the impression that at that date about half the textiles mills were still worked by water.† So long as water was employed as the principal power the textile industry tended to maintain its rural character as it was necessarily scattered along streams or located near water-falls. The adoption of steam meant concentration on the coal areas and the growth of towns, larger works, and increased production. It also meant that industry could practically set up in any spot

Sometimes the cylinders when cast were found to be more than an eighth of an inch wider at one end than the other ; and under such circumstances it was impossible that the engine could act with precision. Yet better work could not be had. . . . There was usually a considerable waste of steam which the expedients of chewed paper and greased hat packed outside the piston were insufficient to remedy and it was not until the invention of automatic machine tools by the mechanical engineers that the manufacture of the steam engine became a matter of comparative ease and certainty." Smiles, " Industrial Biography," p. 180. (1897 Edition.)

*" Lord," op. cit., p. 175. Iron Foundries, collieries, copper mines, and breweries employed steam in 1800. It was also used to pump water for canals and town water supplies.

†The number of steam engines and water wheels in 1835 are given by Ure, " Philosophy of Manufacture," Appendix, as follows :—

	Steam Engines.	Water Wheels.
Scotland	224	214
North of England	21	34
Lancashire	814	340
West Riding	582	526
Cheshire	249	94
North of Ireland	17	23
Staffordshire	13	3
Derbyshire	33	63
	1,953	1,297

and was not confined to regions where there were water falls for power.

There was also a dearth of skilled engineers to make machinery of any sort,* not merely steam engines. It was only after 1820 that the tools were gradually invented with which heavy machinery could be rapidly and accurately made.

The early machine makers were men in quite a small way, and had been as a rule apprenticed to blacksmiths, carpenters or millwrights. Machines were all made by hand, every screw varied and every bolt and nut was a sort of speciality to itself. " As everything depended on the dexterity of hand and correctness of eye of the workmen the work turned out was of very unequal merit besides being exceedingly costly. Even in the construction of comparatively simple machines the expense was so great as to present a formidable obstacle to their introduction and extensive use."† When once the machine was set up there was no guarantee that it would work properly so inaccurate and faulty was it in its parts. " Not fifty years since," says Smiles, writing in 1863‡, " it was the matter of the utmost difficulty to set an engine to work and sometimes of equal difficulty to keep it going. Though fitted by competent observers it often would not go at all. Then the foreman of the factory at which it was made was sent for and he would almost live beside the machine for a month or more ; and after easing her here and screwing her up there, putting in a new part and altering an old one, packing the piston and tightening the valves the machine would at length be got to work."

It is obvious that under these conditions the spread of machinery would not be rapid. The invention of tools for machine making was, therefore, epoch making. Machines could at last be made by unskilled persons who could be easily obtained. The number of machines could be indefinitely multiplied. They were made with such perfect accuracy up to a thousandth part of an inch, that the long process of adjustment after erection could be dispensed with. Machines

*When Brunel invented his block machines considerable time elapsed before he could find competent mechanics to construct them and even after they had been constructed he had great difficulty in finding competent hands to work them. Smiles, " Iron Workers," *p.* 179.

†Smiles, " Industrial Biography," *p.* 212.

‡ " Industrial Biography," *p.* 181.

became much cheaper as well as more reliable and there was far more inducement to adopt them. When these machines to make machines were invented machine making was revolutionized and the adoption of machinery was considerably accelerated.*

Sir William Fairbairn stated that when he came to Manchester in 1814 " with the exception of very imperfect lathes and a few drills the whole of the machinery was executed by hand." Clement's planing machine (1825), his lathes (1827 and 1828), Nasmyth's steam hammer (1839), his machine for cutting key grooves in metal wheels of any diameter (1836), Roberts' punching machine for drilling holes in iron plates (1848) were only a few of the inventions which revolutionized machine production and made iron bridges, railway locomotives and iron steamers become practical possibilities." "The great pioneers of machine tool-making were Maudslay, Murray of Leeds, Clement, Fox of Derby, Nasmyth, Roberts, Whitworth of Manchester, and Sir Peter Fairbairn of Leeds."†

Great Britain had to evolve engineering and create and train a race of engineers before she could carry out a radical transformation of her industrial methods. In the process she trained engineers not merely for herself but for all Europe. Many inventions were also made by Americans and other foreigners but these inventors were obliged to come to Great Britain to get them worked out and then as a rule the inventions were considerably improved upon. No other country had either the coal, the iron-founders or the skilled mechanics, to say nothing of the capital required.

An Englishman, William Wilkinson, had successfully

*Report on the Export of Machinery, 1841, VII., *p* 96. " What used to be called tools were simple instruments as I should call them such as hammers and chissels and files, but those now called tools are in fact machines and very important machines " (the witness instanced planing machines and engines for cutting wheels). " By the production of tools machinery is made by almost labourers and made much better. It required, without those tools, first-rate workmen and the tools that we now have also produce machinery at much less cost than they formerly did. . . . Most of the tools or machines used in machine making are self-acting and go on without the aid of men ; the man who works the planing machine is a labouring man earning his 12s. or 14s. a week."

† Fairbairn Presidential Address to British Association at Manchester, 1861.

started the Creusot Works in France in 1781*. Other
Englishmen had taken over cotton spinning machinery,
both Hargreaves' jenny and Arkwright's water frame, and
had started factories in France.† Although the export of
many kinds of machinery was prohibited up to 1825, machinery
was nevertheless freely smuggled out, the parts being mixed
up with the parts of other machines the export of which was
permitted.‡ A regular insurance business seems to have
existed by which people who smuggled out machinery insured
against the risk of its being stopped or confiscated. Machines
were also erected from English drawings, while English
artisans were to be found in factories and engineering shops
all over the continent and in the United States, in spite of
the fact that the prohibition on emigration was only removed
in 1825, when, with certain reservations, the export of
machinery was also permitted.§

In the second great Commission of 1841, held to consider
the removal of all restrictions on the export of machines,
this tendency to hire English managers and foremen to work
English machines on the continent and train foreign workmen
is even more marked, as machines were more easily obtained
and manufacturers complained bitterly of the way in which

*Ballot, "La révolution technique ; . . dans la metallurgie
française" in Revue des doctrines économiques, 1912.

† Schmidt, Les débuts de l'industrie cotonnière en France, 1760-1806
in Rev. d'histoire économique, 1913.

‡ An amusing account of two attempts by a Frenchman to smuggle
out English cotton machinery in the eighteenth century is told by
C. Schmidt in " Les débuts de l'industrie cotonnière 1760-1806 " in
" Revue d'histoire économique " 1914, pp. 43-44.

That it was not easy to obtain English machines the following
account of an attempt to import them into France in 1798 will show :
" After a perilous voyage pursued by many men-of-war Bauwens
disembarked at Hamburg, where he hid himself at the house of his
brother with the workmen he had brought with him. When these
learned that it was in Paris or Ghent that they were to work, they
fled and denounced Bauwens to the English minister. With five
English who remained faithful Bauwens fled, crossed Belgium, the
fragments of the machines hidden under his carriage. When he arrived
in Paris he put them together." Bauwens finally established a cotton
factory at Ghent and employed over a thousand workmen but was
ruined by the collapse in 1811.

§ In 1825, before the repeal, 15,000 to 20,000 British artisans were
said to be in continental Europe of whom 1,500 to 2,000 were in the
Iron Trade. Report, 1825, V., p. 43. See also an article, Edinburgh
Review, 1819, Vol. 32, on " The Comparative Skill and Industry of
France and England." Also Report, 1825, V. p. 112.

foreigners would tempt away their good workmen.* Many Englishmen also had set up businesses on the continent, notably in France and Germany, to evade the tariffs. They, too, took out English engineers and operatives to train the unskilled continental workmen. When England developed after 1820 the tools for making machines, the continent was said to have gained an enormous advantage in making machinery for itself as then a lower degree of skill in the artisan was required.†

From the foregoing sketch of the slow evolution of the steam engine and machinery it is obvious that had there been no other factors acting as a drag on the wheel yet the pace of the industrial changes could not have been rapid. Not merely was the coming of machinery retarded by the deficiency of machines, their unsatisfactory nature, but the dislike of the hands to work in factories, the possibility of riots and machine breaking by those who thought they would be injured, and the increase of population which provided a larger number of hands always more ready to take up home work than factory work, all worked in the same direction. The soft fibre of wool introduced technical obstacles to the introduction of machinery in the woollen trade and made its adoption a slow process and this was also accentuated by the shortage of raw wool, so that the machinery when set up might not be fully employed. On top of all this came

*The story of the establishment of the flax-spinning industry in France, as told by the witness Marsden, is typical of England's influence on the continental industrial revolution. (Report 1841, VII., *p.* 91.) He went over to Mr. Maberley, an Englishman who was making flax spinning machinery in France, with one hundred English artisans and £6,000 worth of tools and material. The French flax machinery makers who followed in his footsteps were said " to have purchased excellent tools of all descriptions from Manchester and Leeds and purchased one specimen of the best models of machines from the different machine makers in Leeds and other parts of the country and have established works for the purpose of making machines; they have taken over the cleverest Englishmen they could get, not looking to the salary in the first instance but paying very enormous salaries to induce the cleverest men to go and having got the tools together with those clever men and the models of the machines smuggled out of this country they have been able to make machines . . . quite equal to any that I have seen made in England."

†Reports on the Export of Machinery, 1824, V., 1825, V., and 1841, VII.

the risk and expense of building and working a factory, especially when dealing with materials liable to change in fashion. The chief cause acting in the other direction and making for the adoption of machinery was the scarcity of hands which was most strongly felt in such a sparsely populated district as the North, and this partly accounts for the more rapid adoption of machinery there. In other cases machinery seems to have been introduced owing to trouble with the workmen and the inefficiency of domestic work. It is not surprising then, that it took seventy years (1770-1840) to introduce machinery into certain branches of the cotton and woollen trades and that even then there was a large survival of home work.

The distress of the years 1815-1830 is really due to the war and its after effects and only in a small degree to the introduction of machinery, which, if it dispossessed people, slowly, on the one hand, created new occupations on the other. The industries affected were chiefly new trades like iron, cotton and engineering and their expansion increased rather than diminished employment. Even where we find women displaced by spinning machinery there was such an enormous demand for weavers that they found new openings at the hand loom and in the weaving factories.

In no other country was the transition so slow as in this country. Other nations were able to begin where Great Britain left off in her experiments and as they developed later their industrial revolution coincided with the railways which in their case hastened the pace of the change.

Although industrial conditions were eventually radically altered, they were by no means suddenly altered. If a date might be suggested when one might say that England had become " revolutionized " and was something very different from the England of 1750, perhaps the years 1830-1840 might be chosen, though there is much to be said for placing it even a decade later.

By 1840, however, Great Britain had become a typically industrial country instead of an agricultural one. The development between 1820 and 1840 of tools for making machinery had increased the output and use of machinery. Neilson's hot blast, by making the production of iron cheaper, had reacted on the price of machinery, while the permission to export machinery after 1825 had stimulated the engineering trades by increasing the demand. The coal output had increased considerably, the consumption being estimated at 30,000,000 tons in 1829.* By 1835 the cotton manufacture had become definitely a factory industry and wool, linen and silk were beginning to be transformed.†

Thus the decade, 1830 to 1840, witnesses the triumph of the factory system over domestic production and the position which the new engineering, iron and coal trades were beginning to assume as typical features of British industrial life becomes obvious.

V.—ECONOMIC AND SOCIAL EFFECTS OF THE CHANGE.

The term "industrial revolution" is used, not because the process of the change was quick, but because when accomplished the change was fundamental. It involved, in the first place, as we have seen, the rise of new industries. This produced a second change, namely, a radical transformation of commerce. Great Britain became organized for mass production and world exchange. She began to depend as never before on overseas lands for raw materials like cotton and wool and on foreign countries for her markets. During the last half of the nineteenth century she became increasingly dependent on other countries for her food supply, paying for it with coal, manufactures, shipping and financial services. The scale of exchange attained a magnitude hitherto undreamed of and new articles, bulky products, were dealt in to an extent that was not merely new but overshadowed in importance the old historical spices and colonial products. Thirdly, new districts came into prominence. In 1750 the South of England had been the richest part of the country and the Eastern counties of Norfolk and Suffolk and the West of England the chief manufacturing areas. In 1700 the six most populous counties were probably Middlesex, Somerset, Devon, West Riding, Lincolnshire and Norfolk. The

*This was the estimate of a House of Lords Committee held in that year quoted "Galloway, Annals, II.," p. 462. Buddle, the great coal expert and viewer, put the amount at half this, viz., 15,300,000 tons, which shows that there was really no satisfactory basis for an accurate estimate.

†Cf. figures on p. 59 of the numbers of persons employed in factories.

North was a poor, barren district with few roads, a struggling small cotton industry in Lancashire and a developing woollen industry of considerable importance in Yorkshire. There was a small coal export from the Tyne to London and the continent. By 1800 the population had increased to such an extent in Lancashire and the West Riding of Yorkshire that they began to rank as two of the most populous counties. New districts opened out in Scotland in Lanarkshire, in connection with the coal and iron developments, and a cotton industry sprang up round Glasgow and Paisley. South Wales, almost uninhabited in 1750, was transformed by Crawshay, of Cyfartha, into a great iron making centre ; the pottery districts were revolutionized by Wedgwood and cheap transport by canals, while the development of coal, iron and steel made Staffordshire and Warwickshire rank in 1800 along with Lancashire, Yorkshire and Middlesex as the five most populous counties of England. The old textile districts of the South-East and South-West of England became relatively unimportant. Leeds and Huddersfield became the chief woollen centres in place of the West of England while Bradford inherited the worsted trade of Norwich.

The fourth great change was the growth of towns. People massed in numbers on the coal and iron areas, the new canals enabled them to get food and fuel even in regions like the North where the food supply was scanty. As there were no building restrictions houses were run up any fashion, often back to back. There were no regulations to prevent over-crowding or cellar dwellings. There were no arrangements for disposing of the house refuse which always accumulates, ash-pits overflowed and spread " a layer of abomination " about the courts and streets ; there was no system of main drainage and no system of sanitation. An adequate or clean water supply laid on to the houses was rare until after 1850.*

---

*Even in 1850 it was computed that 80,000 houses in London inhabitated by 640,000 persons were unsupplied with water. Jephson, " Sanitary Evolution of London," *p.* 21. The alternative was that " a very large proportion of the people could only obtain water from stand pipes erected in the courts or places and that only at intermittent intervals and not at all on Sundays. The water is then kept in close, ill-ventilated tenements until it is required for use." *Op. cit., p.* 21.

" In Jacob's island (in Bermondsey) there may be seen at any time of the day women dipping water, with pails attached by ropes to the

The wells and pumps were quite insufficient for the numbers who wished to use them and the river and canal water was polluted and disgusting. Towns had always been insanitary places suffering from plague, small-pox and other virulent fevers. In the latter half of the seventeenth century the death rate of the City of London was at the rate of eighty per thousand, in the eighteenth, fifty per thousand.* This enormous death rate was due to the fact that town sanitation was not known. The piling of the population on new areas made an existing evil much worse, it aggravated the filth, congestion and infection, and no machinery existed to grapple with the problem. The Irish who migrated in large numbers to furnish the hands required had a low standard and complicated the problem.

In the new towns, as in the old, typhus and small-pox were chronic, fevers of all kinds were one of the greatest causes of pauperism and the cholera was a frequent visitant. Child mortality seems to have been exceptionally heavy if Preston was a fair sample.†

1ST JULY, 1837—30TH JUNE, 1843.

	Average age at death.		Above 5 years.		Under 5 years.
Gentry ...	47.39 years	...	82.43%	...	17.57%
Tradesmen ...	31.63 ,,	...	61.78	...	38.22
Operatives ...	18.28 ,,	...	44.58	...	55.42

Thus of the total deaths among operatives in these five years more than half occurred among children under five years of age.

On top of the dirt and disease came the great difficulty of the disposal of the dead. The overcrowded state of the little town burial grounds added to the horrors of town life and poisoned the water supply.‡

back of the houses, from a foul, fœtid ditch, its banks coated with a compound of mud and filth, and with offal and carrion—the water to be used for every purpose, culinary ones not excepted." *Op. cit., p.* 23, quoting Report of General Board of Health, 1850.

If this was the standard of the Metropolis it is clear that the early factory towns would not fare better.

*Sir George Newman, Chief Medical Officer, Ministry of Health, Lady Priestly Memorial Lecture, 1920. In 1917 the death rate for London was 13.6 per thousand.

†Report on the State of the Large Towns, XVII., 1844, *p.* 37. Appendix.

‡Of Clerkenwell it was stated positively that " the shallow well water f the parish received the drainage water of Highgate cemetery, of

The fifth change was involved in the break-up of home industry and its replacement by the factory. Instead of manufactures being produced by families working with simple tools, many of whom carried on an agricultural bye-employment, the new unit consisted of large numbers of people massed in one building or on one spot dependent on expensive machinery worked by water or steam power. The new machinery was far too costly for workmen to erect in their own homes, the use of steam or water as power to move the machines necessitated the combination and concentration of the operatives or " hands " in one building and the conditions under which work was performed were radically altered. The various members of the family went out to their work in the morning and returned home at night. They often worked in different factories. The man could no longer combine industry with a little farming. He became an attendant on a machine, his actions governed by that machine and his livelihood dependent on the demand for the product of the machine. Another result was that in many cases the work became more monotonous. " A spinner is required to do but one thing throughout his whole life : to watch a pair of wheels and to walk three steps forward and three steps backward."*

The sixth change consisted in the new relations between capital and labour, *i.e.*, between employer and employed. The new type of industrial production needed considerable command of capital, it became more difficult to rise in the world and as capital became the most prominent feature of nineteenth century industrialism it became increasingly an object of criticism and attack.

numerous burial grounds, and of the innumerable cesspools of the district." Jephson, *p.* 22. The burial ground in Russell Court, off Drury Lane, where the whole ground which by constant burials had been raised several feet, was " a mass of corruption " which " polluted the air the living had to breathe and poisoned the well water which in default of other they often had to drink." *Op. cit.*, *p.* 36. Dickens gives a gruesome account of another such burial ground in Bleak House where Lady Deadlock goes to visit the grave of her lover. Both Martin Chuzzlewitt and Bleak House describe the fevers of the time and there was Sarah Gamp as a nurse to add a new horror to sickness.

*Report, *op. cit.*, 1840, XXIV., *p.* 682.

The man with capital was no new feature in English industrial life. The typical capitalist had been, however, a giver-out-of-orders, and often of raw materials, which people executed in their own homes, in their own way and in their own time. While they often were very dependent on the man who gave them orders the scarcity of labourers and the possibility of agricultural work somewhat mitigated their position and working at their own pace made it seem as if they enjoyed a good deal of personal independence. The workers were so scattered that they had no strong class feeling of resentment against the merchant employer. It was easy for a man to become a little master on his own account and he would not be willing to combine against the position of those to whose ranks he hoped to attain.

With the new f..ctory system the position of the worker changed. He became more dependent on the employer as it became more difficult for him to fall back on agriculture, especially with the increasing growth of large farms. There was a rise of a new moneyed class, often from his own ranks and to them he felt he owed no loyalty. He resented the new factory owner driving in his carriage and pair which seemed to have been unlawfully obtained out of the labour of the worker. The old ties with the village, the squire, the parson or the priest were broken when he had to go to the towns for work. For his social life he was driven back on his fellow operatives. These were vitally interested in their rates of pay and conditions. Massed together they could discuss their grievances and the operative became " class conscious." All through the eighteenth century trade unions had been developing, their object being to secure a standard wage or a limitation of hours. The State was prepared, however, to enforce a system of wages and hours and did not approve of trade unions attempting to do its business. Thus when they became prominent they were prohibited as were the London Tailors in 1721. Trade unions continued to appear and under the scare of the French revolution these prohibitions were renewed in 1799 and 1800.* In this case, however, the State was so completely *laissez-faire* that it instituted no standard of wages and hours and even repealed, in 1813, those that had existed from 1563.† It was, however,

*39 G.III. c.81.   39 & 40 GIII. c.106.   †5 Eliz., c. 4.

impossible for men closely associated in their work not to form some kind of combination for their common interests, and trade unions continued to exist disguised as friendly societies, a form of thrift which the Government encouraged. They were, however, local in action and confined to a group of artisans of a particular town. It was the railways which enabled them to become national societies. Any common action by the men could always be checked and their members prosecuted under the Combination Acts or the common law for Conspiracy. Until the Combination Acts were repealed in 1824-1825 the worker, though beginning to be organized on a local basis* was fairly helpless to effect any betterment in his position as he had no vote and did not therefore count in politics to an extent that would give him a powerful leverage. Rioting was his only chance of expressing discontent and as there were no police the military would be called out to keep order.

The social results of the new methods were not altogether bad though the transition stage gave rise to many abuses and the situation was complicated by the French wars which made raw materials and markets very uncertain at a time when Great Britain was organizing for world sales.

So far as the worker was concerned, once the transition period was over a man was physically better off in a well ventilated factory than when he worked in a home littered with the mess of the family production. He no longer ate, drank and slept with the refuse of his work.

Many Commissions brought out the fact that the weavers' homes were by no means ideal places, though of course they varied.

" In some parts of Bethnal Green and Spitalfields, inhabited by weavers, every house ought long ago to have been condemned and razed to the ground. Ruinous buildings, streets without sewers, overflowing privies and cesspools and open ditches filled with a black putrifying mass of corruption infecting the air for miles round render the district the abode of disease and death, There are streets and alleys from which typhus fever is never absent the year round.

" With regard to health, having seen the domestic weaver

*The shearmen of the West of England seem, however, to have been in close touch with the shearmen of Yorkshire.

in his miserable apartments, and the power loom weaver in the factory, I do not hesitate to say that the advantages are all on the side of the latter. The one, if a steady workman, confines himself to a single room in which he eats, drinks and sleeps and breathes throughout the day an impure air. The other has not only the exercise of walking to and from the factory but when there lives and breathes in a large roomy apartment in which the air is constantly changed. Some of the factories I have visited are models of neatness cleanliness and perfect ventilation ; and there is no reason all should not be the same."*

The description just quoted deals with the older trade of silk, but in the new industry of cotton persons working in their own homes seem to have been in an equally bad position.

" Weaving as a domestic occupation, among the hand loom cotton weavers, is carried on in circumstances more prejudicial to health, and at a greater sacrifice of personal comfort, than weaving in any other branch. The great majority of hand loom cotton weavers work in cellars, sufficiently light to enable them to throw the shuttle, but cheerless because seldom visited by the sun. The reason cellars are chosen is that cotton requires to be woven damp. The air must therefore be cool and moist instead of warm and dry. Unhappily the medium which might be preserved without injury to the constitution and which is preserved in the best power loom factories, the impoverished hand loom cotton weavers are obliged often to disregard. I have seen them working in cellars dug out of an undrained swamp ; the streets formed by their houses without sewers and flooded with rain ; the water therefore running down the bare walls of the cellars and rendering them unfit for the abode of dogs or rats."†

It is scarcely surprising that the hand loom weavers were described by the Commissioner as " never a healthy or strong class of people,"‡ and it is difficult to conceive that the factories would not be better from the point of view of the health of the worker.

It is clear from the report of the Commissioners that weavers in factories were paid in 1840 better wages than

*Hand Loom Weavers' Report, 1840, *p.* 681. Hickson.
†Hand Loom Weavers, XXIV., 1840, *p.* 645. Hickson's Report.
‡ *Ib* *p.* 425.

home workers, and had more regular work. The amount
paid to factory weavers in Gloucestershire was 11s. 9d. a
week in 1840, the out-door master only got 8s. 1½d., and the
journeyman 5s. 7d.* The wages were also more steady
" because a manufacturer finds that it is more difficult to
lower thirty or forty men working under a roof and who are
in communication with each other than to lower the wages
of thirty or forty isolated out-door weavers who, owing to
a surplus of hands, are ever ready to undersell each other."†

In times of slack trade the master who gave out work
was able to withhold work without loss to himself but not
so the employer with a factory. " It is well known that
factory hands when dispersed are not again collected in a
state of efficiency without considerable time, trouble and
loss . . . the power loom manufacturer may, but rarely
does, work short time . . . he never risks the dispersion
of his work-people by leaving them totally without employment."‡

Although the sentimentalists were shocked at the break up
of family life, yet the Commissioner considered that " domestic
happiness is not promoted but impaired by all the members
of a family muddling together and jostling each other constantly in the same room."§ In the opinion of the Commissioner the man was improved morally by working regular
hours,‖ which were said to engender regular habits.

One great advantage of the factory system was that it
became possible for the workers to combine and enforce

*Hand Loom Weavers, 1840, p. 404, cf., also p. 649. " Factory
hands earn about 2s. a week more."

†Report, op. cit., p. 424, " Low as is his (the outdoor weaver's)
pay he would be better off if this amount was regular, if he knew every
week that he would have that certain sum. This is not the case,
his work is extremely irregular, his means of subsistence are consequently precarious, he therefore possesses no certainty of habit or
concentration of his energies."

‡Hand Loom Weavers, 1840, p. 630. Muggeridge's Report. Gaskell
points out, op. cit., p. 43, that employers did not reduce the wages of
their cotton spinners in factories " knowing that the lower the work-
people are reduced in circumstances the less dependence can be placed
on their labour."

§Ib., p. 682.

‖" Mr. Cobden is of opinion that the factory system by its regularity
is working a marked and decided improvement in the character of the
Irish population with which Manchester is inundated." Ib., p. 681.

decent conditions, a standard wage, shorter hours and payment in full in coin and not in truck.

The domestic worker who depended on an employer who gave or withheld orders as he chose was not in a strong position for bargaining when workers were scattered and ignorant of the price usually paid or the state of the market. Although spinners were so scarce one must remember that spinning was always wretchedly paid.

Indeed, domestic industry had always been badly paid because the workers were too scattered to combine. Gaskell, who idealized the system of home work, says that " a family of four adult persons with two children as winders earned at the end of the last and at the commencement of the present century £4 per week when working ten hours per day ; when work was pressed they could, of course, earn more."* He admits that they had animal food very rarely and lived on " meal or rye bread, eggs, cheese, milk and butter, the use of tea being quite unknown."† He mentions that the factory operative had " a limited supply of animal food once a day joined with copious dilutions of weak tea."

The home worker was frequently paid in truck and had to submit to the most arbitrary deductions from his wages.‡ The effect of the coming of machinery was to increase the tendency to pay in truck or make men buy their goods at an employer's shop at enhanced prices. In the country districts where the water factories settled or where mines developed there were no shops and few houses. The result was that the factory or mine owner ran up houses and let

* " Artisans and Machinery," p. 24.
†Op. cit., p. 18. Gaskell was a surgeon.
‡Felkin mentions that in 1847 five hundred frame work knitters earned £194, the deductions were £77, leaving them £117 or 4s. 8d. a week. He says in 1845 " in eight of the villages in our district it was found that of eighteen bagmen only two paid wholly in wages." " History of Machine Wrought Hosiery and Lace " (1867), p. 459.

An extraordinary case of a frame work knitter working at home who was made to pay rent for a machine that was his own is given in the Report on Frame Work Knitters, XV., p. 76.

Samuel Jennings in giving evidence said : " I worked for T. P., of Hinckley, and every time I took my work in he used to tell me if I did not lay the whole of my money out in the shop he could not give me any work, and besides that the frame was my own but I had to pay rent to him for it that he should employ me because I could not get employment anywhere else."

them to the work-people and deducted the rents—often high rents—from their wages.* He also built a shop and in many cases a public house and the hands were forced to deal at the shop. They frequently got credit at the shop in advance ; they became hopelessly bound to their master by debt and could not leave his employ. In cases where they were paid in goods and not in coin they had no money to move and were equally bound. "Around many mills a fixed population has arisen which is as much a part and parcel of the property of the master as his machinery." The following is an account of payment in truck given by a witness before the enquiry in 1845 :†

"In the whole of that time, within a week or two of two years all the money that I received of my employer was 16s. 6d. and 10s. 6d. of that I had of him to pay interest of the pawn-tickets where I had pledged things for necessities for my wife and family when I could get no money in order to buy things he did not sell in cases of sickness. When Saturday night came I had to turn out with a certain quantity of meat and candles or tobacco or ale or whatever I had drawn as wages to dispose of at a serious loss. I used to take a can of ale to the barber to get shaved with and a can of ale to the sweep to sweep my chimney. I was in good receipt of wages and in company with my neighbours I used to take in a newspaper and I was obliged to take a pound of candles at 7d. and leave it for the newspaper the price of which was 4½d. I used to take my beef at 7d. a pound and sell it to the coal woman that I had my coals of for 5d., and any bit of sugar or tea or anything of the kind that my employer did not sell I used to get from the grocer living at the bottom of the yard by swapping soap and starch."

It is worth noticing, however, that the women did not disapprove of this payment in kind. It effectually kept the workman from the public house or beer shop because he had

---

*The cottages were let at a rent which brought in 13½ per cent. "This profitable return is burdened with no drawback ; no rent is lost, every pay night it is deducted from the wages." Gaskell, *op. cit.*, *pp.* 294-302. By building cottages near the mill the employer was able to begin work earlier in the morning, shorten the hours for meals and work late at night.

†Frame Work Knitters Report, 1845, XV., *p.* 77.

no money and his earnings took the form of articles consumed by his wife and children.*

The scandals of the truck system were attacked by the Act of 1831, but it was exceedingly difficult to stop the practice although in many districts the good employers banded themselves together to obtain evidence and prosecute the truck masters.† The growing strength of the trade unions after 1850 did much to put an end to the system but where home work still existed truck continued to flourish.‡ The prevention of truck was placed under the factory inspectors in 1887.

One great advantage of the factory system was the general development of intelligence that seems to have taken place among the artisans collected in the early factories. They formed trade unions and in these early trade unions the man got his training as a member of a group and was educated through industrial to political citizenship.§

* Ib., p. 81.
†" There are five Anti-Truck Associations in South Staffordshire. . . . We have ample funds to carry on prosecutions and to maintain men who in consequence of joining us in our efforts have been thrown out of employment by truck paying masters. The cost of these together has amounted to upwards of £800 in ten months. We have laid between five hundred and six hundred informations and obtained about two hundred and fifty convictions and notwithstanding this I do not think we have reduced the amount of truck five per cent." Report on Mines. Horner, May, 1851, in XXIII. of 1851 p. 463.
‡Report on Truck, 1870.
§Guest, in " A Compendious History of the Cotton Manufacture," published in 1823, devotes a chapter to the " change of the character and manners in the population superinduced by the extension of the Cotton Manufacture."

" The operative workmen being thrown together in great numbers had their faculties sharpened and improved by constant communication. Conversation wandered over a variety of topics not before essayed and the questions of Peace and War which interested them importantly inasmuch as they might produce a rise or fall of wages, became highly interesting and this brought them into the vast field of politics and discussions on the character of their government and the men who composed it. They took a greater interest in the defeats and victories of their country's arms and from being only a few degrees above their cattle in the scale of intellect they became Political Citizens.

" The facility with which the weavers changed their masters, the constant effort to find out and obtain the largest remuneration for their labour, the excitement to ingenuity which the higher wages for fine manufactures and skilful workmanship produced and a conviction

It is true that the worker disliked the regularity and the "tyranny of the factory bell" and the Commissioner recorded "everywhere I observed a visible reluctance to working in factories."* There was strong feeling against working under superintendence and being subject to a foreman's orders. When the family broke up into independent working units and the boys kept their money to themselves and married when they chose and when the girls were mistresses, too, of their own earnings the loss of the family wage tended to lessen the stability of life. The loss of the bye-employment also worked in the same direction. The specialization of workers in a factory and the growth of occupations for boys which led nowhere and merely turned out unskilled labourers were poor alternatives for the all-round industrial training a boy got in the home where industry was carried on. It must, however, be remembered that family work and the family wage often meant that the members of the family were sweated by their parents or the wife by the husband.

On the other hand endless new opportunities were opened out by the new industrial developments. The transport trades expanded and wanted men as drivers and men to manage the barges and load and unload the goods at the wharves and these occupations were by no means monotonous. The building trades were crying out for hands to erect the new towns and factories, coal miners were wanted, so were engineers of all kinds as well as men to work in the blast furnaces, iron works and chemical factories. Many of these new occupations were dangerous or unpleasant but all the

that they depended mainly on their own exertions produced in them that invaluable feeling, a spirit of freedom and independence and that guarantee for good conduct and improvement of manners a consciousness of the value of character and of their own weight and importance."

On the other hand Gaskell laments the decadence in the character of the factory artisan and considers the home worker an infinitely superior person. Gaskell, "Artisans," *p.* 22, 1836 : "That he was inferior in some respects is not denied : he could seldom read freely or write at all, but he went to church or chapel with exemplary punctuality ; he produced comparatively speaking but little work but he was well clothed and well fed ; he knew nothing of clubs for raving politicians, or combinations which could place him in opposition to his employer but he was respectful and attentive to his superiors, and fulfilled his contracts to the letter."

* *Ib., p.* 649.

factory and metallurgical workers seem to have been well paid when compared with the home workers. The factories were not popular at first as they limited a man's independence and lowered his social status since he was no longer a " master," and higher wages were necessary to attract operatives into the factories.

The overwork of children was not new* but it was brought into special prominence in this period and the evils were recognized and combatted.

Conditions of work in the home were by no means ideal and the overwork of children did not begin with the factories. " The creatures were set to work as soon as they could crawl and their parents were the hardest of task-masters,"† was one description of children's work in pre-factory days. This is borne out by a description of the boyhood of Crompton.‡ " I recollect that soon after I was able to walk I was employed in the cotton manufacture. My mother used to bat the cotton wool in a wire riddle. It was then put into a deep brown mug with a strong ley of soap and suds. My mother then tucked up my petticoats about my waist and put me into the tub to tread upon the cotton at the bottom. When a second riddleful was batted I was lifted out, it was placed in the mug and I again trod it down. This process was continued till the mug became so full that I could not longer safely stand in it, when a chair was placed beside it and I held on by the back." We should to-day consider it perfectly preposterous to put a tiny child in petticoats into strong, soapy water, possibly cold, for it to work its little feet up and down cleaning cotton wool. Nor had matters changed much in 1866. A Commission reported in favour of legislation on the following ground : " But more especially would such legislation be a protection and benefit to the great numbers of very young children who in many branches of manufacture are kept at protracted and injurious labour in small, crowded, dirty and ill-ventilated places of work by their parents. It is unhappily to a painful degree apparent through the whole of the evidence that against no persons

* Dunlop, " Apprenticeship and Child Labour," pp. 99-101.
† Cooke Taylor, " Notes of a Tour in the Manufacturing Districts."
‡ French, " Life of Crompton."

do the children of both sexes need so much protection as against their parents."*

The early water frames were worked by pauper apprentices. The machines were low and children could more easily work them than adults. These apprentices were housed and fed by the employer and as they were a peculiarly helpless, friendless class, there were grave abuses, but even here the children seem in some places to have been kindly treated.†

The pauper apprentices seem to have become rare by 1816.‡ They were but a temporary expedient owing to the fact that the early factories were set up in the country near water-falls where the population was so sparse that labour had to be imported. As soon as the newer factories could obtain steam power they set up on the coal fields and in towns and they hired children who lived with their parents. There was then no need for the employer to feed or house the children, who at least enjoyed the shelter of a home. The condition of children's work in factories was very trying. There was the heated atmosphere, often eighty to eighty-five degrees, and there was an enormous amount of dust. In some departments of cotton spinning it was impossible, according to evidence given in 1816, to see across the room.§

* Report, 1866, V., *p.* 24.

† An appalling account of the sufferings of one Blincoe, a pauper sent up to Lancashire from St. Pancras to work in a cotton mill, is given in the " Memoir of Robert Blincoe," 1832 (John Brown). " When Blincoe could not or did not keep pace with the machinery the over-looker used to tie him up by the wrists to a cross beam and keep him suspended over the machinery till his agony was extreme. To avoid the machinery he had to draw up his legs every time it came out or returned. To lift the apprentices up by their ears, shake them violently and then dash them down upon the floor with the utmost fury was one of the many inhuman sports in Litton Mill in which the overseers appeared to take delight." (*p.* 39.)

On the other hand, in the " Life of Kitty Wilkinson," by W. R. Rathbone, we read (*p.* 25): " Long afterwards when Kitty was a woman she used to look longingly back to those Caton days (where she worked in a cotton mill and was lodged in the 'prentice house) and would often say, ' If ever there was a heaven upon earth it was that apprentice house where we were brought up in such ignorance of evil and where Mr. Norton, the manager of the mill, was a father to us all.' Sometimes she was so happy she would sing for joy."

‡ Report on Children in Cotton Mills, 1816.

§ Minutes of Evidence taken before the Select Committee on the State of Children in Factories, 1816, III.

Then there was also the strain of keeping pace with the machinery. The hours worked were, in 1816, anything from twelve to fourteen with only one hour off for dinner. The other meals were apparently eaten while work was going on. The children were all this time on their feet, moving to and fro the whole twelve or fourteen hours. In the case of the water mills they had to work overtime to make up for irregularity in the water supply.

There was abundant evidence that the hours worked were far too long though the Commissioners reported in 1833 that the hours were even longer in the homes where they considered that the work was performed under more unwholesome conditions.

By 1816 very few children under ten seem to have been employed in the factories investigated.

Steam power was far more regular than water and the gradual adoption of the steam engine obviated much overtime pressure.

The tendency of factories to settle in towns and use steam is especially noticeable after 1815.

The employment of steam as power meant that the machinery became larger and more powerful and the machines were no longer suitable for very young children, and their place was taken by older children and women. From the very first attempts were made to regulate the abuses of child labour. In 1784 the Manchester magistrates decided not to apprentice children to any mills where they had to work more than ten hours a day.* In 1802 an Act was passed limiting the hours of apprentices in cotton and woollen factories to twelve and prohibiting night work. It also laid down regulations for their education and attendance at church and was called "The Health and Moral of Apprentices Act." When the factories grew up in towns, ordinary children were employed and were not apprenticed. As the Act providing for compulsory apprenticeship was repealed in 1813 ; they thus escaped legislative protection. In 1819 this was remedied and no child was allowed in a cotton factory till it was nine, and from nine to sixteen, twelve hours were

*Hutchins & Harrison, " History of Factory Legislation," *p.* 9.

the maximum limit to be worked.* The Act did not apply, however, to other textiles like woollen or worsted. These Acts were not really effective till 1833 when inspectors were appointed and the Acts were extended to include not merely cotton but woollen, worsted, flax, silk and linen mills. Thus the abuses of child labour in factories were eventually stopped as it was easier to regulate conditions of employment in a factory than in the home. Indeed, even before regulation the conditions of work seem to have been better in the factories than in the home.† The Factory Acts did not, however, apply to mining where children went down into the pit at a very early age, sometimes at five, to sit in the dark all day and open and shut the trap doors for ventilation. The older children were used in some districts to drag along the coal in trucks or baskets to the pit's mouth. Although to a modern observer the conditions appear terrible, three out of the five Commissioners who investigated the condition of women and children in mines in 1842 reported that the general health of the children was good and ascribed it to the good food they obtained as a result of the high wages.‡ The bad conditions in mines were not tackled till 1842.

*It is difficult to see how long hours could have been worked in winter when artificial light other than lamps and candles scarcely existed. In 1833, in the Report on Children in Factories, some factories in Cheshire were lit by gas, but in the majority of cases where the lighting is mentioned it is usually candles or lamps. Gas seems to have been very unusual.

†" It appears that of all the employments to which children are subjected those carried on in the factories are amongst the least laborious and of all departments of indoor labour amongst the least unwholesome. Hand loom weavers, frame work knitters, lace runners and work people engaged in other lines of domestic manufacture are in most cases worked at an earlier age for longer hours and for less wages than the body of children employed in factories." Report of the Commissioners to enquire into the Condition of Children in Factories, XX., 1833, *p.* 51.

‡Report of the Assistant Commissioners to enquire into the Condition of Women and Children in Mines, 1842, XVI., *pp.* 143, 191-192, 345. " The collier children always look well and the medical evidence abundantly proves their general good health . . . the children after their day's work appear as playful as school boys come out of school." This was not true, however, of the East of Scotland (*p.* 396). " Nevertheless the contrast is most striking between the broad, stalwart frame of the swarthy collier as he stalks home all grime and muscle and the puny, pallid, starveling little weaver with his dirty white apron and feminine look. There cannot be stronger proof that it is

After the passing of the Factory Act of 1833 there were grave difficulties. Of the 40,000 children thrown out of work,* not one in a hundred was sent to school, as their parents, now that the children's earnings had stopped, were less able to pay school fees than before. They tried to get the children employment in factories not under the Act, such as bleaching or dyeing or sent them into collieries. If they could not do this the children ran wild about the streets. They often lost the benefit of the industrial training and got no other. Where they did get taken on in other factories this flood of children lowered the wages of those already working in them. It was quite obvious that an educational system would have to be devised to deal with the child till it was nine. The dame schools in garrets and cellars seem to have been a much worse alternative than the factory.†

The ultimate effect of the factory system as regards children was to make their condition decidedly better. Their exploitation in the home became more difficult when they had to be in school a large part of their time and their overwork in the factories was stopped by the law and the inspectors. Their wretched condition in the mines was brought to light and drastically dealt with in 1842, females not being allowed underground at all and boys only at the age of ten.

not muscular exertion which hurts a man." *p.* 193. "The medical men almost unanimously agree that there is no disease peculiar to collier children." (*p.* 194.)

*Hand Loom Weavers, *ib.*, 1840, *p.* 686.

† "With few exceptions the dame schools are dark and confined. Many are damp and dirty, more than one half of them are used as dwelling, dormitory and school room, accommodating in many cases a family of seven or eight persons ; above forty of them are cellars. Of the common day schools in the poorer districts it is difficult to convey an adequate idea ; so close and offensive is the atmosphere in many of them as to be intolerable to a person entering from the open air, more especially as the hour for quitting school approaches. . . . Mr. Wood . . . notices particularly a school in a garret up three pair of dark, broken stairs with forty children in the compass of ten feet by nine, where on a perch forming a triangle with the corner of the room sat a cock and two hens ; under a stump bed immediately beneath was a dog kennel in the occupation of three black terriers whose barking added to the noise of the children and the cackling of the fowls on the approach of a stranger was almost deafening. There was only one small window at which sat the master obstructing three-fourths of the light." Report on the Sanitary Condition of the Labouring Classes, Appendix. Liverpool, 1844, XVII.

The factory restrictions led eventually to the State education of the child, the scope of which is ever widening as is also the scope of the protective measures designed to help children. Indeed, the twentieth century promises to be the century of the child. We have seen that by 1820 the factory system was not developed to any great extent except in cotton and in the spinning of wool, worsted and linen, therefore the effective enforcement of the Factory Acts accompanied the spread of machinery in the textiles and gradually mitigated its abuses in other trades as machinery spread to them.

Probably the married woman was the most unfavourably affected. Spinning was the first thing to be revolutionized in each textile trade and she was the spinster—the producer of yarn. She had, if married, to choose between remaining in the home and becoming dependent on what the man chose to give her out of his wage or going to work in the factory and leaving the home. The man's wage had not been fixed on the basis of keeping a family. It seems to have been taken for granted that the woman would largely support herself and that the children would largely support themselves.* When the woman's earnings declined the family income was insufficient, the man's wages were not adjusted to meet the deficit and the woman who does not like to skimp the children and dare not skimp the wage-earner is always the one to go short first. The family income became so small when the women's and children's earnings declined that the Poor Law had to step in, especially in the South of England, and wages were subsidized. The reform of 1834, by cutting off the doles, necessitated the readjustment of wages to a point where the man could maintain a family.†

Much of the success of the domestic worker had depended on the fact that he could control a cheap labour supply in

*This facility which once existed for a mother and her children to procure sufficient money to support herself and the family without assistance from the father was naturally productive of early marriages. It was a certain income to the husband and his own wages he employed for his own immediate purposes, be they good or be they dissolute. ." The Commissioner speaks of the wife and children being " now " unable to support themselves "and the husband is obliged to maintain them out of his wages." Hand Loom Weavers' Commission, 1841, X., p. 46.

†On the rise of wages in the nineteenth century, see Bowley, " History of Wages in the Nineteenth Century."

his wife and children or apprentices. This laid the work of wife and children open to considerable sweating and very long hours were worked. It is possible that the dependence of the wife on the husband received a " set-off " in the fact that she had only domestic duties to perform and had not got to earn her living as well. If she did choose to go into a factory it may well have meant a welcome relief and change from the home.

On the other hand, the factory system probably meant more independence for the young unmarried woman.* She could now go into the factory and earn her own living apart from the family and get her own money to herself. Probably neither the single nor married woman regarded it as a hardship to go to the factory. In country districts, however, where there were no factories, the only alternative to domestic production was domestic service or field work and these alternatives to home work were unpopular. Women are social gregarious creatures and factory life seems to attract them. After all " the home " is not always a little six-roomed rose-covered cottage but is often one or two over crowded rooms in a city slum and to go out to work makes a " bit of a change."

One must always guard against imagining that things were specially bad because people complained. It is often the sign of an awakened public conscience when an enquiry is made and searchlights turned on certain trades by Commissions and it does not always mean that conditions have

*" One of the greatest advantages resulting from the progress of manufacturing industry and from severe manual labour being superseded by machinery is its tendency to raise the condition of women. The great drawback to female happiness among the middle and working classes is their complete dependence and almost helplessness in securing the means of subsistence. The want of other employment than the needle cheapens their labour, in ordinary cases, until it is almost valueless. In Lancashire profitable employment for females is abundant. Domestic servants are in consequence so scarce that they can only be obtained from the neighbouring counties. A young woman prudent and careful and living with her parents from the age of 16 to 25 may in that time save £100 as a wedding portion. I believe it to be the interest of the community that every young woman should have this in her power. She is not then driven into an early marriage by the necessity of seeking a home ; and the consciousness of independence, in being able to earn her own living, is favourable to the development of her best moral energies." Hand Loom Weavers, 1840, p. 682.

suddenly become worse. The evils become apparent when an enquiry is held but what one cannot judge is the number of those who are not affected at all or who are affected favourably. People who are prosperous do not often say so for fear of attracting the tax-gatherer or encouraging rivals to set up. It is always difficult to realise that those who are articulate are often a minority. Englishmen are always despondent about their own times, and it would be easy to quote contemporaries in every period so that their testimony would show that we had gone downhill ever since the time of the Norman Conquest. One must remember that there were no police and that when riots occurred the militia would be called out to suppress the rioters. Therefore the calling out of the soldiers has nothing like the significance it would have to-day. Nor had people any other mode of expressing their sense of serious discontent than by rioting. They had no recognized trade unions by which they could bargain for better terms, they had no votes with which to bring political pressure to bear. Acts of violence were, therefore, the only means of drawing attention to their grievances. What is so difficult to measure is the proportion between those who have given expression to their grievances and those who are perfectly satisfied.*

It is this silence of the well-to-do that makes it difficult to judge the extent to which the middle classes were affected by the industrial revolution. They certainly increased in numbers and wealth. Many working men rose from the

*One would imagine, for instance, from reading the figures given in the " Labour Gazette," January, 1914, *p.* 6, that Great Britain was in a state of continual industrial turmoil in 1911, 1912 and 1913

	No. of disputes begun in each year.	Total number of work-people involved.	Duration in working days.
1911 ...	903	961,980	10,319,591
1912 ...	857	1,463,281	40,914,675
1913 ...	1,462	677,254	11,491,000

Yet the official comment is: " There have now been three years of good employment. During these three years of an exceptionally active demand for labour, rates of wages have improved in most industries so that the workers have benefited both by a greater volume of employment and by higher rates of pay."

ranks and recruited the middle classes.* To mention only a few : there were Arkwright, Strutt, Peel and Owen in the cotton industry ; Fairbairn, Nasmyth, and Maudslay among the makers of machinery ; Crawshay, of Cyfartha, and Wilkinson among the iron-founders ; Neilson, the inventor of the hot blast ; Gott and Horsfall among the woollen magnates ; Brindley, the canal engineer, and George Stephenson, of locomotive fame.† There was something dramatic about the rise of these men from mere workmen to great captains of industry. We do not, however, hear much of the great merchants who necessarily existed to dispose of the new manufacturing output,‡ of the builders who must have been in great request to erect the new factories and the new towns. A new race of contractors arose to make the canals, the new docks and the railways of whom Brassey stood out pre-eminent. A new race of coal merchants emerged to furnish the new towns and factories with the new fuel,§ while the shop-keeping class necessarily increased in all towns.‖ Banking also developed with the increase of business.

It is interesting to speculate on the effect of this increase

*Many of the rich manufacturers of Stockport, Hyde and Staleybridge were said to have been " hatters, shoemakers, carters and weavers." Gaskell, " Artisans and Machinery," *p.* 32. *See* also Ashton, Chapter IX., " The Iron Masters," *p.* 210.

† " Dictionary of National Biography," also Smiles' " Invention and Industry," " Industrial Biography," " Lives of the Engineers."

‡For the rise of a new merchant class in Bradford where Scotchmen and Germans migrated to undertake the disposal of Bradford goods, *see* M. Law, " History of Bradford," *p.* 195.

§This was not true of London where the existing coal merchants owned the wharves for unloading the coal ships, all coal being water borne. It was almost impossible for new coal merchants to set up in London till the railway era. The railways provided trucks and storage if required, while daily deliveries of truck loads could be obtained and men with small capital could then undertake coal merchanting and the number of London coal merchants increased rapidly. This information has been kindly supplied to me by Mr. Dale.

‖" The origin of a manufacturing town is this : A manufactory is established, a number of labourers and artisans are collected—these have wants which must be supplied by the corn dealer, the butcher, the builder, the shop keeper—the latter when added to the colony have themselves the need of the draper, the grocer, etc. Fresh multitudes of every variety of trade and business whether conducive to the wants or luxury of the inhabitants are superadded and thus a manufacturing town is formed." Guest, *op. cit.* (1823). *p.* 4.

.n numbers and wealth of the middle class on their women-kind. Among the bigger merchants, manufacturers and shop-keepers the place of business became divorced from the home. The man began to sleep in one place and work in another. When the home was over the shop the wife helped to run the business and marriage was as much a business partnership for the shop-keeping and trading class as it was for the artisan or farmer classes. With her isolation from the business the woman lost touch with affairs, her life became narrowed, if less strenuous. When the children went to a boarding school " to finish," or grew up, she was often con-demned to a tea-drinking, fancy-work, district-visiting existence after a few crowded years of child rearing.

With the growth of capital it became far more difficult for a woman to set up in business for herself, and as she was deprived of any training in business, widowhood often meant ruin. She could not " carry on " when the man died. It is often not realized how prevalent widowhood is. According to the Census of 1901 one woman in every eight over the age of twenty was a widow.*

While many women no doubt lost greatly by being " out of business " others probably greatly preferred the gentility of being a " real lady " with nothing to do. This explains the early Victorian women, the viragos and the sugary nonentities as portrayed for us in Dickens, Thackeray or Miss Austen. The viragos were women who had not sufficient outlet for their energies, and the Esthers and Amelias were so genteel and amiable and futile because they had never been brought into contact with real life.

While there was some advantage as well as loss to the artisan in the change from home work to factory there is little doubt that from the point of view of the work turned out it meant not merely a larger output but a greater efficiency in production. The country weaver was apparently neither a first-class weaver nor a first-rate farmer. It was said in 1840 that the Scottish manufacturers would not put out work with the hand loom weavers in the North of Ireland, although the work was done cheaper there by a third than in Glasgow because the country weavers were so inefficient.†

*B. L. Hutchins, " The Working Life of Women " (1911), *p.* 7.
†" Weaving in Ireland is less an exclusive occupation than in this

The same thing was true of agriculture. Gaskell, after expatiating on the great advantages to the weaver of the occupation of a few acres of land, says, "It cannot indeed be denied that his farming was too often slovenly and conducted at times as a subordinate occupation and that the land yielded but a small proportion of what under a better system of culture it was capable of producing."*

Great Britain was outgrowing her own food supply ; import in the absence of railways was very difficult and expensive and bad farming was a national danger. Nor had the spinners been able to cope with the growing demand for yarn.

As the output of yarn increased the price dropped. It was said in 1840 that the cost of spinning had declined from 1s. 2d. a pound to 1d. per pound, and that the machine yarn was better than the hand spun.† This in its turn meant a cheaper cloth, larger sales and wider markets.

It must not be forgotten that the effect of the industrial revolution was to enable Great Britain to survive a war lasting twenty-two years (1793-1815) in which she not merely bore the brunt of the sea fighting but subsidized a large part of Europe against the French. Napoleon's difficulty lay in the fact that the English were so un-get-at-able once the French sea power was broken. He therefore set out to destroy the source of Great Britain's prosperity, her overseas trade. Her increased production by machinery made it absolutely essential that she should have large markets ; she was organized for world production and it was the capital made in this wide-spread trade that enabled her to buy allies and focus the opposition against the French Emperor. Napoleon did not merely wish to destroy the sources of British economic strength, he wished to set France up in her place and to make France the great industrial nation of Europe as she had been prior to 1789.

French industry had been ruined by the collapse of credit

country. The vast majority are small occupiers of land and the weaver who is one day driving his pig to market, another digging his potatoes and a third getting a job at harvest work can neither become an expert at the loom nor as careful a workman as one living in a large manufacturing town and giving his whole attention to the same operation." Hudson's Report, Hand Loom Weavers, 1840. p. 647.

*" Artisans and Machinery," p. 12.

†Hand Loom Weavers, 1840, p. 370.

and economic life generally during the ten years following the Revolution. Napoleon had to reconstruct France industrially ; he was anxious to start machine production partly to beat the English at their own game, partly to combat unemployment in France. If goods were to be produced by the French in quantities Napoleon needed markets. To keep British goods out of Europe would give the reviving French industry an enormous opportunity. By his " continental system " Napoleon tried to exclude British goods from Europe, hoping thereby to ruin his antagonist and make France her industrial heir. The struggle between 1806 and 1812 was largely a struggle by France to wrest the industrial supremacy from Great Britain.

It was the development of the cotton industry that enabled Great Britain to find compensating markets in tropical and semi-tropical countries for her products when she partly lost the European market for woollen goods. There was always a large smuggling trade with Europe carried on in British goods but the risks of capture and the detours of routes* so greatly enhanced the price that if Great Britain had not been able to produce cheap yarns by machinery and therefore cheap cloth it would have been almost impossible for the continent to have bought her goods. They must have bought French goods instead. Therefore instead of trying to keep goods out of Europe, as was the object of the blockade in the war of 1914-1918, Great Britain tried to thrust them in. Every sale she made tended to limit the markets for French goods.

Napoleon failed to accomplish the ruin of Great Britain, so strong was her economic position, and she emerged in 1815 the work-shop of the world, the forge of the world, the banker of the world and the world's greatest carrier. She had fallen heir to the shipping supremacy of the Dutch and the industrial leadership of the French.

The advantages of the new methods may be summarised as follows :

Mechanical production increased Great Britain's wealth and power and tended to make her the foremost industrial power in the world. It enabled her to hold her own against Napoleon and in saving herself she saved Europe. It helped

*English cloth was at one time smuggled into Germany *via* the Balkans and sugar was sold at Mayence that had been landed at Riga.

her to lift the great burden of the war debt and lighten the effect of the crushing taxation to pay the interest. It provided more and not less employment, steadier work and better pay. The British artisan became unrivalled as a skilled worker and helped to train Europe. The price of textiles fell which enabled the British both to push markets abroad and provide cheaper, cleaner and healthier clothing at home for cotton was pre-eminently washable as wool was not.

In the transition stage there was an increase of truck, the pressure of children's work was probably increased, especially in the busy times. The horrible insanitary condition of London was repeated over and over again in the new factory towns or in towns like Glasgow and Liverpool that were rapidly increasing in size. Where the competition of home work entered into competition with machines the position of the home worker became desperate. Although the younger men could go into the factories the older men were probably thrown on to the scrap heap of the Poor Law.

After the transition was over conditions improved generally through the limitation of children's work by the Factory Acts and the development of sanitary reforms. Truck was gradually checked by Truck Acts and trade unions, and wages rose. Very few people can doubt that the home was not a suitable work-place and that to separate the home and the workshop was a great sanitary gain. The general type of artisan improved in skill and intelligence.

On the other hand, there were certain permanent disadvantages in the change. There was a great increase in the monotony of the work, there was a loss of independence, many employments such as iron-founding, chemicals and coal mining became more arduous and more dangerous. There was a large yearly toll of industrial accidents. There also grew up a number of unskilled jobs for mere boys that led nowhere, " blind alley " employments, which tended to manufacture paupers, not because they were " too old at forty," but because they were " no good at twenty-five." We have seen that there were two types of home workers, those living in the country carrying on a combination of agriculture and industry and those who worked either in the towns or in the country and were specialized in industry. The first class suffered a considerable loss in stability. They could no longer fall back upon their farm or garden in bad times ; the pure artisan probably gained, however, from the steadier work and higher wages offered by the factory.   But there was a general loss in stability owing to the loss of the family earnings which constituted a sort of family insurance. Finally, for good or for evil, Great Britain changed her character.   Instead of being an agricultural country feeding herself and exporting corn with one great industrial export, wool, for which she grew the bulk of her raw material herself and for the product of which there was a steady demand all over the world,* she became a country of engineers, iron works, coal mines, factories and chemical works.   She imported the bulk of her raw materials, both wool and cotton, and after 1870 she imported hæmatite iron for acid steel made by the Bessemer process.   She imported after 1846 the bulk of her food stuffs and relied chiefly upon the export of her manufactured goods to pay for the food imports.   The old self-sufficing stability of the country had disappeared and she relied during the nineteenth century to an ever-increasing extent upon foreign trade and exchange and upon the ability of new countries to grow her food and provide raw material.

Very little was done to improve conditions in England till the thirties, when a great era of reform took place.   The brain of the country was absorbed by the French wars and their after effects.   No one knew whether France might not break out again.   Europe was seething with revolution and these political events absorbed the attention of statesmen to the exclusion of economic reform.   One must remember that the changes had been going on for about a hundred years if one dates them as beginning with Newcomen's steam engine in 1712, and they probably did not appear to contemporaries as revolutionary as they do to us.   Nor had statesmen, any more than economic or industrial experts, any idea what to do to improve matters.   The art and science of engineering were only beginning, hence preventive measures against accidents in both mines and factories took a long time to develop.   The huge mass of reports and inquiries with which the period between 1815 and 1840 abounds bears witness to the desire

---

*There were fluctuations in particular places owing to wars.   Norwich suffered owing to the wars with Spain in the eighteenth century and also felt severely the loss of the American colonies.

of the ruling classes to find out the real state of affairs and if possible provide remedies. One is always struck by the fact that the remedies proposed were so futile. They did not really know what to do. Sanitary engineering, too, was in its infancy and the modern iron drain-pipe had not yet been evolved. It did not seem to be possible to make arrangements for an adequate water supply or for the wholesale reconstruction of houses and streets involved in a system of main drainage. Besides, no one believed that it would do any good if all this were done and until the municipal Reform Act of 1835 it was impossible to get town authorities vigorous enough to take the matter in hand. Before any sanitary reform could be really effective three things were necessary. They had to know what to do ; they had to have the appliances, such as drain pipes, to carry out the schemes and they needed special powers to carry them out on a wholesale scale. It was difficult to get all three things to combine. The doctors began to have some idea how to fight the fevers ; the difficulties of the engineering problem and the state of the municipalities held the reforms back. Nevertheless the population both in town and country increased rapidly, though as one reads the sanitary reports it seems difficult to conceive how anybody was left alive at all.* Only in 1848 was a beginning made by the appointment of a Board of Health to tackle the problem and it almost at once fell foul of the engineers. Little was really done in the matter of effective sanitation till after the Crimean war. The appalling loss of life and the reforms instituted by Miss Nightingale advertised the fact that sanitary measures could really do some good and the science of preventive medicine developed hand in hand with sanitary engineering. In this way Great Britain invented for the world the way to live healthily in the mass in towns. It was difficult, however, to convince people that " something ought to be done." They were intensely interested in the state of their own and other people's souls, and they were by no means sure that hard work, disappointment and privation were not designed by an all-wise Providence to wean people from desiring to remain in this

*On the declining death-rate see an article in *The Edinburgh Review*, 1827: " Rise, Progress, and Present State of the Cotton Manufacture."

Vale of Tears. It would be " made up " to them hereafter when they lay safe in Abraham's bosom.

It was thought to be good discipline for people to work long hours. " Satan finds some mischief still for idle hands to do," was not merely embroidered on samplers but was a warning to be acted on while the " busy bee " improving " each shining hour " seemed to be a model person, and no one reflected that her end was to be smothered in a straw skep so that someone else might enjoy the honey. Illness equally was regarded as good for one and a discipline from the Lord that it would be impious to prevent. Parents also honestly believed in the dictum of Solomon that " to spare the rod was to spoil the child," and that whipping and disappointment were a good preparation for bliss hereafter. People were very hard on children generally, often from the kindest motives.* The evils of overwork, crowding and child labour were not new only a little more obvious, and after all " people had lived through it before." There was a fatalistic attitude abroad. The economists taught that *laissez faire* was the only possible plan that would be fair to all alike. Each man would then seek out his own welfare and attain it if left untrammelled and the result would be the greatest happiness of the greatest number. This was believed to be the ideal quite as firmly in Revolutionary France as in Conservative England. In fact, no one had the knowledge at first to cope with the new conditions. Only gradually did English statesmen work out for the United Kingdom and for other countries the way to combat the new evils through inspection, Factory Acts, Health Authorities and main drainage.

Thus the modern system of drainage which has made the modern system of living together in masses a healthy possibility took some decades to devise. The requisite appliances had to be invented and made, local authorities needed to be stirred to action involving so much expense and reconstruction and people had to be educated to appreciate its importance.

The same difficulties were encountered in mining. When the mine owners of Northumberland and Durham got Sir

* *Cf.* The boyhood of Sir Walter Scott in Lockhart's " Life." The first six children born to Sir Walter Scott's parents all died, which shows that child mortality was not confined to the poor as Sir Walter Scott's father was a Writer to the Signet.

Humphrey Davey to invent a safety lamp it proved a great boon but did not involve much capital outlay. Other matters such as ventilating and double shafts involved large capital expenditure and were a doubtful venture. The knowledge of mining engineering was very defective ; experts contradicted one another and a Committee which sat in 1835 to enquire into the dangers in mining and to propose remedies frankly stated that they did not know any remedies to recommend.* In factories, too, when it was so difficult to get engineers to make the machines it was equally difficult to get them to make appliances to safeguard machines. The theory was honestly held that the longer hours people worked the larger would be the output and that the profit was made in the last hour. It took a long time to make people see that long hours did not pay in the long run while "industrial fatigue" and its effects have only been scientifically studied during the recent war. As a matter of fact what actually happened was that the power of dealing with inanimate objects by mechanical means increased, man's power over Nature increased, his power as a productive animal enormously increased but his new powers outstripped his power to remedy the evils that the changes brought. His capacity for production increased out of proportion to his capacity for securing the welfare of the human instrument engaged in that production. It is the problem of the twentieth century to invent a social mechanism to promote human welfare which shall correspond in power to the industrial mechanism of last century. But it must be remembered that machinery and mechanical transport have enabled the working classes to command a variety of food, a standard of clothing and a possibility of change such as would not have been possible for even princes and nobles three centuries ago.

*1835 V., p. VIII. " On a review of their labours your Committee cannot but feel apprehensive that they have in great measure failed in devising adequate remedies for the painful calamities they have had to investigate ; they entertain notwithstanding a sanguine expectation that the attention of the public will be availingly turned to this interesting subject . . . The great dissimilarity of the mineral stratifications of the kingdom, the constantly varying circumstances of particular mines render it in their opinion impossible at present to lay down any precise directions or to form any rules of universal application."

SUMMARY

The general features of the industrial revolution in every country have been a decline of handwork and domestic production relatively to machine work in factories, the separation of agriculture and industry as two separate pursuits, the growth of towns, the cleavage of social classes, the rise of huge businesses and monster impersonal corporations, the growth of education under the supervision of the State and the authoritative regulation of industrial conditions so as to fix a minimum standard.

Although the same features are encountered in each country where the industrial revolution takes place, the sufferings of the readjustment were probably more severe in England than in any other country. In the first place by the time other countries had started England had worked out methods of dealing with the most obvious of the evils. In other countries, too, the transition did not take place in the middle of a great war lasting twenty-two years (1793-1815) with only a short break. The after effects of the struggle in high taxation and an exhausted continent unable to purchase except on a very restricted scale were felt in the United Kingdom till 1840.

It was Great Britain that experimented with machines and bore the burden of the failures. Other countries were able to begin where she left off. It was Great Britain who worked out through her own doctors and sanitary engineers the way to live healthily in the mass in towns. It was Great Britain who invented the factory inspector,* the great instrument of State regulation and control.

For the first three-quarters of the nineteenth century although France was the second great industrial nation and a rival of whom the English manufacturers were always afraid, the British were the great world manufacturers and traders. Great Britain continued to influence the whole of the economic development throughout the world during the nineteenth century by her inventions in the technique of manufacturing and transport. The raw material producing countries were drawn upon to provide material for the workshops of Northern Europe, the food producing countries were opened up to feed the population grouped on the coal and iron line of Europe. This great producing area looked in

*Inspectors of the *quality* of the goods had existed in France from the time of Louis XIV., but not of the social condition of the worker.

its turn to the rest of the world for its market. America, Asia and Africa became focussed upon Europe and the result when the mechanical transport developed by Great Britain had provided swift and rapid connection was to make the world one great market.

The sixteenth century was the century when Spain swayed the economic destinies of Europe, the Indies and the Americas. The seventeenth century belongs to Holland with her vast exchange business and shipping, the eighteenth is the century of France with her great industrial, commercial and colonial development, and her leadership in ideas, but the nineteenth century is the century of the predominance and world-wide influence of this tiny island on the outskirts of Europe.

PART III

INDUSTRIAL AND COMMERCIAL POLICY IN GREAT BRITAIN DURING THE NINETEENTH CENTURY

SYNOPSIS

I.— LAISSEZ-FAIRE AND THE REACTION

1—1793–1815. Period of the French Wars.
2—1815–1830. Period of social distress due to after effects of the wars. Heavy taxation to pay the interest on the National Debt. Spread of machinery.
3—1830–1850. Period of Reforms.

    (a) Government regulation of industrial conditions. Tory Humanitarians led by Shaftesbury.
        1—Factory Act, 1833. Creation of inspectors.
        2—Truck Act, 1831. Payment of Wages regulated.
        3—Coal Mines Act, 1842. Prohibition of female labour under ground. Minimum age of 10 for boy miners.
        4—1844
            1847 } Factory Acts included women. Hours fixed.
            1850
        5—1848. Board of Health created.

    (b) Repeal of restrictions—Laissez-faire—The Benthamites.
        1—1825. Trade unions permitted with certain reservations.
        2—1825. Emigration free.
        3—1825
          and } Export of machinery permitted.
         1843
        4—1834. Reform of the Poor Law. Free movement.
        5—1835. Reform of the municipalities.
        6—1826. Joint stock banks other than the Bank of England.
        7—1855
          1862 } Joint Stock Companies with limited liability.
        8—1833
          1854 } Repeal of Usury Laws.
        9—1822–24 } Reduction of the tariff and the abolition
          1842–46 } of corn and meat duties.
          1853 } Establishment of " free trade." Tariff for
          1860 } revenue only.
      10—1822–25 } Repeal of shipping restrictions and conse-
          1849 } quent freeing of the colonial trade.
          1854 }

In Great Britain the free traders were the manufacturers and the land-owners were the protectionists : on the continent the reverse was the case.

The years 1844–1846 mark the crucial period of Government regulation of industry and free trade in commerce.

(c) Growth of labour movement. Trades union movement, 1834. Chartist movement 1838–1853. Both unsuccessful.

(d) The coming of the railways.
1—Great increase of employment.
2—Stimulated large businesses.
3—Stimulated agriculture.
4—Stimulated iron and steel trades.

4—1850–1873.   The Good Years.

(a) Great expansion of trade especially metallurgical industries. Large foreign demand for rails and machines. The victory of the iron steamship. Increased demand for coal.

(b) Rise of prices and greater rise of wages. Formation of trade unions on a national basis. Factory Act, 1867, included work-shops and non-textiles. Successful development of co-operative movement, 1844.

(c) Great Britain's monopolistic position due to inability of France to compete in the metallurgical trades and to wars on the continent and the Civil War in the United States. Great Britain supplied the belligerents and financed constructional works in all parts of the world with the result that orders for equipment and renewals came to this country.

(d) The development of municipal activities. Four possible correctives to the power of capital emerged during this period—the co-operative movement, the trade union movement, increased factory legislation, municipal action.

5—1873–1886.   The Great Depression.   General fall in prices.   Currency changes.

(a) Agriculture—depression due to wheat and meat imports which were facilitated by railway and steamship.

(b) Shipping—depression due to competition between steamship and sailing ship, the Suez Canal, new types of steamships.

(c) Iron and Steel trades—depression due to Bessemer process with reconstruction of the iron industry.

(d) Distributing trades. Alteration of trade routes due to the Suez Canal and the railways. Direct shipments to the continent.

(e) The new German rivalry.

6—1886–1914.   Reaction from Laissez-faire.
1—New Colonial policy—constructive imperialism.
2—New era of labour legislation.

3—New policy for children.
4—Extended powers for trade unions, 1871-76 and 1906.
5—Rise of trusts.
6—New agricultural policy in Ireland and Great Britain.
7—Transport. Fixing of railway rates, 1888-1894.
8—Commerce. Commercial education ; provision of commercial intelligence ; consular service reorganised.
9—Increase of municipal activity : tramways, electricity, housing.

Causes of the growth of State action :

1—The great depression.
2—German influence.
3—Intensified competition due to railways.
4—The labour movement.
5—Influence of railway amalgamations.
6—Science—health movement—scientific investigation.

Increased expenditure met by Income Tax and Death Duties.

II.—CAUSES OF THE SUPREMACY OF GREAT BRITAIN DURING THE NINETEENTH CENTURY

1—Pioneer of the new industrial technique—long start.
2—Abundant supplies of coal and geographical situation of the coal fields.
3—The development of shipping.
4—Financial connections all over the world.
5—British investments abroad.

III.—GROWTH IN WELFARE OF THE WORKING CLASSES DURING THE CENTURY

IV.—THE CONTRAST BETWEEN THE INDIVIDUALISM OF GREAT BRITAIN AND FRENCH AND GERMAN PATERNALISM

British economic development in the nineteenth century owed nothing to State aid. British individualism and French and German paternalism date from the seventeenth century. The survival of serfdom and prevalence of feudalism made for paternal government in France and Germany. Their early disappearance in Great Britain made for individual action. The influence of Puritanism worked in the same direction. The poverty of the English monarchs made it difficult for them to adopt a constructive policy involving the expenditure of money. The tradition of laissez-faire was reproduced in the United States. Different evolution in Ireland.

V.—THE ECONOMIC POSITION OF GREAT BRITAIN IN 1815 AND 1914.

I.—LAISSEZ-FAIRE AND THE REACTION.

IN looking at British economic development during the nineteenth century we can discern two main threads running through it.

In the first place the nineteenth century is remarkable for

the great growth of the power of capital. Starting with ordinary partnerships or one-man firms, capital became mobilized in joint stock banks and in joint stock companies with limited liability. These companies were able to undertake works on a scale hitherto undreamed of ; they were able to look ahead and sink money and wait for returns ; they had behind them great reserves of capital and could usually get more from their shareholders or by new issues.*
Then with the coming of mechanical transport and the shrinkage of the world as a trading area through the development of rapid communications afforded by the telegraph, railway and steamship, the companies began to amalgamate and combine, businesses grew larger and were carried on in several countries, the power of capital was enormously increased, it became international in scope and outgrew the boundaries of national systems.

To this growing dominance of capital there developed certain correctives or limitations. In the first place trade unions increased in strength, the national societies coincided with the growth of joint stock companies and played, as far as Great Britain was concerned, an ever larger part in politics.†
They were instrumental in obtaining a minimum wage, shorter hours and the proper enforcement of the law as it existed with regard to truck and sanitation. They also provided a mechanism for bargaining.

The co-operative distributive societies, which had been the subject of unsuccessful experiments during the earlier years of the century, were started successfully upon what is known as the Rochdale system in 1844. They aimed at eliminating " profit on price " by selling at current rates and by returning the profits periodically to the purchasers in proportion to their purchases. With the extension of these societies, each of which enjoys complete local automony, the necessity for a central purchasing agency arose. They therefore formed a federation of societies for this purpose and the Co-operative Wholesale Societies were started in

*The number of companies carrying on business under the Companies' Acts, was stated in 1906, to be more than 40,000 and the capital exceeded £2000 millions. Cd. 3052, 1906.

†In 1918 the members in 1,200 unions were said to be 6,620,000, of which 5,400,000 were men and 1,220,000 were women. Labour Gazette, January, 1920, p. 7.

1864, which returned the profit to the shops purchasing from them in proportion to their purchases. The surplus the shops gained by having the dividend returned to them by the Wholesale Society was handed on again to the customers at the shops. From purchasing the Wholesale Society started production in their own factories and farms. The underlying idea is to eliminate the capitalist and establish a new system based on co-operation. This is an instance of the mobilisation of capital by consumers as opposed to the mobilisation of capital by producers.*

The development of municipal activities after the reform of the municipalities in 1835 provided another possible corrective or limitation on capital by setting up a monopoly of certain services, the profits on which were supposed to be given back to the rate-payers in cheaper services or in reduction of the rates.

The State itself could not see this growing power of capital without imposing certain restrictions on its exercise and thus there developed during the nineteenth century a great body of Company Law laying down conditions under which companies should work so as to ensure publicity and honesty. A great industrial code was devised setting up a minimum below which no worker should fall in hours, sanitation, or safety, the scope of which was gradually enlarged to include wages.

During the nineteenth century it has been the policy of the central government to restore something of the security of life imperilled by the industrial changes, hence old age pensions, compensation for accidents, sickness insurance, labour exchanges to dovetail work if possible, and unemployment insurance. As apprenticeship, the old technical education, became less and less important with the coming of machinery, a new system of general and technical education had to be devised to take its place. Except in connection with the Post Office and Savings Bank the State did not, before

---

*The number of members in the 1,362 distributive trading societies was 3,547,567 in 1916; their sales were £125,363,364, the surplus on the year's working was £16,650,576 and the dividends returned to purchasers were £13,394,854. The Co-operative Wholesale Societies had 2,106 members in 1916, to whom their sales were £66,732,485 and their surplus on the year's working was £2,365,141. (Royal Commission on Income Tax, Cmd. 288-5. Fifth Instalment.)

1914, engage in national trading although the municipalities had started various forms of municipal trading.

The second great line of thought running through the nineteenth century is to be found in the fact that at the beginning of that century the general theory was that laissez-faire, involving free trade and free contract, was the only possible policy for a government to adopt or for a free people to live under.   It was held to be unfair to deprive the worker of the right to sell his labour as he liked.   The growing class of manufacturers had become more important than the old landed interest and it was undesirable to hamper them in their activities as they would have to pay a large proportion of the taxation after the war.   Between 1815 and 1840 trade was stagnant.   It was to the new captains of industry that men looked to pioneer a revival of trade and an increase of employment and nothing must be allowed to hinder the free exercise of their energies in production, in the opening of new markets and in the expansion of the old. The economists seemed to teach that wages could not be raised either by any action of the government or by any combination on the part of the workers themselves.   The capital in existence was a result of saving in the past ; that capital was divided between plant, labour and raw materials. If labour temporarily got more than its share, then plant or raw materials would suffer, fewer people would be employed, wages would decline and the " Wages Fund " automatically sink back to its original sum.   As the amount could not be increased, if either by legislation or combination one class got more, then some other class must get less.   It was robbing Peter to pay Paul to interfere.   Under the circumstances no government felt that it could rightly interfere and the conditions which mitigated the doctrine of the iron law of wages were not obvious at the time.

On the other hand it was also generally believed that if wages were increased or if the price of corn went down, more marriages took place, more children were born and there was a greater struggle for existence because more people existed to scramble for a share of the " Wages Fund." It made people afraid to do anything in the way of philanthropy lest they should increase that " devastating torrent of children."

What with the economists, who seemed to say that whatever

was done was bound to be futile, with the obvious misery of large families, the shortage in the food supply and the over-crowding that everyone saw around them and which again the experts seemed to say would only be aggravated by anyone doing anything to relieve it, one may pardon the early nineteenth century legislators for feeling that laissez-faire was the only possible line that would be fair to all. Meanwhile the growing commerce was hampered by an antiquated system of customs, so elaborate as to be absolutely unintelligible. Great Britain was making a bid for the trade of the world and to that end the greatest possible freedom of export and import seemed desirable. Her system of taxation had not been designed to reach the new rich and therefore a readjustment of taxation was in any case desirable. The substitution of the Income Tax and the Death Duties for the revenue obtained under the old protective system cleared the way for free trade. Everyone, so it was thought, benefited by exchange, since nobody parted with goods unless they wanted what they got more than what they gave away. The greatest happiness of the greatest number would be obtained by allowing free play for each person to attain his goal.

Certain exceptions were, however, admitted. A child was not a " free agent," therefore it should be protected by law. Women were poor, weak, helpless creatures ; they, too, must be protected and so there developed the beginning of an industrial code. Only very slowly and reluctantly did the legislature begin to include men. It was obvious throughout the century that the country had not been ruined by the restrictions on the work of women and children, that the most regulated industries, cotton and coal, had become the two leading export industries. It became clear that the limitation of hours had meant greater efficiency and that sweating was nationally improvident since the persons sweated were thrown early on to the rubbish heap of the Poor Law. It was obvious that the employer or joint stock company could afford to wait, while the workman could not wait any length of time, that he was ignorant of the state of the market for his labour and was at a disadvantage in bargaining. He had no knowledge of safety appliances and could not by himself insist on their being installed. He was himself often most reluctant to use them when provided.   There were always good factories and good employers and it did not pay to let them be driven out of the trade by bad employers.* Besides, it is the business of the State to think in generations and not in terms of momentary individual profit, therefore it must take long views and secure a healthy efficient population by preventive medicine and education.   As these things are costly and need experts it is impossible for individuals to provide them in sufficient numbers or on a large enough scale. Hence the State must step in.   Trade unions were given special powers to safeguard the interests of men under the Acts of 1875 and 1906, while the Factory Acts envisaged women and children.   After the depression of 1886 and the new rivalry with Germany the State began to legislate more and more for the welfare of the industrial classes.   Laissez-faire was gradually abandoned not merely in industry but in agriculture, in colonial relations, in Irish matters and in transport.   Even in commerce free trade was challenged.

There is a striking contrast between the belief in the efficacy of laissez-faire which prevailed for the first three-quarters of the nineteenth century and the growth of State intervention and State control which have been the characteristic of the last fifty years.†   Thus the growth of capital and its correctives and the complete change of national policy are two features which seem to explain the nineteenth century as far as Great Britain's internal development is concerned.

But one would get one's perspective wrong if one concentrated solely on national policy or the Labour movement or the organization of capital as the chief characteristic of the nineteenth century.   The dominant commercial position of Great Britain during the century is one of its most salient features.   During that period the world's trade pivoted on Great Britain.   She was the exchange place of the nations, the financier of the world, the developer of the undeveloped countries.   Her ships, sailing or steam, were to be met with at all times within the seven seas, she girded the earth with her rails and moved bulky commodities and food stuffs with her locomotives and steam engines.   She not merely

*Webb, S. & B., " Industrial Democracy " *Ch.* on the Economists and the Higgling of the Market.

†Dicey, " Law and Public Opinion in the Nineteenth Century."

manufactured for the world but she helped to open up continents, to move millions of people to new countries to grow new products and food.   She helped to make man the dominant power over nature.   She stood to the world for power, enterprise, constructive ability, capacity and financia' strength.   Incidentally during the century she became the centre of two great Empires, one containing coloured races and mainly situated in tropical or semi-tropical areas and the other consisting of people of her own race reproducing her own type of institutions, the first requiring paternal government, the other a common economic, legal and fiscal agreement on the basis of free allied peoples of one race and type.

The nineteenth century as far as Great Britain is concerned is a century of achievement.   Although no doubt her turn will come to decline relatively to that of some other Power she will have left an indelible impression on the world even greater than that of Rome.

(1) During the nineteenth century there are certain well-marked periods in English development.*   From the outbreak of the French Revolutionary war in 1793 to 1815 England was largely occupied by wars with France, yet during the whole period great internal changes were going on.   An agricultural revolution was being accomplished leading to the ousting of the peasantry and the first phase of the industrial revolution, viz., the adoption of machinery in cotton spinning and the accelerated development of coal mining, iron smelting and engineering, was proceeding at such a rate as to make cotton and iron goods the principal branches of English trade.   From about 1800 onwards machinery began to be increasingly used for spinning wool as well as cotton. Cotton was a new trade and comparatively few people were

---

*They may be summarized as follows:

i.—1793–1815.	The period of the French Wars. Beginnings of the industrial revolution.
ii.—1816–1830.	Depression and readjustment after the War. Laissez-faire.
iii.—1830–1850.	Period of Reforms.
iv.—1850–1873.	The good years.
v.—1873–1886.	The great depression.
vi. 1886–1914.	Abandonment of laissez-faire. The new rivalry.

ousted by machinery but with wool it was different. This was a trade that was spread all over the country and women's earnings were particularly affected. From 1815 to 1830 there was a great deal of unemployment. The war demand had stopped, the continent was exhausted and could not purchase largely, returned soldiers found a difficulty in obtaining work, many of them took to weaving and increased the number of weavers struggling for orders. Accordingly sweated rates* were paid for piece-work to the domestic workers such as weavers and frame work knitters and this tended to prevent the rapid spread of machinery. Labour was so cheap that it scarcely paid to put in machines. Cases are mentioned by the Commissioners enquiring into the condition of hand loom weavers in 1834 where employers gave up machinery and went back to hand labour as it was cheaper.

The period from 1815 to 1830 was one of deep depression. Taxation was crushing since the interest on the huge war debt had to be paid.

DEBT IN 1816.

	British.	Irish.
Funded - -	£772,764,937	£23,435,254
Unfunded -	£ 44,463,300	£ 5,304,992
Annuities and Interest funded -	£ 30,731,555	£ 1,323,795
TOTAL -	£847,959,792	£30,064,041†

The total debt amounted, therefore, to £878,023,833 in 1816. As the Irish could not possibly meet the interest on their portion of the debt it was added to the British debt in 1817 when the two Exchequers were merged. Thus the greater part of the interest on the Irish as well as the British debt

*The following shows the decline in the weekly earnings of home weavers given in the Report of the Select Committee on Hand Loom Weaving, 1835, xiii., p. 355-56.

				Lbs. of food.
1797–1804	-	26s. 8d.	-	281
1804–1818	-	14s. 7d.	-	131
1818–1825	-	8s. 9d.	-	108
1825–1832	-	6s. 4d.	-	83
1832–1834	-	5s. 6d.	-	83

†" Public Income and Expenditure, 1869," XXXV., p. 306.

had to be borne by Great Britain whose population in 1811 was 10,164,256 in England and Wales, 1,805,864 in Scotland. The Irish population was probably about 6 millions.

Say, the French economist, writing on " England and the English people " in 1815, says :* " We shall not be far from the truth in asserting that the government consumes one-half of the income produced by the soil, the capital and the industry of the English people." He goes on to point out that the enormous taxation made it impossible for the people to live without working.†

" The English nation in general," he says, " with the exception of those favourites of fortune (the great land-holders and rich capitalists) is compelled to perpetual labour. She cannot rest. One never meets in England professed idlers ; the moment a man appears unoccupied and looks about him he is stared at. There are no coffee houses, no billiard rooms filled with idlers from morning till night, and the public walks are deserted every day but Sunday. There everybody runs, absorbed in his own affairs. Those who allow themselves the smallest relaxation from their labours are promptly overtaken by ruin." He considered that the English were relapsing into barbarism in consequence.

Taxation after the War fell heavily on articles of consumption because the Income Tax was repealed in 1815, and with that went fourteen millions of revenue that had to be made up somehow. Hence articles were taxed and retaxed until there was an intolerable confusion in the Customs which gave a point to the agitation for free trade.

People were afraid to do anything for fear of making the lot of the working classes worse. They were up against the so-called law of the " Wages Fund " which seemed to say that capital at any time was fixed in amount, that labour's share of that capital was automatically fixed and that any gain by one class must be at the expense of another class. It would therefore, be clearly unfair for the government to help one class of workers at the expense of another. More-over, people believed that there was a law of population, the discovery of which was ascribed to Malthus, by which any rise in the standard of comfort would result in earlier

*p. 21 of translation (1816).
†p. 23.

marriages and in more children being born. These, it was held, would compete in their turn for work, wages would fall and the result would be an intolerable struggle for employment, only intensified by the temporary improvement in conditions.

(2) The fifteen years after the peace, 1815-1830, were years of great social difficulty. The Poor Law was strained by the Allowance system of doles in aid of wages, the working men were not allowed till 1825 to help themselves by forming combinations.* Trade was stagnant and a great agricultural depression set in which finally ruined the small farmers, already in difficulties, and accentuated the problem of unemployment by adding them to the numbers of those seeking work.

Lowe, writing in 1822,† reckoned that two to three hundred thousand men had returned from the war to be re-absorbed in industry; he also considered that no less than 100,000 domestic manufacturers had lost their work through the cessation of the demand for army clothing and armaments. There was no need, as has been observed before, for manufacturers to put in machinery, labour was too plentiful and cheap to make it worth while to adopt machines. This abundance of very cheap labour, however disadvantageous morally, did help the manufacturers to maintain the manufacturing and commercial supremacy of Great Britain by cheap production, as it had also helped them during the Napoleonic Wars. They could produce so cheaply that the cloth had been able to stand the extra expense of smuggling it inside the continental blockade set up by Napoleon, and the high tariffs after the Peace could not keep it out.

"After the Napoleonic wars," says Professor J. Shield Nicholson, "this country escaped bankruptcy and national ruin and eventually entered on a period of unprecedented prosperity. Amongst the causes and conditions favourable to this escape and recovery were :—The lead taken by Great Britain first in the great industrial revolution (and later in the great revolution of transport) ; the favourable balance of trade ; the low cost of labour ; retrenchment of further

---

*This was not peculiar to England. It was also a feature in democratic revolutionary France.

†Joseph Lowe, " State of England," *p.* 62.

expenditure ; the reduction of taxation of property and income ; security against social revolution ; the check to inflation ; and the speedy return to normal foreign exchanges after the war."*

But the labourer paid the price because it was he who was chiefly hit by the retrenchment of public expenditure which meant less employment. The abolition of the Income Tax only meant an increase of indirect taxation which it is admitted falls more heavily, in proportion, on poor people than on the rich, the security of the social order meant severe repressive measures, with the military frequently called out since there were no organized police forces. The check to inflation by the return to cash payments in 1819 did, however, assist the labourer considerably by helping to lower prices. On the other hand it is certain that national bankruptcy would have meant even greater unemployment and a slower recovery, since credit facilities, which enabled a continual extension of business, would have failed ; transport would have remained unimproved and the railways, which did more than anything else to increase wealth, exchange and employment, would probably have not developed so early.

It must again be repeated that to any one who studies this period closely it is not the coming of machinery but the after effects of the war that caused the social trouble of the decade after 1815. Not till 1840 did the burden really grow lighter, partly as a consequence of the reforms and partly as a result of increased employment, itself a result of increased business due to the railways and the recovery of Europe which began to buy machines and textiles in increasing quantities.

It is interesting to notice, however, that bad as the condition of the British domestic and agricultural workers seem to us to have been, it compared very favourably with that of the corresponding classes on the continent. A great enquiry was held into the condition of foreign labourers in 1834† and the verdict was that wages in England were nearly double those paid on the continent, that fuel was cheaper, clothing

*" Recovery after the Napoleonic Wars," *Glasgow Herald*, 1st October, 1919.

†Report on the Poor Laws, 1834, Appendix F., Part II., Vol. XXXIX.

less expensive and mortality lower in Great Britain than elsewhere.* Although food in this country was said to cost more than on the continent, the Englishman is represented as enjoying a much more generous diet as the continental labourer very rarely seems to have seen meat. " In the North of Europe the usual food seems to be potatoes and oatmeal or rye bread accompanied frequently by fish but only occasionally by meat. . . . The French returns almost exclude fresh meat."† In 1839 Symons, an expert observer who had been one of the Commissioners appointed to investigate the condition of the hand loom weavers here in 1835, remarked frequently on the inferior condition of French housing.‡ Nor do children seem to have been in a more favourable position on the continent under either the domestic or the factory system. We find from the answers sent in to the British Government in 1834 in reply to its questionnaire sent out for the purpose of ascertaining the condition of foreign labourers that it was quite usual for a child of five to contribute to the family income.

The following budget speaks for itself. Annual earnings of a labourer's wife and children in France :§

			£	s.
Wife, 120 francs -	-	-	4	16
Eldest boy, 80 francs	-	-	3	4
Child, 11 years, 50 francs -	-	-	2	0
,, 8 ,, 30 ,,	-	-	1	4
,, 5 ,, 20 ,,	-	-	0	16

The employment of young children, especially orphans, was also a feature of the development of cotton spinning in France. Nor were conditions in the early factories any better there. A French doctor reporting favourably in the early nineteenth century on the " paternal tenderness " of a certain employer mentions incidentally that the children got up at 5 o'clock, went to work at 5.30, at 9 had some bread and a half-hour's rest, at 2 o'clock they dined and had one hour off and then they continued at work until 8.‖

*Ib., p. cii.
†Ib., p. cii.
‡ " Arts and Artisans at Home and Abroad."
§Quoted Symons, op. cit., p. 54.
‖Schmidt, C. : " L'industrie cotonnière en France " in Revue d'histoire économique, 1914, p. 48.

The fact that these conditions were almost normal everywhere, even where machinery had not as yet penetrated, makes the efforts of the English reformers all the more remarkable. They were constantly confronted with the bugbear of this cheap foreign labour which might take to manufacturing and undercut English prices, and their opponents could correctly state that the English labourer was better off here than the corresponding labourer anywhere else. It was a powerful argument for laissez-faire. On the other hand, the discontent with and determination to improve the relatively favourable state of things here does give some idea of the strength of the humanitarian feeling of the time especially when it is remembered that no one could know then that the country would not be ruined by the experiment and that the economic experts predicted disaster. A great deal of the reforming impetus arose from sheer religious conviction. It was felt that with incessant work, people had no time to devote to spiritual things. Even the first Factory Act of 1801, regulating the labour of pauper apprentices, made statutory provision for the attendance of the children at church. Nothing is more striking than the way in which the early Commissioners always enquire into the " moral welfare " of the classes whose condition they are investigating.* It was not always the cruelty to the body but

*The following extract from the report presented by Commissioner Leifchild who enquired into the condition of women and children in mines in Northumberland and Durham (1842, XVI., *p.* 523) illustrates this point :

" It struck me as an astounding fact when more than one boy both in pits and iron works, being closely pressed upon the subject, confessed that their sole knowledge of sacred and awful terms was derived from their daily desecration at the works. The case of witness No. 586, aged 14, who states that he never heard of hell except when he has heard men swearing about it, was neither solitary nor altogether uncommon."

He continues : " The unbroken monotony of the duty in conjunction with its duration, and the darkness, solitude and other peculiarities of the scene of its performance, must at least blunt the feelings and deaden the intellect so as to diminish the capabilities of receiving instruction."

To us it seems quaint that people should consider a boy ought to have shorter hours to learn properly about hell but there was this strong feeling among sincere people at the time and it made them press on the reforms.

Compare also the Committee on Children's Work in 1816.

the starving of the soul that moved many people to embark on the risky path of regulating the hours of labour.

(3) The third period, from 1830 to 1850, witnesses a long series of reforms proceeding from two opposing quarters. One group of Tory reformers headed by Lord Shaftesbury believed in making experiments to see if some of the visible evils could not be stopped. Lord Shaftesbury himself was a deeply religious man and a pillar of the evangelical party, and many persons felt themselves quite safe in following his lead. He had no axe to grind, he was a good man and being a lord would not wish for equality as did those " mistaken persons," the Jacobins in France. He seemed, therefore, to many persons to be the safest of guides and this gave him a very strong position.* The result was the first effective Factory Act in 1833 the novelty of which lay in the fact that it created four inspectors. Children under nine had already been prohibited from entering cotton factories in 1819 but the law was evaded as there was no one whose business it was to see that it was kept. For instance, how can anyone tell that a child who is alleged to be nine, really is nine, in the absence of any birth certificate ? It was also useless to keep a child out of a cotton factory till it was nine if it could go into a mine or into a printing or bleaching factory adjacent at the age of five. Other industries would necessarily have to be brought into line and better records would have to be kept if the law were to be effective. The starting of the inspectorate is therefore epoch-making, it only meant at first State control of certain industries to prevent breaches of the law, it gradually extended its scope until in the United Kingdom the State regulates the conditions of work in all factories and workshops and to a greater or lesser degree in all industries and that not merely for children but for women and men.

It was not the least important part of the work of the inspectors to stimulate the adoption of inventions which should make for safety. They could insist by getting fresh legislation that the inventions utilized in the best mines and factories should be adopted by others and that a minimum standard should be set up below which no business should fall.

---

*Robert Owen, the pioneer of factory reform in 1819, was considered in 1830 to be a dangerous revolutionary, and Oastler, another advocate of reform, was imprisoned for debt.

As the inspectors were officials of the central government they were able to ensure a standard of uniformity which would not have been possible had they been appointed by local authorities. This type of legislation was copied in every country starting the new industrial methods. Abandoning the ideal of free competition and free contract, so dear to the English mind, the efforts of the Shaftesbury party further secured in 1831 an Act against truck in the textile, the metallurgical, and the mining industries, *i.e.*, against the payment of wages in things other than coin, then a prevalent abuse.

If the humanitarians prevented children going to a factory till they were nine they were equally impelled to assist in establishing some sort of education for a child till it reached that age. It is no accident that the first effective Factory Act and the first State grant for schools come in the same year. A small sum of £33,000 was to be advanced by the State in 1833 to help on the erection of school buildings in poor areas. Just as the Factory Acts were the beginning of the industrial code and State regulation of industrial conditions, so this was the beginning of the intervention of the State in education.*

As the mines had got deeper with the ever increasing demand for coal, conditions had become worse, explosions were frequent and ventilation and other safeguards were very deficient. With the great demand for coal and the scarcity of workers, women and very young children had been increasingly employed in the pits. Lord Shaftesbury went on to attack this evil also with the result that women and children were prohibited from going underground by an Act of 1842— a drastic interference with the liberty of adults to choose their occupation—while from 1850, inspectors were appointed to ensure the greater safety of men. Women were included in the Factory Acts and their hours limited and fixed within a definite period by the Statutes of 1844, 1847 and 1850. In the forties the appallingly insanitary condition of the towns was brought to light by a series of enquiries, and the Board of Health was instituted in 1848 with Lord Shaftesbury as a member. Its object was to provide the elements of a decent

---

*The State had since 1563 supervised the conditions of apprenticeship which was the technical education of its day, but compulsory apprenticeship was abolished in 1813.

life in the shape of a water supply, street scavenging and main drainage. The health activity soon spread so as to include the prevention of infectious diseases and fevers, and after 1872 concentrated on trying to keep people well and not merely cure them when they fell ill. Every hour of the day we can now witness the engineering miracle of the great volume of clean water flowing into a town and the great mass of dirty fluid being drawn off.

The second group of reformers believed that the ideal to be aimed at was the greatest happiness of the greatest number. This, they held, would be attained by leaving each man free to work out his own salvation and make his own contracts unhindered. If free, he would always make the best bargain for himself. In trade, if each man were left alone to exchange goods freely the best result, they held, for the nation would also be obtained as each man would act advantageously for himself and the sum total would be greater wealth. To ensure perfect freedom any laws hindering free movement, free contract, free choice of an occupation or free sale should be swept away. Each man would strive, unhampered, to obtain happiness and wealth in his own way and the sum total of happiness and prosperity would be greater than if he were cribbed, cabined, and confined by legislation.

There was a religious motive at the back of this freedom movement also. The laissez-faire party believed that the Lord had endowed certain peoples with certain aptitudes and that mere man had no right to try and hinder them in the exercise of their faculties by putting man-made restrictions in the way of the exchange of goods or utilization of their opportunities. It was clearly opposing the will of the Almighty, who had endowed France with the climate to produce wine and England with the climate and skill to produce cotton, to place tariff barriers in the way of an exchange which was bound to prepare the way for the brotherhood of man and universal peace by making each nation dependent on the other.

This point of view is well brought out in the following passage which also shows how Great Britain considered that the natural division of labour was for her to manufacture and other people to grow raw materials for her use :

" It is clearly seen that to our beloved land Great Britain

has been assigned the high mission of manufacturing for her sister nations. Our kin beyond the sea shall send to us in our ships their cotton from the Mississippi valley. India shall contribute its jute, Russia its hemp and flax, Australia its finer wools and we with our supplies of coal and ironstone for our factories and workshops, our skilled mechanics and artificers and our vast capital, shall invent and construct the necessary machinery and weave these materials into fine cloth for the nations ; all shall be fashioned by us and made fit for the use of men. Our ships, which reach us laden with raw materials, shall return to all parts of the earth laden with these our higher products made from the crude. This exchange of raw for finished products under the decrees of nature makes each nation the servant of the other and proclaims the brotherhood of man. Peace and goodwill shall reign upon the earth, one nation after another must follow our example and free exchange of commodities shall everywhere prevail. Their ports shall open wide for the reception of our finished products as ours are open for their raw material."

Not merely ought there to be a general repeal of all laws hindering liberty in any form but after their abolition these reformers held that the State ought to impose no more restrictions and intervene as little as possible. This group was following the ideas of Jeremy Bentham and was necessarily opposed to the tenets of the Shaftesbury party which believed in imposing definite restrictions, especially for persons unable to bargain freely for themselves.

The outcome of the work of the Benthamite party was a series of repeals of laws that had stood in the way of free contract or free trade.

Workmen were not allowed to form trade unions under the Combination Laws of 1799-1800. In 1825 these laws were modified and men were allowed to combine for certain specified purposes. The legislators were, however, careful to provide that this liberty should not endanger the liberty of other people by undue pressure on the part of a group. Thus the right of combination was hedged in with certain conditions designed to secure the liberty of other people. Englishmen had been prohibited from emigrating to any place not within the Empire. This restriction was also removed at the same

time and they obtained the right to choose their domicile
either in this country or out of it. A further step in the
direction of the free choice of an occupation or domicile was
found in the reform of the Poor Law in 1834. When the
system of doles was stopped a man was no longer bound
to his parish. He had been legally able to move previously
but if he fell into pauperism he would have been removed to
his original parish for relief under the Settlement Laws.
Hence the poor person often preferred to stay where he got
regular relief rather than move to another parish to be moved
back again as a pauper. There was no incentive to move and
get employment elsewhere. With the abolition of doles an
able-bodied man was offered the work-house or had to seek
for work. He got no more out-door relief. The result was
that a large number migrated and were absorbed in building
the railways. The laws of settlement were also modified in
1850, severing the last tie that bound a man to one particular
spot. In this way the legal hindrances to industrial liberty
were removed.

The municipalities were reformed in 1835 so as to break
up the close monopoly which had governed the towns. The
extension of the franchise revitalized all municipal government, and the Benthamites unconsciously prepared the way
for the efficient working of the health reforms of the Shaftesbury party by providing a local government which could
carry those measures out. The recasting of municipal
government which followed these reforms paved the way
for the development of municipal trading—a proceeding
which would have filled the Benthamites with horror, as
they disapproved both of officialdom and monopoly.

Further reforms in the direction of commercial liberty
followed.

The export of machinery was permitted to a large extent
in 1825. In many cases a license had to be obtained first.
It was completely freed in 1843. The Bank of England was
deprived of its monopolistic position as the one great Joint
Stock Bank in 1826, and a great banking development ensued.
The promotion of joint stock companies was facilitated so as
to allow of the much easier concentration of capital. Previously companies had to get a charter from Parliament;
from 1825 they were allowed to develop freely after complying

with certain legal directions intended to prevent fraud.

The process of company formation was rendered easier in 1844 ; in 1855, with certain exceptions such as banks, joint stock companies were allowed to limit the liability of their subscribers. In 1862 this privilege was extended to all trades.*

The Usury Laws were partly repealed in 1833 and finally in 1854, and there ceased to be any hindrance on the employment of capital by limitations on the rate of interest. In general the greatest possible freedom was allowed to the accumulation of capital and the conduct of business. This was facilitated by the introduction of the penny postage in 1838, which opened the way for a vast increase of communications.

The duties on raw materials, lowered between 1822 and 1824, were finally swept away between 1842 and 1846 ; the duties on half-manufactured articles were substantially reduced, in some cases abolished ; the duties on manufactured goods were also reduced or repealed. The duties on corn and meat went at the same time and only a few other food stuffs and manufactures remained to be dealt with by Gladstone in 1853 and 1860. The principle was then adopted that the tariff should henceforward be for revenue only.

Shipping had been deprived of much of its protection by

*The expansion of this form of business may be seen from the following figures :

		Number of Companies Registered.		Nominal Share Capital Thousand £.
1862	-	165	-	57,007
1863	-	790	-	139,988
1864	-	997	-	237,237
1865	-	,034	-	205,392
1866	-	762	-	76,825
1867	-	479	-	31,465
1868	-	461	-	36,528
1869	-	475	-	141,274†
1870	-	595	-	38,252
1871	-	821	-	69,528
1872	-	1,116	-	133.041
1873	...	1,234	-	152,057

†A company was registered with a capital of £100,000,000, but its paid-up capital did not exceed £200 ; the actual amount of this year would therefore be about £41,274,000.

a series of laws between 1822 and 1825, and by the reciprocity treaties concluded with foreign countries after that date.* It was completely thrown open to foreign competition by the abolition of the Navigation Acts in 1849. Merchants were now free to charter the cheapest ship, whether English or American. With the Navigation Act went the remainder of the restrictions on colonial trade under " the old colonial system." The colonists could from henceforward trade freely with foreign countries, employ ships of any nationality they chose, even in the inter-imperial trade, and admit foreigners freely to their own markets. The coasting trade of the United Kingdom was also thrown open to all comers in 1854.

Free Trade was victorious by 1850, but it must be remembered that the tariff changes were only a part of the general movement for the abolition of all restrictions.

Although there was a free trade movement on the continent as well as in Great Britain the two movements differed fundamentally. The English free traders were the manufacturers. They feared no competition and they wished to be able to get raw materials under the cheapest possible conditions. They also wished for the free import of corn, partly because they thought that foreign countries could not pay for manufactures unless England would open her ports freely to their food stuffs. They also desired to stop a demand for a rise of wages on account of dear food and they thought that free imports would lower the price of food. They considered that if the continent could sell the surplus corn to Great Britain, the price of food would rise on the continent and fall here, thus the great advantage of the continent in cheap labour based on cheap food might be neutralized. Hence the manufacturers financed the movement for the repeal of the Corn Laws. The Tory agricultural party opposed it. They had no wish to be subjected to the effect of foreign imports and urged the risks of the ruin of English agriculture and the dangers of depending on a foreign food supply in time of war.

As Great Britain had become a manufacturing State the free trade party were victorious. Their victory meant a complete reconstruction of British finance. If the revenue derived from the Customs were diminished some other form

*For further details *see* Part V., Division IV.

of revenue must be found. Hence the Income Tax was revived in 1842 and in 1853 the Death Duties were developed to cover the deficit created by Gladstone's tariff changes. Thus direct taxation was substituted for many of the old Customs and Excise Duties.

On the continent, on the other hand, the backbone of the protectionist party was formed by the manufacturers who feared the English imports ; the free trade party were the agriculturists who wanted cheap manufactures. Thus in Germany the Agrarians or Junkers were free trade, so were the great land owners of Russia, the wine producers of France and the cotton growers of the South in the United States. They were all exporters and wanted markets abroad and were willing to take manufactures in return.

When the United Kingdom went over to " Free Trade " in the forties it was a turning-point in her economic develop- ment. She deliberately abandoned her national policy of self-sufficiency and made a bid for the world's trade.   She relied on importing food and paying for it with her manu- factures. She then definitely adopted a world economy when other countries such as the United States, Germany and Italy were only struggling to attain to national unity and a national economy.

France had already attained to a national economy but the French Revolution had put her back, commercially speaking, for forty years. Not till 1830 did her exports and imports reach the same figure that they had attained in 1788. She was not in a position to make a bid for world trade although Napoleon III made the attempt when he took France and Europe over to a system of low duties by the treaties he concluded in Europe between 1860 and 1870.

It is interesting to see how both the Humanitarians and the Benthamites attained their greatest success almost within two years of each other. In 1846 the Corn Laws were repealed and if one were not going to safeguard the nation's food supplies there seemed to be no valid argument for any sort of protection. It was the great triumph of the freedom party. In 1844 the first Factory Act included adults—*i.e.*, women, and this meant that men and women when working together would usually stop at the same time and thus the Act indirectly limited the hours of men working with women.

In 1847 there was secured a ten-hour day for women and children and incidentally for men also where they worked with women. In the first place, by the repeal of the Corn Laws the way was opened for freedom in commerce while by the Factory Acts of 1844 and 1847, the way was prepared for the great labour code of the nineteenth century and the government regulation of industry. Generally speaking, it was the views of the free trade party that were predominant up to 1870. The Factory Acts and Mines Acts were considered to be exceptional measures to meet the exceptional case of two classes, women and children, who were unable to look after themselves. The general aim was to ensure liberty of action for the individual. " The best government is that which governs least " was the maxim. Laissez-faire was the ideal.

During this period there were two working-class upheavals. The trade unions had combined in 1834 to form great federations with the purpose of rapidly overturning by a general strike the existing system of industry, which was henceforward to be carried on by the workers themselves, the capitalist being eliminated. Their views would now be called " Syndicalism," but the term had not then been invented.

The second movement, known as the " Chartist " movement, aimed at securing political rather than economic power for the workers. The Chartists thought that once they had the political power in their hands they could transform society. In this they were the forerunners of Marxian Socialism. Chartism was equally an outcome of the new economic conditions. The new Poor Law with its " bastilles " —the work-houses—was very unpopular, while the spread of machinery and consequent dislocation of employment and the conditions in the mines, all created a favourable soil for revolutionary propaganda. Both these movements failed to achieve any success.

It is interesting to notice, however, that in this phase of the Labour movement its adherents all expected that the complete transformation of society would be rapidly accomplished. With the French Revolution before their minds they expected a quick and sudden change ;—evolution, compromise or gradual amelioration were far from their thoughts. The new Utopia was to be accomplished in the twinkling of an eye. Hence the swift collapse of the movement when the objects were not promptly attained. The disillusionment of the worker made him accept after the fifties the capitalist organization of society and support the practical proposals and definite aims of the new trade union movement for amelioration of existing conditions. A complete *bouleversement* of the existing state of things is alien to the British temperament. As a matter of fact there is no English word to express a *bouleversement*. There is a passionate attachment to precedent in this country and the British never do more than " tinker " when changes are required but they keep on tinkering until the patches finally become a new pot. Although its shape and size may have altered entirely in the process of tinkering it always retains traces of the original pot. France, however, breaks up the pot and throws away the sherds and begins to make a new pot of entirely different material. The general result is that any change in France is an intellectual one, the result of an intellectually conceived plan thought out beforehand, and the French always seem willing to start afresh from the very beginning. In England a change though ultimately fundamental, perhaps, is the result of slow experimentation ; a definite problem is tackled and then the new problem arising out of that is tackled again in its turn. This is the history not merely of factory legislation but of all the other industrial legislation of the nineteenth century in the United Kingdom.

The failure of this revolutionary Utopian movement was not merely due to temperament but also to the development of the railways. After 1830 they began to extend rapidly and provided new fields for employment. In 1848 it was calculated by one of the greatest statisticians of the day that 188,000 navvies were employed on constructing the railways. Others were engaged in making the iron rails, preparing the stone, brick and cement for erecting the stations, building the carriages and waggons and cutting the sleepers and altogether a new employment was found for 300,000 workmen who with their families would make up a million persons dependent on the railways. As the numbers of men on construction declined the numbers engaged in working the railways increased.* The same authors considered that

*Tooke & Newmarch, " History of Prices," Vol. V., *p.* 357.

during the five years, 1846 to 1850, 600,000 people found employment in the railway works, which were as many as were employed in the whole of the factories of the United Kingdom, and that this "mitigated the disastrous effects on the working classes of the commercial and political convulsions of 1847, 1848 and 1849."* Perhaps this partly accounts for the fact that the United Kingdom was the only European country, except Norway and Sweden and autocratic Russia, that escaped a revolution in 1848.

The railways did more than provide employment, they made enormous demands on capital for their construction.† They gave an impetus to all large-scale enterprises by providing increased facilities for the transport of raw materials and goods. They were a brilliant example of the success of joint stock companies ; they gave a striking example of the size to which large businesses might attain and by their amalgamations, especially after 1850, they became the pioneers of the movement towards monopolistic combinations in business. They afforded a great stimulus to agriculture by providing better markets for all produce. By conveying cattle rapidly so that they did not lose weight by the way as was the case when they had to be driven for days to reach the towns, they stimulated stock breeding. The whole of the iron and engineering trades were quickened by the enormous demand for rails and locomotives. The creation of this new channel for investment at home and abroad almost amounted to a financial revolution.

The following table shows the stagnation of the export trade after the war. The figures after 1835 show the revival :‡

*Tooke & Newmarch, *op. cit.*, Vol. V., *pp.* 368-369.

†Leone Levi gives the capital of the new joint stock companies of the years 1834-1836 as £135 millions, of which £69,666,000 were for railways. "History of British Commerce," *p.* 220. Tooke & Newmarch consider that this capital was the result of saving by the middle classes.

‡Value of Exports of Produce and Manufactures of United Kingdom, Customs Tariffs of the United Kingdom, 1800-1897, C. 8706, 1897, *p.* 51.

£ millions.		£ millions.		£ millions.	
1815	- 51.6	1828	- 36.8	1841	- 51.6
1816	- 41.6	1829	- 35.8	1842	- 47.3
1817	- 46.4	1830	- 38.2	1843	- 52.2
1818	-	1831	- 37.1	1844	- 58.5
1819	- 35.2	1832	- 36.4	1845	- 60.1
1820	- 36.4	1833	- 39.6	1846	- 57.7
1821	- 36.6	1834	- 41.6	1847	- 58.8
1822	- 36.9	1835	- 47.3	1848	- 52.8
1823	- 35.3	1836	- 53.2	1849	- 63.8
1824	- 38.4	1837	- 42.0	1850	- 71.3
1825	- 38.8	1838	- 50.0	1851	- 74.4
1826	- 31.5	1839	- 53.2	1852	- 78.0
1827	- 37.1	1840	- 51.4	1853	- 98.9

(4) In the fourth period, from 1850 to 1873, the effect of the two great series of reforms was felt. The worst evils were checked and freedom was given for expansion. Towns became healthier places to live in. The conditions of work improved, employment was brisk and these twenty-three years are sometimes termed the " golden age." The improvement in trade and employment was no doubt largely due to the railways which facilitated exchange not merely in England but all over Europe, where tariffs were being generally lowered after 1860. A further stimulus was due to the gold discoveries in Australia and California, which, by increasing the world's stock of gold, raised prices and made it worth while to embark on new undertakings. Engineering knowledge had developed and the mechanism of sanitation was beginning to be grasped. Factory and mine inspectors saw that the new knowledge was applied and the extension of the industrial revolution which took place in this period proceeded under proper safeguards.

There was a great expansion of trade owing to the fact that the Continent had got over the effects of the Napoleonic wars and was proving an increasingly good customer for British manufactures. But above all the great improvement arose from the railways.

Once the great constructional work for the railways was over in England, there always remained the foreign demand for rails and locomotives, and a railway once built begins to deteriorate with use. The iron rails wore out in about seven to ten years and there were increasing demands for renewals. Machinery was spreading rapidly with better transport facilities for the production and distribution of masses of goods and this made fresh demands on the engineering and iron trades, to say nothing of the increasing continental demand for British machinery. Machines like rails and locomotives wear out and need replacing every ten to fifteen years.* A new type of ship was coming in, the iron ship propelled by steam, and this created fresh demands on the iron trades. Railways, machinery and steamers all required coal as a motive power, the blast furnaces were one of the largest consumers of coal and this demand reacted on coal mining, the estimated production of which was 64,666,000 tons in 1854 and 110,431,000 in 1870. It is interesting to speculate where all the new " hands " came from. Agriculture probably supplied the bulk of them. Machinery was coming into agriculture after 1850, a good deal of labour was rendered superfluous, and the agricultural labourers or their sons found new outlets in the police, the railways, in coal mining and the expanding engineering trades. Numbers of Irish came over to build the English railways and formed a large part of the personnel of the English factories. Wages rose, especially in the metallurgical industries, and although prices were rising wages outstripped prices and the working classes were better off.† There was no longer any question of a social revolution.

Employment was steady for all classes of workers. They began to form trade unions of one particular trade extending over the whole nation. These no longer had in view revolutionary changes but devoted themselves successfully to obtaining higher wages, shorter hours and greater safety appliances by negotiation. They provided a mechanism for bargaining and an insurance that bargains would be kept and thus helped to promote the " industrial peace " of the period. A great Factory Act was passed in 1867 which extended the provisions of the earlier Acts to all industries— textile and non-textile and included workshops as well as factories in its scope. A flourishing movement had developed among the workers for the co-operative purchase of goods.

*See Schedule of agreed rates of depreciation for Income Tax purposes on p. 69 (Appendix 7) of the Royal Commission on the Income Tax, 1919, First instalment of Minutes of Evidence.
†Bowley, " Wages in the United Kingdom," Diagrams, pp. 130-133.

In 1844, the Rochdale pioneers had started a shop which gave back the profits to the members in proportion to their purchases. This movement spread in the fifties and provided a training for the operatives in self-government, self-dependence and thrift. They combined for wholesale purchasing as well as sale in 1864. The good times extended to agriculture and the agricultural population proved to be excellent customers for the manufacturers. England held at the time an almost monopolistic position in many branches of manufacture, notably iron goods. France was unable to compete as far as the engineering trades were concerned owing to the difficulty of obtaining cheap coal and iron, while industry in Germany and the United States had hardly begun to develop. Germany was absorbed by three wars, one with Denmark (1864), one with Austria (1866) and with France (1870). Austria and Italy were engaged in war or preparing for the renewal of the struggle between 1848 and 1870. Russia was freeing her serfs after the Crimean War and reconstructing generally. The United States was also occupied with the Civil War (1861-1865) and the consequent readjustment. To all belligerents Great Britain was able to supply equipment for soldiers, iron for armaments, and the services of shipping and capital, while they themselves were hindered from competing by the drain of men and money. A striking instance of the monopoly conferred by war is the way in which American shipping, very important before the Civil War, declined during that war, leaving the ocean supremacy to Great Britain. As most of the continent had adopted a system of low duties between 1860 and 1870 there was little or no barrier to the entry of British goods. The opening of the trade with China and the acquisition of Hong Kong in 1842, the further opening of new Treaty ports in 1858, the opening of the trade with Japan in the same year and Siam in 1857 stimulated trade with the Far East.

It was during this period that England began to take such an active part in financing and constructing railways and similar enterprises all over the world. Brassey, the great contractor, not only built many of the railways of France but built them also in Italy, Holland, Denmark, Norway, Poland, Austria, Hungary, Switzerland, Mauritius, India, Argentina (the Central Argentine Railway) and Canada (the Grand Trunk). He built no less than thirty foreign railway lines between 1850 and 1870,* another instance of the reaction of English technique and trained engineers on the continent.

In the twenty-three years from 1850 to 1873 Great Britain was the forge of the world, the world's carrier, the world's ship-builder, the world's banker, the world's workshop, the world's clearing house, the world's entrepôt. The trade of the world during this period pivoted on Great Britain.† She was organized for a world economy when other countries such as Italy, Germany and the United States, were only feeling their way to a national system. Italy up to 1859 consisted of eight States with eight tariff barriers ; Germany had only formed the beginnings of her internal Customs Union in 1834 and it was still uncertain in the sixties whether it would hold together. The United States was divided into three economic areas, the North, the South and the Middle West and it was doubtful until after 1865 whether there would be one, two or three nations within that area. It had already undergone two sharp crises which had threatened to break up the union. The brilliant success of the English seemed at the time still further to prove the value of individual initiative and enterprise and the wisdom of free trade measures.

*Table given by Sir A. Helps in " Life of Brassey," Bohn Edition, *p.* 84. As to the financing of French railways by Englishmen *see* Sir Edward Blount's Memoirs. He raised £600,000 in England to construct the railway from Paris to Rouen opened in May, 1843. " I had a concession together with M. Charles Lafitte in the construction, in 1845, of the line from Paris to Boulogne by way of Abbeville and Neufchatel. Subsequently, in 1852-1853 I was interested as Director Administrator in the line from Lyons to Avignon and also in that between Lyons, Macon and Geneva. I practically financed the Western of France Railway. I was chairman of the line for thirty years and relinquished my position only in 1894." *p.* 61.

†Felkin, " The Exhibition of 1851 of the Products and Industry of all Nations : Its probable influence upon Labour and Commerce." *p.* 22 : " The silk of China, for example, is woven in Coventry and sold wholesale in New York, retailed amongst a thousand other articles in New Orleans and is consumed by a neighbouring planter's wife as a ribbon attached to her dress. That American planter grows cotton wool which is exported and woven into cloth in Manchester. This cloth finds its way into the interior of Bengal and is retailed by a trader who probably gives two seasons' credit upon the sale, and may be paid for it at least partly in produce which will be sold for food in the English market ten thousand miles off. A halfpennyworth of meal from America, a halfpennyworth of coffee from Jamaica, a halfpenny-

There were during this period only two sets-back. The Civil War in the United States led to a great cotton famine (1861-1863) and much unemployment in Lancashire, where 500,000 persons were said to be in receipt of relief at Christmas, 1862.* After 1860 the silk industry declined owing, so it was said, to the Cobden Treaty with France by which all duties on French silks had been abolished.†

An interesting development which took place during this period was the great enlargement of the functions of municipalities. Under the stimulus of the health movement which had created the Board of Health in 1848 some of them began to provide a water supply. They were all obliged to take steps to develop the main drainage system and the removal of refuse. To accomplish this adequately they began to take over the maintenance and repair of urban roads. Municipal gas works also followed in certain cases. Thus at the very period when joint stock companies were rapidly increasing, when banking was growing and capital becoming more and more powerful, certain alternatives or correctives emerged. The Co-operative movement eliminated profit by returning the profit to the consumer in proportion to his purchases. The action of certain municipalities in creating a monopoly of certain services, viz., water and gas, was intended to benefit all the inhabitants of the town either by selling cheap or by allocating the profits to the reduction of the rates. At the same time the abuse of the power of capital was being

worth of sugar from Brazil are sold at the same humble counter to the occupant of a neighbouring garret in St. Giles'. A chandler's shop in the dirtiest, darkest thoroughfare of the outskirts of London or Limerick cannot exist without supplies from every quarter of the globe."

p. 28. " The ways by which the inhabitants of the earth may come together upon this occasion have been opened up and made plain in an equally extraordinary manner. Twenty-five years ago, not one-tenth of our expected visitors could have travelled hither from want of time or money, steamboats and railways have strangely diminished cost, fatigue and time. . . . The finances of the world have been heavily taxed to supply these means for locomotion but with wonderful results and amongst the things that have been rendered possible by the agency of steam this gathering is the most wonderful of all.

*The price of cotton had been 6*d.* and 7*d.* a pound in 1860. It rose in December, 1863, to 29½*d.*

†Rawlley, " The Silk Industry in Great Britain."

held in check by the trade union movement on the one hand and by more stringent factory legislation on the other.

The growth in the foreign trade of the United Kingdom during the period can be gauged by the following figures:

Annual Average.	Imports. million £	Re-exports. million £	Exports of U.K. produce. million £
1855–1859	169	23	116
1860–1864	235	42	133
1865–1869	286	49	181
1870–1874	346	55	235*

The expansion in the exports of iron and steel, textiles and coal may be illustrated from the increase from decade to decade in the value of the exports.

	Exports of Iron and Steel. £000 omitted.	Machinery and Millwork. £000 omitted.
1830	1,079	209
1840	2,525	593
1850	5,350	1,042
1860	12,138	3,838
1870	23,538	5,293†

TEXTILES.

	Cotton Goods and Yarn. £000	Woollen Goods and Yarn. £000	Silk. £000	Apparel. £000
1830	19,429	4,851	521	983
1840	24,669	5,781	793	1,290
1850	28,257	10,040	1,256	2,535
1860	52,012	16,000	2,413	2,474
1870	71,416	26,658	2,605	3,881‡

COAL.

	Exports. £000		Production. tons 000
1830	184		
1840	577		
1850	1,284	1854	64,666
1860	3,316	1860	80,043
1870	5,638	1870	110,431

(5) The fifth period, 1873 to 1886, was characterized by a great depression. The effects were world-wide. In the United Kingdom three great industries suffered especially

*Fiscal Blue Book, 1909, Cd. 4954, *p.* 18.
†" Commerce and Industry," Vol. II., ed. Page, *p.* 137, taken from Accounts and Papers and Statistical Abstract.
‡" Commerce and Industry," Vol. II., ed. Page, *p.* 133.

severely : agriculture, shipping and the iron and steel trades. There was, however, a general drop in prices in all commodities owing to currency changes, though some things were affected more than others.*

In agriculture, apart from the currency complication, the depression was partly due to the opening up of the " Middle West " in the United States. After the Civil War a great era of railway building set in which opened up the prairie lands. The railways were built as speculations and competed against one another at cut throat rates to carry the grain and these low rates acted as a sort of bounty on export. The competition of the steamer and the sailing vessel lowered freights at sea and a great flood of American exports began to pour into Western Europe. The continental powers dyked up against this flood with tariffs and it was diverted into the great free trade market of the United Kingdom and produced a great depression in agriculture.

On top of the wheat imports came the meat. The railway and the refrigerator car enabled meat to be moved as it had never been moved before. Cold storage steamers facilitated the transfer of frozen or chilled meat by sea when the railways brought it to the coast. The American meat combines started and made such profit out of the sale of bye-products like hides, bristles, horns, etc., that they were able to dispose of the actual meat at very low prices. American beef and pork were soon reinforced by Australian and Argentine mutton, and British cattle growers, as well as wheat producers, suffered severely from competition as far as second and third rate meat was concerned.

With the developments of transport, " free trade " became really effective, and the result was a great depression and a reconstruction of British and Irish agriculture.

In shipping the depression was due to the competition between the old sailing ship and the new steamer. The new steamer was capable of doing far more work, i.e., of putting in more voyages than the sailing ship and so the effective tonnage of the world was enormously increased. This was greatly added to by the opening of the Suez Canal. As the journey was so much quicker by the Canal than by the Cape again more tonnage was liberated, which increased the shipping

*Layton, " An Introduction to the Study of Prices," p. 68.

competing for employment. On top of this, new and more efficient types of ships were being rapidly evolved, which in their turn displaced the earlier steamship. The compound engine with its great economy in the use of coal was being displaced by the triple expansion engine with a still greater economy in fuel. In the eighties the iron ship was being superseded by the steel ship. There was therefore a continuous scrapping of ships going on. This " over-production " of shipping, as it was termed, was still further accentuated by the attempts of foreign governments after 1880 to start their own steamship lines with some form of State assistance. There seemed, therefore, to be no end to the violent competition and drop in freights until the ship-owners settled the matter for themselves by combining into rings in the eighties and nineties.*

The iron trade was suffering from the great change from iron to steel. The Bessemer process had made steel cheap. In 1880 steel plates had cost £11 14s. 6d., in 1886 they were £6 2s. 6d.† Hence there was a general substitution of steel for iron. This meant the scrapping on a wholesale scale of the greatest iron industry in the world. Nearly all the British iron industry based on puddling had to be reconstructed for acid steel. Moreover, British iron ores were not pure enough for the new Bessemer process which needed an iron free from phosphorus, and a large importation of iron ore took place from Bilbao and Sweden, ousting the British product. England, instead of being dependent for her own supplies on her own production, now became largely dependent on import for steel-making as hæmatite ores were only found in Cumberland. The proximity of her coal fields to the sea in South Wales, Western Scotland and the North of England facilitated import and enabled Great Britain still to maintain a position as a great engineering workshop ; but her monopoly was gone. Just at this time the great demand for rails, etc., for railways fell off as the main work of construction had been finished in England and as steel lasted longer than iron there was a lessened demand for renewals.‡

*Report on Shipping Rings, 1909, XLVII.
†Depression of Trade Commission, 1886, XXI., XXII., XXIII. 2nd Report, p. 332.
‡Sir Lowthian Bell giving evidence before the Depression of Trade Commission said that the cost of making a steel rail was less than

In addition to this, other countries started making steel. They were able to begin where Great Britain left off and could start an iron industry without such an appalling waste of capital as was involved in the British reconstruction. The Gilchrist-Thomas process enabled Germany to use the great resources of minette ores in Lorraine to make basic steel after 1880.* The railways joined the vast iron deposits of Lake Superior to the Pittsburgh coal fields and the world saw an unparalleled production of steel which temporarily outstripped the demand. The price for steel rails fell from £12 1s. 1d. per ton in 1874 to £5 7s. 6d. in 1883. Iron rails declined from £9 18s. 2d. to £5 per ton.† Cleveland pig iron declined from £4 17s. 1d. in 1872 to £1 12s. 10d. in 1885.

Other causes were operating to produce a depression here. As the railways were developing on the continent a good deal of the traffic that had previously gone by sea now went by land : the trade routes altered with the opening of the Suez Canal and that brought the Mediterranean ports like Marseilles, Genoa and Odessa into prominence. The monopoly of the world's shipping trade passing over London or Liverpool was impinged upon. Continental countries were beginning to start an industry of their own and no longer relied upon Great Britain to supply them with engineering tools, machines

an iron one because it required less iron ore and less coal. The steel was put in a liquid state from the blast furnace into the converter, thus saving one re-heating, and was rolled while still hot. He thought himself that a steel rail would last twice as long as an iron rail but actual experiments on a certain railway line showed that a steel rail lasted ten years as against the seven years' life of an iron rail. "This arose from the fact that an iron rail split up and became useless long before the actual wear as measured by diminution of weight rendered it unsafe. This often happened when the loss of weight did not exceed four per cent. of the original weight. Steel rails, on the other hand, go on losing weight until they are ten to twenty per cent. lighter than they were when laid down." Depression of Trade Commission, Second Report, *pp.* 4 and 51. In actual practice the steel rail proved to have a far longer life than ten years. It varies according to user but twenty-five years on an average would probably be nearer the mark.

*The increase in English pig iron production was thirty-one per cent. greater in 1884 than in 1870, that of foreign countries was 138 per cent. Final Report of Royal Commission on Depression of Trade, VIII. 1886, C. 4893.

†Evidence of Sir Lowthian Bell. Second Report Depression of Trade Commission, *p.* 43.

and manufactures to the extent they previously had done. As they began to require increasing quantities of raw material they caused it in many cases to be brought direct to their own ports, whereas formerly when they only required small quantities or parcels of raw material they had bought them at the great entrepôt, England. This growing direct trade affected the English entrepôt and distributing trade adversely. The competition of Germany was felt even in neutral markets ; the United States had barred herself in behind her tariff wall and to a large extent supplied herself. The result was that the export trade of England to the United States ceased to grow, although the population there increased rapidly.* England had expanded her own production to fill the gap created by the two belligerents in the Franco-Prussian War and found in the eighties that two of her largest markets, Germany and the United States, were increasingly inclined to cater for themselves.

The great depression marks the period when England's world predominance was challenged and she encountered a foreign competition unknown during the nineteenth century. It also marks the decline of the French rivalry and the rise of the German. The French had been the only industrial power of which the English captains of industry had hitherto been afraid. After Sedan, however, Europe found that it " had exchanged a mistress for a master." The Germans, successful in war, applied their great capacity for organization to commerce. They had acquired a belief in their own powers, they had developed an admirable system of technical education which was backed up by an excellently organized system of transport. Their banking system was developed in such a manner that the banks not merely supplied the capital but took a share in the management of the businesses to which they supplied the money. Thus technical skill and financial knowledge were combined. A system of cheap railway rates was worked out so as to promote export and Germany laid herself out to become the great distributor of Europe in virtue of her central position. Science was applied to assist the growing industries of Germany and the result was that new trades such as the electrical industries and the chemical dyes were developed in Germany with conspicuous success.

*p. 196 n.

As soon as the Gilchrist-Thomas process made it possible to utilize the phosphoric ores, the iron of Lorraine was brought to the coal fields of the Ruhr district and a great iron and steel industry was the result. The cotton makers of Alsace, transferred in 1870, reinforced the small German cotton industry and in this way the German victories over France stimulated a textile and metallurgical rival for Great Britain. France with her shortage of coal and her great artistic skill had not produced along the same lines as Great Britain. Germany was, however, preparing to compete along England's own lines and in England's own preserves.

In 1886 the Commission on the Depression of Trade recognized this new rivalry and reported that : " A reference to the reports from abroad will show that in every quarter of the world the perseverance and enterprise of the Germans are making themselves felt. In actual production of commodities we have now few, if any, advantages over them, and in a knowledge of the markets of the world, a desire to accommodate themselves to local tastes or idiosyncracies, a determination to obtain a footing wherever they can and a tenacity in maintaining it, they appear to be gaining ground upon us."*

Ever since the sixteenth century England has been moulding her policy on that of some real or supposed rival. In the sixteenth century Spain had impelled the English to develop some of their home resources, such as mining, as a means of defence. It was partly to curb the power of Spain that England went to the New World. In the seventeenth century England moulded her policy on that of the Dutch. She followed them to the Spice Islands and only fell back on the mainland of India when she was unable to make her footing good against the Dutch in the islands. She copied their fishing, their shipping, their finance and their agriculture. From 1660 onwards the great model was France, as reconstructed by Colbert. To build up English industry to the French level was the ambition of British statesmen throughout the eighteenth century, and so greatly was Great Britain influenced by France in colonial expansion and in industrial development that the eighteenth century has been termed the period of Parliamentary Colbertism. After the French Revolution and the development of the English industrial

*Final Report, *p.* XX.

revolution we still find English merchants and pamphleteers talking of French progress and French rivalry, but it is half-hearted. In three-quarters of a century Great Britain could really find no rival of whom she was seriously afraid. After 1880, however, we get a new model—the German, and as Germany stood for State guidance and State intervention her reaction on this country was marked.

The following figures will illustrate the decline in values of the exports of the produce and manufacture of the United Kingdom during the depression (excluding ships) :

		Exports to	
		foreign countries.	British Possessions.
Annual Average.	£ million	£ million	£ million
1870–1874	235	175	60
1875–1879	202	135	67
1880–1884	234	153	81
1885–1889	226	147	79
1890–1894 (revival)	234	156	78

During the period the value of the imports rose.

		Imports from	
		foreign countries.	British Possessions.
Annual Average.	£ million	£ million	£ million
1870–1874	346	270	70
1875–1879	375	292	83
1880–1884	408	312	96
1885–1889	380	293	87
1890–1894	419	323	96*

(6.) The result of the depression was that the period from 1886 to 1914 witnessed a great change in English policy. It is the period of the abandonment of laissez-faire in colonization, commerce, industry and agriculture. Great Britain began to modify her cosmopolitan ideas of free trade and laissez-faire, and to concentrate on developing trade within the British Empire. It is no accident that in 1887 representatives of the Dominions appeared as part of the pageant at Queen Victoria's Jubilee ; it was repeated in 1897 and became a regular institution. The United Kingdom denounced her treaties with Germany and Belgium in 1897 which had prevented her giving preferences to or accepting them from her

*Fiscal Blue Book, 1909, pp. 22, 24.

colonies. She gave up her pure free trade attitude for the sake of the West Indies and joined the Sugar Convention in 1902 and prohibited the import of bounty-fed sugar from Russia, Denmark, Spain and the Argentine. She thus enabled Germany, France and Austria to suspend their sugar bounties. As the United Kingdom was the great market for sugar and as she agreed to prohibit sugar from bounty-giving countries, it was not worth while for Germany, France and Austria to continue the bounty system once they could not send the surplus sugar to this country. The United Kingdom had gained in the past by the cheap bounty-fed sugar. She had reared up the jam, confectionery, ærated water and biscuit industries, all based on sugar. Was it worth while to run the risk of making them pay dearer for sugar and so possibly cripple their output for the sake of some scraps of tropical islands inhabited chiefly by coloured races ? Was it worth while to run the risk of a possible rise in the price of sugar to the forty-seven million consumers of the United Kingdom for the sake of British Guiana and Jamaica ? It was a turning-point in British trade policy when she decided that it was worth while. From that time onwards laissez-faire in trade has been abandoned and the economic ties of the Empire have been knit closer in numberless ways.

With the coming of Chamberlain to the Colonial Office in 1895 a new constructive imperialism was organized. The self-governing dominions were not merely linked up with each other and the mother country by a regular conference system but inter-imperial cables were subsidized, postal facilities enlarged and commercial attachés, known as Imperial Trade Commissioners, were appointed to give information and assist the development of inter-imperial trade. Many colonial government and railway stocks were put within the magic circle of safe investment known as Trustee stocks. They thus received the imprimatur of the Imperial Government and the colonies were able to borrow at rates of interest otherwise unobtainable by new countries. The United Kingdom thus gave her colonies a substantial financial preference in return for preferences on British goods in the colonies developed after 1897. This financial preference tended to lower the value of British Consols as it extended the market for first-class securities.

In the tropical dominions constructive imperialism took the form of subsidizing the building of railways in such regions as Uganda, West Africa and the Soudan. Life was made healthier in the tropics by the encouragement given to tropical medicine both in Great Britain and in the tropical areas. Sleeping sickness and malaria were investigated and something was done to stop their ravages.

In addition, the Government encouraged and subsidized the study of scientific agriculture in the tropics. The sugar cane was improved, experiments were carried out on cotton in Egypt, the West Indies and Uganda ; insect and fungoid pests were studied and in some places successfully combated. The tropical and semi-tropical regions are now studded with agricultural experimental stations and laboratories subsidized by the home or colonial governments. The new colonization is colonization by railway and science. This could not fail to react in its turn on the United Kingdom. Men could not be trained in India and elsewhere to carry out a policy of constructive development and not wish to apply it in their own country. All this new imperialism was bound to weaken the belief that the State could do no good thing. The new competition seemed to change popular opinion, which had hitherto regarded colonial possessions as undesirable, and one result of the great depression was to prepare the way for a new colonial policy.

The effect of the depression was no less striking in the way in which it reacted on industrial legislation and increased the State control of industry and social conditions. The depression had resulted in wide-spread unemployment.* A commission was held to enquire into Sweating ; a great Labour Commission sat and reported on all phases of labour in 1892 ; new labour legislation was framed and a period of activity in social reforms ensued only comparable with that of the thirties and forties. A Shop Hours Bill was passed in 1893 and an Early Closing Bill in 1904. So far as the conditions of work in factories and workshops were concerned the chief change was in the elaborate regulations framed in 1891 to prevent the special dangers in certain trades such as pottery or wool sorting, and power was given at the same time to the Home Secretary to extend the list of trades

*Webb, " History of Trade Unionism," p. 346, 378.

L

labelled "dangerous," with the result that the number is constantly being added to.

After the nineties, however, there is a different spirit in industrial legislation. Up to that time there had been an attempt to tackle the most flagrant evils one by one as the factory inspectors brought them to light. The definite aim was to stop abuses. After this date the aim was far wider, namely, to restore or create the security of life, much of which had been lost by the break-up of family work and the loss of the agricultural bye-employment when the factory system became predominant. The family wage and the little farm, where this had existed, had been a sort of insurance. Although an employer had been liable since 1880 for accidents caused not only by his own negligence, but also in certain employments by the negligence of a foreman, he was in 1897, owing to Chamberlain's initiative, made liable for accidents arising out of and in the course of the employment whether caused by negligence or not in the case of the larger and n ore dangerous industries. Thus accidents were made part of the cost of production. The class of persons to whom the Workmen's Compensation Act applied was considerably extended in 1906.

But why stop at accidents? Old age and sickness are part of every-day life. Why not provide against known misfortunes? In this spirit Old Age Pensions were given in 1908 without any necessity for contribution on the part of the recipient, and compulsory insurance, to which the worker and employer both contribute, was instituted in 1911 for sickness. There was so much unemployment during the depression that it was felt to be unfair to force the able-bodied into the workhouse. Hence after 1886 it became the practice of the Local Government Board to favour the starting of relief works by municipalities, and the Unemployed Workmen's Act, of 1905, carried this still further and tried to organize and improve upon these local efforts. Compulsory Unemployment insurance for certain seasonal trades was also inaugurated in 1911. Labour exchanges or State employment agencies were set up in 1909, the idea being that the State should bring employers and employed together without cost to themselves, so that jobs should be dovetailed, unemployment minimized and labour made more mobile. The remuneration of work also attracted attention. The Truck Acts were elaborated and the industries to which they applied were extended so as to prevent unfair deductions which should amount to a lower wage than that agreed on. They were placed under the supervision of the factory inspectors in 1887, and fines and deductions regulated in 1897. As the existence of sweated trades was held to be injurious to the nation, the Trade Boards Act of 1909 made arrangements for fixing a minimum wage in the badly-paid domestic trades such as chain-making, lace-mending and finishing, ready-made tailoring and box-making. In 1914 it was extended to other trades such as food-preserving, sugar and confectionery, linen and cotton embroidery, shirt making and hollow ware, and has since been greatly extended in scope including in 1921 even retail trades like grocery. The miners in 1908 secured an eight-hour day by law and to them, a well-paid trade, the principle of fixing a minimum wage was extended in 1912.

The Education Acts were extended after 1902 so as to include a technical and not merely general training for the child. The State thus began to take into account the training of the child for industry or commerce in addition to equipping him with the knowledge necessary for his development as a human being. As it was a pity for the education of a child to be wasted through underfeeding, meals for necessitous school children were made possible by the Education (Provision of Meals) Act, 1906, and the Act was largely taken advantage of by local authorities.

It was desirable that the child should be diverted into promising avenues of employment and not enter blind alley forms of labour. Hence school advisory committees were set up and Juvenile Labour Exchanges instituted to try and help the child into a suitable post.

State regulation went still further when it took account of children's work after school hours and prohibited certain forms of street trading and night trading by children under the Children's Act of 1903. By another Statute passed in 1908 the State did still more towards protecting the child from the irresponsibility of parents. No child under fourteen is permitted to enter a public house nor is he allowed to smoke till he is sixteen years of age, while State regulation goes so far as to make parents purchase fireguards to save the children from being burnt.

Such elaborate details show how far we have travelled from the period when it was considered almost revolutionary to prohibit a child under nine from going into a cotton factory. It also shows how legislation has begun to invade the home.

The general trend of development has been for the regulation to begin with a child on the ground that the child could not help itself, that it was not really " a free agent " and to extend the regulations to women on the ground of protecting the mothers of the future. In actual practice during the nineteenth century they have not been able to protect themselves as efficiently as men. Men have formed trade unions, women have done so only to a very limited degree. They marry and leave their work. It is therefore not worth while for them to subscribe to something from which they may never reap a benefit. They have not as much time as men to go to meetings of an evening. They have to wash their clothes, or make a blouse, or mend stockings, or help to wash dishes, or cook supper. Moreover there was a strongly-rooted early Victorian idea that " modesty was the ornament of the female character," and women felt a certain dislike to putting themselves forward and speaking at meetings. The lower wages of women also precluded anything like a substantial subscription. Hence the Factory Acts envisaged women and children, while the trade unions have helped to protect men. The unions were the masculine side of the Factory Acts and were given greatly increased powers and wider scope in the period 1871-1876. Their funds were made immune in respect of possible actions arising out of conduct amounting to what the law calls a " tort " by the Trade Disputes Act of 1906. Thus while a Bank or Railway Company guilty of tortious conduct may be condemned in damages, a trade union under similar circumstances is not liable. They were thus placed in a privileged position.

The very severity of the struggle to get business during the Depression led to the formation of combinations and amalgamations to avoid cut throat competition, with the result that free competition tended to disappear and prices were increasingly fixed by rings and agreements. Trade unions were then faced with great Employers' Federations and were driven to rely on legislative action rather than collective bargaining. This partly accounts for the rise of the Labour party as an important group in the House of Commons, and the tendency to press for increased industrial protection by legislation still further quickened the growth of the power of the State in industrial matters.

The employers' combines began to include not merely national trades or branches of a trade but they extended their scope to include foreign concerns. These international combines made " free trade " of little or no effect where they existed. National trusts also neutralized the effect of free competition as the chief protection of the consumer and made it necessary for the State to step in to prevent abuses.

The growing power of the State in agriculture is also marked. The Great Depression had vitally affected English agriculture. While the continent had dyked up against the American imports with tariffs the United Kingdom still adhered to free trade. The flood of imports was disastrous for the Irish peasantry, already over-rented and with little or no capital or staying power. A new policy had to be adopted in Ireland which meant an ever-growing extension of State influence. Fair rents were fixed for a term of fifteen years after 1880, to be reconsidered and refixed at the expiration of that period. The result was that neither tenant nor landowner was free to ask or give what price they liked for the hire of land when once government intervention was invoked to settle the rent.

As the fixing of " fair rents " gave rise to great friction and led to bad farming by the tenant in order to get a rent reduction, the next step, started in 1885 and developed in 1903, was for the Irish tenant to be assisted by the financial resources of the Government to buy his landlord out. The process is for the State to fix the purchase price through its Land or Estate Commissioners, advance the money, and the purchaser has to pay it back in instalments over a certain period of years. Thus one of the greatest land transfers of modern times was being carried out by the Government of the United Kingdom prior to the outbreak of war in 1914.

It is not, however, sufficient to change tenures ; the Government having assisted a man to get a farm must make him a farmer able to pay his way. The Department of Agriculture and Technical Education was set up in Ireland in 1899 with the express purpose of inducing the Irish peasant to adopt improved methods. For the outlying poverty-stricken regions of the West a special body was created in the Congested Districts Board to deal paternally with the region and raise its general economic condition by State aid.

In Great Britain the reaction from laissez-faire in agriculture has taken two forms. In the first place the Board of Agriculture was created in 1889 to help to stamp out cattle diseases and to assist the farmer by giving information and protecting him from forms of unfair competition. It therefore administers Acts intended to safeguard the farmer from adulterated fertilizers and food stuffs or from such unfair forms of competition as margarine passing as butter. Secondly, an elaborate system of agricultural education has been developed and subsidized, scientific research has been organized, and England was divided into great agricultural administrative areas, each of which contained experts who would give advice on matters pertaining to cattle, crops and forestry.

The reaction is still more striking, however, when we come to the small holdings movement. The peasantry were allowed to disappear as the typical features of English agriculture in the first quarter of the nineteenth century.* At the beginning of the twentieth century great attempts were made to re-establish a peasantry. The County Councils were obliged by the Government in 1907† to acquire land and sell or re-let it to small tenants. Should the County authorities decline to act the Government could act in default and land might be and was compulsorily acquired for the purpose. The Government advanced money for the preliminary expenses and subsidized co-operation which is the chief means by which the small farmer can hope to circumvent the disadvantages of small scale production. The general result is that public authorities have become one of the largest landowners in the country.

The increase of the power of the State over railway transport is also noticeable. The great depression raised the question as to whether the whole of the rates of carriage of goods were not too high. The result was the fixing of railway

*The farms under 50 acres were still more numerous than those over 50 acres, being 292,720 as against 143,166 in 1913, but the former only occupied sixteen per cent. of the total acreage. Agricultural Statistics, 1913, Cd. 6597. †7 Ed. vii. c. 54.

rates by the Government between 1888 and 1894, while the amalgamations at the beginning of the twentieth century raised the whole question of the acquisition of the railways by the State.

In commerce the amount of State intervention has been less. The action of the Government has been chiefly confined to fighting unfair forms of competition. Thus Merchandise Marks Acts have been passed since 1887 which are intended to secure English makers against the fraudulent imitation of their trade marks by foreigners or against fraudulent descriptions which should lead the buyer to think the goods English when they were really foreign.* Greatly increased attention has been paid to adult training for commerce and economics have become part of the curriculum of every University. In the nineties one Professor was supposed to be able to deal with the whole of Economics; the London School of Economics, founded in 1895, showed that it was as useful to have one Professor of Economics as it would be to have one Professor of Science. Hence the enormous expansion of economic training and teaching, much of it subsidized by Government grants.

The Board of Trade started a Commercial Intelligence Department in 1900 to give information to traders; it began to increase the equipment of consuls and started a new service, that of the Trade Commissioner to the Colonies and Dependencies. It published a monthly† journal of trade intelligence, *The Board of Trade Journal*, in addition to the elaborate series of Consular Reports and special reports, to keep the British trader *au fait* with tariffs and other regulations in foreign countries.

The enormous extension of State activity which is characteristic of the period after 1870 could not have been carried out in England had not the Civil Service been reformed. In that year the appointment of Civil Servants by competitive examination was instituted; a very high standard of probity and efficiency was established and has been maintained. It would have been hopeless to try to carry through all this constant regulation of every side of life with a dishonest or bribeable officialdom.

During this period there was a great extension of the

*As to the Patent Act, *see pp.* 197-8.   †Now weekly

activities of municipalities and not merely of the State. To water and gas many municipalities now added tramways, and after 1880 electrical undertakings, thereby increasing the area of municipal trading.* The health movement again enlarged their activities in the matter of the inspection of food, which became one of their more important functions. The same movement caused many of them to undertake large schemes of municipal housing and their powers were enlarged by the Housing Act of 1890, while the London authorities obtained a considerable extension of their powers in this respect under the Act of 1894.

This reaction from laissez-faire was not merely due to the great Depression creating circumstances which forced the Government to act. The railways had induced an intensified form of competition in the world. To assist her own people Germany gave them all the support that the State could bring to bear. It was hopeless for individuals to fight powerful governments with the State resources behind them. Hence it quickened a movement in other countries towards State intervention. If Germany gave bounties on the export of beet sugar, Austria and France had to follow suit or see their subjects injured by the action of a foreign government. No country could afford to be outdone. It was obvious that Germany was flourishing, hence it could be argued that State intervention and protective tariffs did not necessarily produce inefficiency as had previously been maintained.

Regulation in industry had not produced unfavourable results on the output or the efficiency of the worker and this, too, made people willing to abandon laissez-faire. The English coal mining and cotton industries were highly regulated by the nineties and their products constituted the two most valuable exports. Nor had the labour organizations brought the country to ruin as had been predicted. The Depression of Trade Commission considered that the depression could not be ascribed to either trade unions or legislation.

" At the present moment there is, as we have already pointed out, a good deal of distress owing to the want of regular

---

*The indebtedness of local authorities in England and Wales increased from £192 million in 1887-1888 to £215 million in 1892-1893. Statistical Abstract for United Kingdom, 1896, *p.* 40, and 1906, *p.* 48.

work, but there can be no question that the workman in this country is, when fully employed, in almost every respect in a better position than his competitor in foreign countries and we think that no diminution in our productive capacity has resulted from this improvement in his position. We may add that in our opinion the unfavourable elements in the existing condition of trade and industry cannot with any justice be attributed to the action of trades unions and similar combinations."*

The Labour movement was growing in every country and was forcing governments to interest themselves more actively in working-class conditions and as what had already been done had not proved injurious, the various States were willing to go on trying experiments and each borrowed from the other.† In Great Britain the creation of the Trade Boards to fix minimum wages on which both employers and employed were represented, had a great influence in stimulating the formation of trade unions among unskilled workers who had not joined unions before. The National Health Insurance Act had the same effect. People had to join an " approved society " and many workers joined their trade union which was recognized as an " approved society," because they would get other benefits as well. This great growth of trade unionism strengthened the Labour party.

The growth of huge railway amalgamations worked in the same direction in England as in the United States. People became afraid of the power of the railways to throttle trade, and the amalgamations stimulated the growing control of railway rates by the State. To all these causes making for the abandonment of individualism came Science. It became obvious that in matters of health, no man could live to himself, hence the increased regulations intended to prevent disease and create a wholesome standard of life accustomed people to interference even in the home. Moreover, as scientific investigation became more elaborate ordinary individuals could rarely afford the expense of carrying out

*Final Report, Depression of Trade Commission, *p.* XXI.

†The social insurance system of Germany has been copied, for instance, in a modified form in the United Kingdom. *Cf.* Comparison of Labour Laws in different countries, in 24th Annual Report, United States Labour Statistics.

the necessary enquiries. Thus there was a growing tendency to insist that the State should subsidize scientific investigation for purposes of agriculture or industry as it was already doing for industrial diseases or for the sake of the general health of the community.

Although the reaction was due to a series of economic causes perhaps Mr. Joseph Chamberlain did more than any other personality, both on the social and colonial sides, to shape the general course of events.

All these new social services assumed by the State necessarily meant increased expenditure. The larger part of the extra money required in the United Kingdom came from an increase of the Death Duties and the Income Tax. The latter was so graduated after 1894 as to bear more heavily on the larger incomes, to which a super tax was added in 1909. Unearned incomes were less favourably treated than earned incomes after 1907. Measures were also taken in 1909 to relieve the man with children whose income was less than £500 a year. These allowances have undergone a rapid and continuous development since that date.*

Thus even in collecting its revenue, the State has begun to carry out a social policy of assisting or relieving certain classes and has met the deficit by taxing the rich.† In the process the majority of the incomes of the community come under review every year in the claims for repayment, abatement or exemption.

The following tables illustrate the position of the United Kingdom as compared with France, Germany and the United States at the beginning of the twentieth century and the comparison with twenty years earlier.

FISCAL BLUE BOOK, 1909, CD. 4954.

I—POPULATION. ANNUAL AVERAGE. THOUSANDS.

	United Kingdom	France	Germany	United States
1880–1884	35.188	37.728	45.185	52.514
1900–1904	41.966	39.052	57.984	79.015

*Royal Commission on the Income Tax, 1919. Minutes of Evidence, Vol. I., *pp.* 51-60.

†During the war this policy was carried further in that manufacturers and others were not allowed to keep more than one-fifth of the profits hey made over the pre-war standard.

	United Kingdom	France	Germany	United States

2—BIRTH RATE PER THOUSAND.

	United Kingdom	France	Germany	United States
1880–1884	32.4	24.7	37.6	
1900–1904	27.9	21.4	34.7	

3—DEATH RATE PER THOUSAND.*

	United Kingdom	France	Germany	United States
1880–1884	19.5	22.4	26.1	
1900–1904	16.8	20.0	20.2	

4—FOREIGN TRADE AGGREGATE IMPORTS, NET, ALL ARTICLES MILLION £.†

	United Kingdom	France	Germany	United States
1880–1884	343.6	190.9	151.8	140.1
1900–1904	466.0	182.1	287.0	186.0

5—IMPORTS MANUFACTURES.‡

	United Kingdom	France	Germany	United States
1880–1884	64.7	28.2	42.8	65.8
1900–1904	113.4	32.4	57.0	78.6

6—EXPORTS, ALL ARTICLES, DOMESTIC. MILLION £.§

	United Kingdom	France	Germany	United States
1880–1884	234.3	138.3	152.8	165.4
1900–1904	282.7	168.6	235.6	292.3

7—EXPORTS. MANUFACTURES ONLY.‖

	United Kingdom	France	Germany	United States
1880–1884	206.4	73.1	91.9	30.6
1900–1904	224.7	94.6	154.2	99.8

8—FOREIGN TRADE PER HEAD OF THE POPULATION. §§ IMPORTS, NET. ALL ARTICLES.

	United Kingdom			France			Germany			United States		
	£	s.	d.	£	s.	d.	£	s.	d.	£	s.	d.
1880–1884	9	15	3	5	1	2	3	7	2	2	13	4
1900–1904	11	2	2	4	13	3	4	19	0	2	7	1

*Note the favourable death rate of the United Kingdom.

†Note the enormous extent of the import trade of the United Kingdom as compared with the three other great powers.

‡Note the small proportion of the import of manufactured goods in the figures of the United Kingdom's trade.

§Note that the imports of France, Germany and the United Kingdom in 1900-1904 exceed the exports. In the United States the exports exceed the imports.

‖Note the enormous preponderence of the United Kingdom when it is a question of the export of manufactures. Note also the rapid increase of Germany's export of manufactured goods, a rise of £62 million as against £18 million for the United Kingdom, also the £69 million rise of the United States.

§§These figures show the preponderant position of the United Kingdom in another way.

	United Kingdom	France	Germany	United States

EXPORTS, DOMESTIC. ALL ARTICLES.

	£ s. d.	£ s. d.	£ s. d.	£ s. d.
1880–1884	6 13 2	3 13 4	3 7 8	3 3 0
1900–1904	6 14 9	4 6 4	4 1 3	3 14 0

9—EXPORTS, IRON AND STEEL. MILLION £.*

1880–1884	27.6	0.9	11.5	1.5
1900–1904	33.3	3.8	22.8	10.7

10—MACHINERY (EXPORTS). MILLION £.

1880–1884	11.5	1.1	2.7	2.6
1900–1904	19.5	2.3	10.1	14.7

11—COTTON YARNS AND MANUFACTURES (EXPORTS). MILLION £.†

1880–1884	75.9	3.7	5.0	2.6
1900–1904	74.7	7.5	14.9	5.0

12—WOOLLEN MANUFACTURES (EXPORTS).

1880–1884	18.5	14.7	11.3	0.07
1900–1904	15.8	8.7	11.5	0.13

13—(a) SHIPPING TONNAGE ENTERED AND CLEARED AT SEAPORTS. THOUSAND TONS.§

1880–1884	61.482	25.960	14.519	28.538
1900–1904	101.384	39.087	31.432	48.652

(b) SHIPPING TONNAGE ON REGISTER. THOUSAND TONS NET

				Gross tonnage
1880–1884	6.937	0.971	1.233	sea (a) 1.317
				lakes (b) 4.160
1900–1904	9.958	1.190	2.183	sea (a) 0.877
				lakes and rivers (b) 5.773

(c) TONNAGE SOLD TO FOREIGNERS. THOUSAND TONS.

1880–1884	103	14	22	28
1900–1904	249	22	47	12

*The preponderant position of the United Kingdom in the export of manufactured iron goods is striking when one compares the fact that the United States surpasses her in the output of pig-iron (*see* 18) and in steel output (19). It is interesting to observe the rapid increase of the export of machinery from the United States in twenty years by £12 million, while the United Kingdom increased by £8 million.

†Here again the predominant position of the United Kingdom is obvious. Note the rapid increase of German cotton exports.

§The business of the British seaports is obvious from these figures. The sea tonnage on the British register is more than twice that of the other three powers put together. Great Britain also built largely for other nations.

	United Kingdom	France	Germany	United States
	14—RAILWAYS—MILES IN OPERATION.*			
1880–1884	18.422	15.938	21.719	111.564
1900–1904	22.231	24.027	32.232	208.408
	PASSENGERS, MILLIONS.			
1880–1884	652	192	243	...
1900–1904	1,179	431	922	649
	GOODS, MILLION TONS.			
1880–1884	253	84	187	...
1900–1904	434	124	370	1,069
	GROSS RECEIPTS, MILLION £.			
1880–1884	68.6	44.1	47.9	...
1900–1904	108.7	59.7	103.4	361.6
	15—COTTON CONSUMED, MILLION CWTS.†			
1880–1884	12.9	2.0	2.9	8.3
1900–1904	14.3	3.6	6.8	18.2
	SPINDLES, THOUSANDS.			
1880–1884	41.170	3.887	4.900	12.087
1900–1904	46.640	5.940	8.450	21.403
	16—WOOL CONSUMED, MILLION LBS.‡			
1880–1884	336.0	409.5	219.8	343.1
1900–1904	450.8	517.2	345.3	448.4
	17—COAL OUTPUT, MILLION TONS.§			
1880–1884	156.4	19.3	51.3	88.7
1900–1904	226.8	31.8	110.7	281.0
	18—PIG-IRON OUTPUT, MILLION TONS.‖			
1880–1884	8.1	1.9	3.2	4.2
1900–1904	8.6	2.6	8.9	16.4

*Note the huge mileage and goods traffic of the United States and the large passenger traffic of the United Kingdom.

†It is interesting to note the large consumption of cotton by the United States as compared with the value of the export of cotton manufactures (11). Note the predominance of spindles in the United Kingdom.

‡It is interesting to notice that France consumed more wool than the United Kingdom both in the period 1880-1884 and 1900-1904.

§The coal output of the U.S.A. passed that of Great Britain for the first time in the quinquennial period, 1900-4.

‖The pig-iron output of the U.S.A. surpassed that of the United Kingdom in the period 1890-1894.

	United Kingdom	France	Germany	United States
	19—Steel Output, Million Tons.*			
1880–1884	1.8	0.4	0.8	1.6
1900–1904	4.9	1.7	7.3	13.4

The following table shows that Great Britain was main-taining in 1913 her enormous lead in trade and bears witness to the strength and vitality of British industry as a whole.† It also shows the growing importance of German trade.

INCREASE in Million £ on Average of 1910-1913 over Average 1895-1899.

	United Kingdom	Germany	France	United States
Net Imports - -	218	260	155	188
Imports of manufactured and partly manufactured goods - -	72	48	37	81
Total Exports of domestic produce and manufactures - -	230	244	115	221
Exports of manufactured and partly manufactured goods - -	177	170	71	140

II.—Causes of the Supremacy of Great Britain during the Nineteenth Century.

The supremacy of Great Britain between 1789 and 1914 was the result of a combination of several factors.

In the first place she had a long start and although that meant that she had to bear the burden of the experiments and that other countries could begin where she left off, it did mean that she had evolved a race of skilled and trained workers such as no other country in the world possessed and this enabled her to improve upon or adapt machines invented

*These figures illustrate the growth of the steel output and the important position occupied by the United States.

†Report on Commercial and Industrial Policy after the War. Cd. 9035, 1918.

elsewhere.* Although English machines were exported in large numbers after 1825, foreigners could not work them to anything like the same advantage as the English.† This highly developed skill in engineering enabled her to acquire and develop the new trade of iron ships so that she became the world's ship builder. It is well known that the Lancashire cotton spinner could work more spindles than any cotton operative in the world and that English fine yarns are unsurpassed. During the nineteenth century British goods invariably stood for good quality and workmanship and great was the difficulty Great Britain experienced in preventing other nations putting her trade marks on their inferior goods.‡

Moreover, as Great Britain was the first to develop machine industries on a large scale the subsidiary industries had grown up around the principal industries and acted as their feeders.

If the long start had given the British a race of trained workers it had also helped to develop a race of bankers, merchants and manufacturers who were able to take large views, who were highly specialized in finance, exchange and manufacturing and who were able to visualize the world as one market. The development of rapid communications first of all in this country tended to foster the sense of statesmanship of commerce in the British directors of companies, general managers, captains of industry and merchant princes. The result was that an extraordinarily high level of specialized

---

*" It is admitted by everyone that our skill is unrivalled, the industry and power of our people unequalled : their ingenuity as displayed in the continual improvement of machinery and production of commodities without parallel." Report, 1825, V., *p.* 16. Speaking of inventions brought here from the United States the Report of 1841 says : " Those machines have been subsequently improved upon by the great skill of the mechanics of this country and by the greater aptitude which the artisan here has in suggesting or carrying out improvements on account of the knowledge which the great quantity of machinery constructed in this country has given to the machine maker." Report, 1841, *p.* 112.

†Report on the Export of Machinery, 1841, VII.
It was said that the English artisan earning 50s. was more profitable than the foreign artisan earning 20 *frs.* Men brought from Staffordshire working at the same piece-work rates as Belgians, earned £3 - £5 a week, while the Belgians earned 18 - 20 *fr.* Report, *p.* 30. This was also Brassey's experience. A. Helps, " Life of Brassey."

‡Reports on the Merchandise Marks Acts.

skill in business was developed.* Great Britain has also gained enormously from having been first in the field and having all the agencies established. Once a trading connection has been set up it is very difficult to break it. People get used to the kind of things they have been accustomed to purchase, merchants get used to trading with certain firms and the presumption is strongly in favour of the continuance of the historical line of connection.†

It is impossible to overestimate the importance to this island during the nineteenth century of the abundant and accessible supplies of coal. It gave her a cheap motive power from the very beginning and was especially valuable as her water power is limited.

It is only when one sees how Britain's great industrial rival, France, was hampered by the cost of coal all through the century that one realizes the enormous bounty bestowed by nature upon this country. Not merely was coal abundant for power but it was the right kind of coal for iron smelting. Durham coal is probably the best coking coal in the world.‡ Cheap and good coke joined to the existence of skilled artisans enabled Great Britain to make cheap machines, cheap locomotives, steamers and engines, and she was thus able to become the world's construction shop and forge. Cheap coal enabled her to work her steamships cheaply and the general demand for English coal gave the ships that went out to fetch corn and raw materials an outward as well as an inward freight and this made for cheap rates of transport and lower prices. The abundance of coal thus enabled this country to buy her food and raw materials cheaply as well as to manufacture cheaply.§

Not merely were there vast deposits of coal but the geographical situation of the coal fields was most favourable, i.e., the coal was not merely there, it was get-at-able and

*Whelpley, " Trade of the World."

†" When the (cotton) mills were started (in Moscow) about forty-five years ago, British carders, weavers and spinners, as also managers and assistant managers were brought over from the United Kingdom to set things going and organize the cotton industry on a stable basis. The managers and foremen in many mills are still Lancashire men, and the result has been that the machinery ordered has been and still is almost exclusively of British manufacture, the small balance coming from Alsace."   Consular Report, Moscow, for 1910, published 1911.

‡Cd. 9084. Appendix I., p. 79.

§In 1913 the export of coal was worth £53,659,660. Cd. 9093, p. 29.

transportable. When the revolution took place in the steel industry after 1870, it was necessary to import iron ores from Spain and Sweden for Bessemer steel as the English ores were not pure enough. It was the situation of the English coal fields on the coast in Wales and the North that enabled the ore to reach the coal from Spain and Sweden without an expensive inland haul by railway. Great Britain was thus able to continue to be one of the great iron producers of the world. In an economic civilisation which during the nineteenth century was based on coal England was the largest coal producer. It was only in 1900 that she was surpassed by the United States,* but she continued to be the largest coal exporter up to 1914. Here again the geographical situation of the coal fields on the coast helped the export. German coal exports to the Mediterranean, for instance, were hampered by the long railway haul. It was easier for Great Britain to reach the Southern countries by sea. The cost of coal production in the United States is considerably less than in the United Kingdom but as far as export is concerned the extra haulage from the United States mines to the ports and the longer sea passage neutralized the cost of production.

The magnitude of her coal exports has caused Great Britain to provide docks and other accommodation in foreign ports for bunkering ships. " Most of the docks and coal handling facilities in South American ports are controlled by English companies who have long had the business of bunkering

*PRODUCTION OF COAL.
MILLION TON.

Annual Average.	United Kingdom	France	Germany	United States
1855–1859	66.0	7.5	—	12.4
1860–1864	84.9	9.8	15.4	16.7
1865–1869	103.0	12.4	23.5	26.7
1870–1874	120.7	15.1	31.8	43.1
1875–1879	133.3	16.3	38.4	52.2
1880–1884	156.4	19.3	51.3	88.7
1885–1889	165.2	20.7	60.9	115.3
1890–1894	180.3	25.4	72.0	153.3
1895–1899	201.9	29.6	89.3	189.1
1900–1904	226.8	31.8	110.7	281.0
1905–1908	254.1	34.0	135.3	380.2
		Fiscal Blue Book, *pp.* 166-167, Cd. 4954.		
1913	287.4	40.1	187.0	508.9

M.

English ships at call all over the world."* It is to this control of the facilities for handling coal that the American Federal Trade Commission are inclined to attribute the success of the English in supplying South America with coal.

The magnitude of her shipping connections has been another cause of the expansion of British industry and trade. Her ubiquitous tramp has given her facilities for the receipt and despatch of goods which were unrivalled by any other country before 1914. It is scarcely realized, however, what an important asset the English ship captain has been in pushing English trade. As he goes all over the world it is his business to get freights, and he is willing to carry for anyone who will charter him, but he wishes above all to get back to England, and will work towards that end in getting cargoes.† He is one of the best agents for British trade and he is found everywhere.

As has already been observed, the United Kingdom was organized for world trade when other countries, France excepted, were only developing national unity. The result is that during the last half of the nineteenth century English financial connections were established all over the world. The bill on London has become the international currency of commerce. Purchases in the United Kingdom have been stimulated because financial settlement has been so easy. Great Britain " has provided all the financial facilities needed for its exporters and importers to do business with other men anywhere on the globe, and conversely through British agencies, merchants in the most remote regions have been able to transact business not only with British subjects but with the merchants of any other country. . . .   In short, wherever

---

*U.S.A. Federal Trade Commission, Co-operation in American Export Trade, I., p. 340.

†" There is no merchant that is so good a drummer for trade as the ship-owner. You get a fellow with a ship in the Far East, and he has no cargo for it ; anything short of stealing that fellow will resort to to get a cargo for his ship, so we often get very low rates of freight which is a great advantage to the Pacific coast, for instance. If I was living in London do you think I would be pulling for the Pacific coast ? Not at all. I would be pulling for London . . . every ship-owner, wherever you find him, is working to try and get his ships back home to his home port . . ." Witness before the Federal Trade Commission, U.S.A., 1916. Report on Co-operation in the American Export Trade, I., p. 35.

British imports are bought or British exports sold there is either a local bank intimately connected with London or there is a British bank for the accommodation of British commerce." No other nation has any such comprehensive financial organization for foreign business.*

Great Britain has been above all the financier of great constructional works—railways, docks, electrical light and power works, water works, electric tramways, telegraphs, telephones, cables, all over the world, to say nothing of mining and plantation companies.† The materials for those works were generally ordered in this country. The engineers were British and preferred British goods in the use of which they had been trained at home, the directorate of most of the companies was domiciled in England, they had interests in other British concerns and the orders were placed in the United Kingdom.‡ The orders for renewals, always such an important consideration, would also come here.

Thus the early start, the abundant coal supplies, the skill of the British artisan, the ubiquity of British shipping, the universality of the British financial organization, the magnitude of British investments abroad combined with the excellence of the manufactures of the United Kingdom have all combined to ensure her predominance during the past century.

III.—GROWTH IN THE WELFARE OF THE WORKING CLASSES.

This great industrial and commercial development and consequent creation of wealth has not been carried out at the

*Federal Trade Commission, I., *pp.* 40, 44.

†The Federal Trade Commission, II., *pp.* 537-574, covers thirty-seven pages with the enumeration of mere names and approximate capitalization of British Companies abroad. In round numbers about £4,000 millions of British capital were invested abroad in 1913.

‡Federal Trade Commission, *p.* 66, *p.* 281.

The importance of training engineers who get used to certain types of mechanism and do not readily take to others and therefore act as commercial agents for the country where they are trained, was recognized by the Germans. " Promising young Chinese were educated as engineers in the schools and universities in Germany at the expense of German business organizations as well as the establishment of an engineering school in Shanghai with German equipment and German instructors for the training of young Chinese in China in German engineering standards, methods, equipment, etc." The Report of the Federal Trade Commission, *op. cit* , I., *p.* 112.

expense of the working classes.   Not merely have their wages risen but they get more for the money, *i.e.*, real wages have risen as well as nominal wages.   The following table drawn up by Professor Bowley* brings out that fact :

	Nominal Wages.	*Prices.*	*Real Wages.*
1790–1810	Rising fast.	Rising very fast	Falling slowly
1810–1830	Falling	Falling fast	Rising slowly
1830–1852	Nearly stationary	Falling slowly	Rising slowly
1852–1870	Rising fast	Rising	Rising considerably in the whole period
1870–1873	Rising very fast	Rising fast	Rising fast
1873–1879	Falling fast	Falling fast	Rising fast
1879–1887	Nearly stationary	Falling	Rising
1887–1892	Rising	Rising and falling	Rising
1892–1897	Nearly stationary	Falling	Rising
1897–1900	Rising fast	Rising	Rising
1900–1914	Falling a little	Falling and rising	Stationary

* "Dictionary of Political Economy," ed. Palgrave, 1908.   **Appendix,** *p.* 801.

Professor Bowley shows the rise of wages by an index number as follows :— *

	Index Number of Nominal Wages.	Real Wages.
1850–1854	55	50
1855–1859	60	50
1860–1864	62	50
1865–1869	67	55
1870–1874	78	60
1875–1879	80	65
1880–1884	77	65
1885–1889	79	75
1890–1894	87	85
1895–189	92	95
1900–1904	100	100

This table shows that real wages doubled in the last half century. Education has been free since 1891, illiteracy has declined, food is more varied, transport has developed opportunities for holidays and change, pauperism has lessened in proportion to the population,† serious crime has diminished.‡ Thrift, as shown by the accumulations of friendly societies and in savings banks showed a marked increase.

FUNDS.§

	1877 £	1905 £
Ordinary friendly societies	5,211,052	18,056,640
Branches of registered orders	7,752,050	23,888,491

	1850	1907
Deposits in Post Office and Trustee Savings Banks	£29 millions. £1 1s. 0d. per head of population.	£209½ millions. £4 15s. 1d. per head.

It seems clear that the Englishman was better off in 1834 than the continental labourer and as a consequence of the

*He says : " The result is not to be regarded as final in any sense but rather as showing the direction of the effect of the change of prices, *i.e.*, the nature of the numerical relation between nominal and real wages."

†Statistical Memoranda and Charts relating to Public Health and Social Conditions. Cd. 4671, 1909. Charts *p.* 52, *ib. p.* 104.

‡Statistical Memoranda Chart 3 and 5.   Section VI.

§" Statistical Memoranda," *p.* 103-104.

tariff reform agitation, another great enquiry was undertaken in 1909 into the cost of living in France, Belgium, Germany and the United States. It is almost impossible to make an exact comparison of the condition of workmen in two countries. It is clear that if an Englishman had gone to France in 1909 and wanted to live in just the same way as he had lived in England he would have been worse off. His tea would have been dearer, his coal would have been dearer, his jam would have cost him more, his house would have been a very different type of thing, probably a flat for which he would have had to pay a higher rent if he wanted as many rooms as he had in his English house. Nor would his wife have found the water " laid on." She would have had to fetch her water from the pipe in the court. On the other hand if he were willing to drink wine instead of beer he would have found it cheaper than in England, he would have found his milk, eggs and poultry cheaper. His meat diet would have been more varied and he would have eaten beef, veal, mutton, bacon, charcuterie, pork and poultry and if he had adapted himself to French ways, he would have eaten far more fruit and vegetables. In different countries people have quite different wants and habits and it is difficult to compare them. Bread in France, though bearing the same name as in England, is a totally different thing, a house is a different thing, wages are differently paid and occupations are differently planned. In France the wage was still largely a family wage, women habitually added to the family income in 1909. In Roanne, in the budgets obtained, 97.5 of the wives worked, 82 per cent. worked at Grenoble, the centre of the glove industry, and 81 per cent. at Fougères, a boot and shoe making town.

Wages of the French workmen as compared with the English were said to be as 75 is to 100, the corresponding ratio for Germany being as 83 is to 100 ; the hours of labour of the French artisan were from 13 to 23 per cent. higher, yet who can measure the intensity of effort put into the two periods—a man may take a long period in a leisurely fashion or put in very strenuous work in shorter hours with more resultant fatigue.*

*Cd. 4512 (France) 1909, Cd. 4032 (Germany) 1908.

IV.—THE CONTRAST BETWEEN THE INDIVIDUALISM OF GREAT BRITAIN AND THE PATERNALISM OF FRANCE AND GERMANY.

A peculiarity of British industrial development in the nineteenth century lay in the fact that it owed practically nothing to State aid. The English inventor received neither capital nor encouragement from the authorities, whereas in both France and Germany the State afforded substantial assistance.

The energy of the Prussian Government in bringing the knowledge of English machines to the very doors of its people is remarkable. This is an account given by a witness before the Royal Commission on the export of machinery in 1841 :

" I found at Berlin the most enterprising and systematic exertions made on the part of the Government to obtain a command of the manufacture of machinery. I found no expense spared for that purpose and the exertions quite astonished me. There is one very important institution at Berlin called the Gewerbe Institut which is a large establishment for practical education combining design with almost every branch of manufacture into which science and mechanics enter. In going through the room of this institution with the Professor I saw suites of apartments completely filled with models of English machines. The Professor informed me that they had in it models of every machine in use in Great Britain for the manufacture of cotton, flax, silk and wool and likewise a number from America and Germany, that by these means they were enabled to have our recent improvements but what was a matter of importance which we cannot command that they were enabled frequently to combine in the same machine two distinct English patents.

" The system, he told me, was that this Machinery as soon as produced in England was immediately imported at the expense of the Government and set up at the Gewerbe Institut ; that it was proved, that a working model was immediately made from it to be deposited in the institution and that the original was presented as an honorary prize by the Government to some manufacturer in Prussia who had distinguished himself in the peculiar branch to which it was applicable. In the Institut likewise the pupils were taught to make the machinery themselves, they were supplied with the tools and they were permitted to carry away the machines which they themselves had constructed."*

While on the one hand this extract shows the influence of English technique on the country destined to become her great industrial rival at a later date, on the other it is a striking instance of the paternalism of the Prussian government and its direction of industry and technical education especially when contrasted with English laissez-faire methods.

Not merely was the introduction of machinery into the United Kingdom and the consequent remodelling of industrial life due to the efforts of private individuals working with their own or borrowed capital, but they were equally free from any government restrictions as to wages or conditions under which the employees should work. All these had been swept away finally in 1813 though obsolete long before. Only when it was a question of compulsorily acquiring property for canals or railways had the sanction of Parliament to be obtained. To this scope for individual enterprise and initiative the manufacturers were wont to ascribe their success and were enthusiastic advocates of laissez-faire.

"The freedom which under our government every man has to use his capital, his labour and his talents, in the manner most conducive to his interests are inestimable advantages ; canals are cut and railroads constructed by the voluntary association of persons whose local knowledge enables them to place them in the most desirable situations and these great advantages cannot exist under less free governments. These circumstances when taken together, give such a decided superiority to our people, that no injurious rivalry either in the construction of machinery or the manufacture of commodities can reasonably be anticipated."†

The reconstruction of the industrial life of the nation by individuals in Great Britain is in striking contrast with the developments of machine industry in France.‡ Ever since

*1841, VII., *p.* 87.
†Report on Export of Machinery, 1825, *p.* 16.
‡*Edinburgh Review*, 1820, " State of Science in England and France,"
p. 416 :
" In France, too, the Government is the great protector and promoter of science ; and not merely urges on, but even directs the pursuit of the learned. This likewise has been much extolled. . . . In England the Government does less because the subject does more.

the days of Henry IV. the manufacturers were accustomed
to look to the government for assistance. The collapse of
French industry had been so great at the Revolution that it
was only by Napoleon's personal effort and financing that
machine industry could be restarted but it was quite in
accordance with French tradition that the State should help
manufacturers and above all provide the means of good
transport in roads and canals.

In Germany the assistance afforded the manufacturer
by the government arose from much the same reason. The
destruction of economic life in the Thirty Years' War was so
complete that it could only revive again with State help.
Foremost among the princes at the work of economic reconstruction were the Kings of Prussia and the people became
accustomed to look to the monarch to do things for them.

Apart from this in all the countries of Europe the agricultural population, a large proportion of whom were still
serfs in 1780, had looked to the feudal lord for justice, guidance,
employment and maintenance. This habit of depending on
authority was not lightly broken and there is, as has been
already shown, a much stronger tradition of paternalism and
State control in France and Germany than in Great Britain.
People cannot be suddenly deprived of guidance, as the
English have found in India.

Lack of capital, destruction of economic life, poverty of
initiative and the traditions inherited from the days of feudalism and serfdom caused the governments of Prussia, France
and Russia to take the lead in encouraging, introducing or
organizing the new methods of either industry or transport
in the nineteenth century.

Apart from the non-existence or early disappearance of
serfdom in Great Britain, the reason for her different evolution
is to be found first of all in the poverty of the English monarchs.

In free governments, it is not so much the function of the rulers to
enlighten the governed as of the governed to enlighten them. In
order that the people may be wise wisdom must be a demand of the
people. The only knowledge which men truly appreciate is that of
which they feel the value ; not that which they are told is excellent
or which is pointed to as glorious. The enlightened state of the wealthy
British population and the efforts of those who would become both
enlightened and wealthy spare our government from all solicitude
upon science."

Queen Elizabeth had a royal income which rarely exceeded £300,000 a year and with that she had to carry on the whole government with an occasional dole, every four or five years, from Parliament for an emergency. Charles I.'s income was about £550,000 a year and he was always crippled for money, as his annual expenditure ran to about £850,000 in the period between 1630 and 1642. The English monarchs might bless an enterprise but they could take very little share in it. Thus the expansion of English foreign trade and colonisation was carried out by merchants associated in Chartered Companies. Even the Pilgrim Fathers were a joint stock company. What the English merchant or manufacturer wanted he had to do for himself. This independence of government was stimulated by the coming of Puritanism. The same spirit which led men to refuse to acknowledge an established church or to set up any priest between themselves and God, led them to be as independent in matters of trade as they were in religion. They resented authority in both. Many of them considered that to be poor was to prove that the Lord had turned His countenance away ; those whom the Lord had blessed prospered. The incentive to " get on " was very strong when the acquisition of wealth seemed to be an earnest of ease and comfort both here and hereafter. Moreover, there were very few things on which the righteous Puritan could really spend his money without sin and he became an excellent vehicle for the accumulation of capital.*

This tradition of independence naturally went with the colonists to the New World, especially as so many of them were Puritans, and it is no accident that the Americans should follow Great Britain so closely in leaving so much to private enterprise and coming so late to that government intervention which seems to be the characteristic of the modern world. Ireland was a conquered country and had to depend on what the conqueror would permit. She has therefore inevitably had to look to the government as the power which could give or withhold. She was also a poor country and for that reason looked to her richer neighbour for financial assistance†

---

*On the influence of Puritanism, Levy, " Economic Liberalism," Schulze-Gævernitz, " Englischer Imperialismus," Cunningham, " Christianity and Economics."

†" Of the £119,421,373 advanced to the United Kingdom up to March, 1893, the sum of £52,283,698 or 43.78 per cent. of the whole

which meant State intervention. She has been in an especially privileged position since 1880 but the assistance she has received has been assistance to agriculture and not to industry.

In the district where the conquerors are the predominant element and where the Puritan tradition is strong, viz., in Ulster, such assistance has not been necessary. A great ship-building industry has been developed during the nineteenth century by private initiative and the linen industry has been successfully re-modelled along the lines of machinery. It is also easy to import coal from Scotland to the North of Ireland* which no doubt assisted the development of industry in the North.

France received something like the Puritan stimulus in 1789 with her faith in " liberty, equality and fraternity " which counteracted the effects of late emancipation from feudalism. She learnt to believe in the principles of the " glorious Revolution " and once capital was accumulated again and political life had become fairly settled, French industry has owed little to the State except good transport facilities. Machine industry

had been advanced for Irish purposes and of that no less than £10,718,095 or more than a fifth of the amount advanced to Ireland has been remitted or written off, been treated as a free grant, while only £1,154,514 or one fifty-eighth part of the amount advanced in Great Britain has been so dealt with. This seems to be due to two main causes, viz. (1), the difficulty or supposed difficulty of raising capital in the open market for Irish purposes and on the credit of the Irish local authorities without the intervention of the credit of the State. (2) The special social and political circumstances which have led to a large expenditure in Ireland upon public works and relief of distress." Financial Relations Commission, *pp.* 160-161, C. 8262, 1896.

This illustrates that where the country is poor or dependent State intervention becomes necessary even when the government professes laissez-faire principles as the English government did between 1815 and 1886.

*The characteristic difference between Ulster and the rest of Ireland is brought out in the following passage : " Only the other day a Royal Commission investigated the needs of Irish harbours. The Southern ports dwelt much on the improvements necessary and the impossibility of carrying them out without assistance from the Treasury. ' Belfast explained that it itself had made its harbour, and would itself determine and carry out all improvements required. It has already commenced these improvements which will cost it ultimately millions of pounds." Strahan, " A Tale of Two Cities," in *Blackwood's Magazine* August, 1919, *pp.* 152-153.

required State aid to restart, after that, it developed along the lines of laissez-faire. German industry, on the other hand, has continuously owed a great deal to the State, especially in the matter of technical encouragement. The extract quoted above shows how early and vigorously this tendency manifested itself in the nineteenth century. The leading characteristics of French, English and German industrial development can, however, be traced from the middle of the seventeenth century. The year 1649 saw the execution of Charles I. in England, and this meant the destruction of the King as the guiding power in economic life. After that Parliament was supreme, and Parliament, being then an unwieldy body of several hundred country gentlemen, allowed English industry to work itself out in its own way unhampered internally by restrictions on enterprise. It merely saw to it that there was a high protective tariff and wide overseas markets. A year before the King's death in 1649, the Peace of Westphalia ended, in 1648, the Thirty Years' war and the collapse of Germany was complete, throwing her, so to say, into the arms of her princes for reconstruction. Almost within a year the King was dispensed with in one country and made a necessity in another. In 1642 Richelieu died, having completed the destruction of the political power of the nobility, having created a great economic administrative instrument of control for the King in the intendents, and having made the royal power the greatest thing in France. It was therefore inevitable that there would be a direction of industry and trade from above by the monarch, especially as the French king was a very wealthy potentate, who could give subventions and other financial assistance when he chose. Colbert, after 1660, reorganized the economic life of France along the lines of royal autocratic control and the whole of French industry looked to the King for encouragement, regulation, direction and inspection. The Revolution destroyed the tradition of State assistance but it was revived by Napoleon and the two conflicting ideas of freedom and control are active at different times and in different ways during the nineteenth century in France.

In the eighteenth century France had been the great industrial country, England came second; in the nineteenth century the positions were reversed. The interesting thing

is to notice that both Great Britain and the United States*
have tended to develop in the direction of increasing State
control as well as France and Germany, autocratic by tradition.

v.—The Economic Position of Great Britain in 1815 and 1914.

There is a striking contrast between the rural nature of
Great Britain at the beginning and the urban character of
the country at the end of the nineteenth century. In 1815
she may still be termed an agricultural country, as four out
of every ten of the male workers in England and Wales were
engaged in agriculture. She provided her own corn and other
food supplies. To her old woollen industry she had added
two new trades—cotton and engineering. She was weighted
down by a huge debt left from the war and the heavy taxation
and stagnation of trade combined with the agricultural revolution, the inflation of prices by the over-issue of paper, the bad
Poor Law, the insanitary condition of the growing towns and
the friction caused by the introduction of machinery made
the condition of her people very miserable though it seems
to have been better than that of the continental labourer, and
real wages were rising all the century after 1810. The State
did not intervene for many reasons. It was afraid of injuring
the manufacturers on whose efforts it relied to increase
production and lift the burden of the war ; it believed that
State regulation was a bad thing in itself and contrary to
" natural liberty " ; and in any case it knew no adequate
remedies.

By the beginning of the twentieth century Great Britain
had become a great food importing nation ; only one in ten
of her male workers were still in agriculture, seventy-seven
per cent. of the population was massed in urban areas in
1901.† Her manufactures had developed in all directions ;

---

*Hadley : " Undercurrents in American Politics."

†For every 100 persons living in 1851 in

London, there were approximately	203	in 1908.
84 large urban areas do.	282	,,
14 rural counties do.	95	,,
Rest of England and Wales	184	,,

[Cd. 4671 (1909)]

the debt had been partly paid off by sinking funds and terminable annuities but a debt of £651 millions (1913-1914) distributed among the 45,221,615 people of the 1911 census was a very different thing from the £878 millions (1815-1816) borne by the 18½ million people who comprised the United Kingdom in 1811. Taxation had been entirely recast, free trade was substituted for the old protectionist tariffs, the Poor Law had been reformed and this had cleared the way for the tackling of the problems of special classes of poor ; the towns were becoming increasingly healthier, the death rate was the most favourable in Europe, having fallen from 21.3 per thousand in 1870 to 15.4 in 1908* and certain fevers such as typhus and small-pox had disappeared while all epidemics had diminished. Britain's great industrial rival, France, had ceased to be so important after 1870 but a new and vigorous nation, Germany, had emerged and was becoming a formidable competitor ; the United States had become an industrial nation but she was not as yet an important rival. She exported food stuffs and raw material such as cotton, but the exports of her manufactures were small in comparison.

A great industrial code had been built up and with its inspectors and trained civil servants and the great increase in scientific and engineering knowledge the State could attempt to do many things in 1914 that were impossible in 1815. Perhaps nothing is more striking than the contrast between the belief in laissez-faire in the early part of the nineteenth century and the wide field of government and municipal action at the beginning of the twentieth century.

The contrast between the extent and value of the colonies and dominions making up the British Empire is no less remarkable. In 1783 England had lost the continental strip of coast line with its three-and-a-half to four million inhabitants which has now expanded into the United States. She still possessed in 1815 a northern region, " our lady of the snows," which seemed to offer little hope of successful development, partly owing to climate and partly owing to the feuds between the two races, French and English, inhabiting Upper and Lower Canada. At the Cape there was the same difficulty of two races but here the question was accentuated by the

*For England and Wales it was 13.7 in 1913.

English desire to abolish slavery on which the agriculture and livelihood of the Dutch depended. Australia seemed to be a waterless region without native products and was chiefly used as a dumping ground for convicts. Until about 1830 it was not apparent to people in this country that there was the possibility of the lands of the Antipodes being great wool producing countries. The West Indies were vitiated in English eyes by the fact that their economic existence was based on slavery.

The Empire between 1815 and 1850 seemed unpromising, and Englishmen in general, organized as they were for world-wide enterprise, expected little from it in the way of economic benefit while they did anticipate great disadvantages in the complications that might arise with foreign governments over the colonies.

The developments of mechanical transport altered that point of view. Railways penetrated interiors and helped the inner settlement of continents—the wheat producing prairie belt of Canada was opened up as was also the interior of West Africa stimulating the export of oils and cocoa. The railway assisted the cotton production of Egypt and made the multifarious raw products of India available in enormous masses. Its carriage provided increased employment for shipping while cold storage steamers showed the value of Australasia for mutton as well as wool. The dominions and dependencies became increasingly good markets for manu- factures ; after 1880 India was our largest customer for manufactured goods.

Instead of being, as old colonies were, either islands or a fringe of people along a coast-line or river, the new Empire consisted of developed continental areas, and included a quarter of the inhabitants of the world linked up by railway and rapid steam communications with the mother country and each other.

A medal was struck in 1670, probably for the Royal African Company, bearing the inscription, " Britannus diffusus in orbe."* What was then a prophecy is now an accomplished fact.

---

*Reproduced in Cunningham, " Growth," Vol. II. Frontispiece.

PART IV

THE COMMERCIAL REVOLUTION CAUSED BY MECHANICAL TRANSPORT

SYNOPSIS

The coming of the railway and the steamship meant the substitution of a world economy for a national economy—the general results of which were world interdependence and world rivalry. Mechanical transport created a revolution in the commercial and industrial importance of States. It created a new mobility of goods and persons. National policies were affected by the need to build and control railways.

I.—THE REVOLUTION IN THE IMPORTANCE OF CONTINENTAL AREAS

The penetration of interiors by railways resulted in the development of North and South America from populated coast lines to populated continents. Asia and Africa were also brought into the world economy.

1—In Europe Germany was enabled to become the Central European land distributor and a Mediterranean Power ; Russia was able to get an outlet by railway during the winter when the ports were frozen. Both countries were able to become iron and steel producing nations owing to the development of railways which brought the ore and coal together and facilitated distribution of the finished product.

2—The interior of the United States was opened up by railways and she began to ship her grain and meat to Europe. The railways developed her great iron and steel works.

3— The British Empire.
    (a) Great Britain.
        1—As pioneer of railways Great Britain was the first to experience the speeding up of production caused by railways but she suffered from having to make the experiments.
        2—They confirmed her in the position of the workshop of the world, 1850-1873.
        3—The railways created new rivals for Great Britain in the new industrial Empires.
        4—A new industry arose—iron and steel ship building, which made Great Britain by far the greatest ship builder and carrier of the world.
    (b) The Dominions and Dependencies.
        1—Mechanical transport provided a medium of rapid communication between all parts of the Empire and linked it together.

2—It made colonies increasingly valuable as markets
and sources of raw materials when the interiors
were opened up.

3—It created a new desire to dominate colonial areas
among the Great Powers. Great Britain had to
increase her Empire to preserve the " open door."

4—New and intensified international rivalry due to the new accessibility.

II.—THE REVOLUTION IN COMMERCIAL STAPLES AND COMMERCIAL ORGANIZATION

The importance of commodities changed.

1—Bulky commodities, coal, machinery, food stuffs and raw material
took the place of spices and colonial products as the principal
articles of commerce.

2—There was a new demand for iron for construction and renewals
of railways and rolling stock.

3—Disappearance of fairs.

4—Growth of combinations to eliminate the competition made easy
by rapid communications :

      Vertical combinations.

      Horizontal Combinations.

      International Combinations.

      Combinations to purchase raw material.

5—Growth of co-ordinated organization in business :

      Multiple shops.

      Allied industries under one management.

6—The difficulty of controlling international combines by national
machinery.

III.—THE CREATION OF A NEW FINANCIAL ERA

1—National Finance affected by the building and working of State
railways.

2—New and extended fields for investment.

3—Problems of taxation of capital invested in several countries.

4—New financial mechanism.

IV.—SOCIAL EFFECTS OF THE COMMERCIAL REVOLUTION

1—New personal mobility led to the growth of towns.

2—Growth of new industrial class of transport workers.

3—Increase of shop-keeping and trading classes.

4—Effect on the position of women.

5—Emigration

      (a) of Europeans.

      (b) of Asiatics.

6—Government policy with regard to immigration and emigration.

N

ECONOMIC development has passed through three stages. There is first the stage of local economy when the manor and later the town with its surrounding district aimed at being self-sufficing and there was very little intercourse between one part of the country and another. This is the characteristic of the thousand years known as the Middle Ages. After 1492, with the discovery of the sea routes to India and America a national economy began when the nations, supplemented by their new colonies, also aimed at being self-sufficing but within larger areas. Sailing ships, boats on rivers and riding and pack animals were the chief means of communication. The roads were earthen tracks suitable for animals. Towards the end of the period, *i.e.*, in the late eighteenth century, the roads began to be metalled, canals were built and rivers were improved. The capacity of transport was enlarged by two new features : the canal barge drawn by a horse and the cart or other wheeled vehicle which could now be used on the new roads with made surfaces. This period came to an end with the general development of the railway and the steamship and a world economy took the place of national economy. Instead of each nation being obliged to live to itself within a tariff wall of isolation all parts of the world are linked together into a common economic system and no country can remain isolated. Nations have developed into great land or sea empires each owning or dominating financially large portions of the globe. This period of world economy which means world production, world distribution, world interdependence and world rivalry may be held to date from 1870, by which time railways and steamships were developed in England, France, Germany and the United States to a point where their means of communication were revolutionized and this was also the case in Russia during the nineties.

Between 1850 and 1890 the Great Powers were in turn confronted by problems arising from the new methods of distribution in addition to those arising from the new methods of production in mines, factories and blast furnaces. The Powers were, however, chiefly occupied at first with getting railways built, subsidizing steamships to carry national goods and controlling the grosser and more obvious abuses of railway monopoly or corruption. The revolution which the

ra'lways and steamships were going to create in world commodities began to be apparent in the eighties when the new
facilities for food imports seriously affected European agriculture.

As in the case of machinery, the steam engine and the
metallurgical industries, the new transport inventions
emanated from England and are another instance of her
technique reacting powerfully on the economic life of other
countries as well as her own. If French ideas helped to
transform the status of persons, English inventions helped
man to control the forces of nature which had kept him in
subjection hitherto.

The combined effect of railways and steamships was to
introduce some wholly new factors into economic life, viz.,
speed, safety, regularity, cheap transport, and the power
to move bulky and weighty goods in large quantities over
great distances. Mechanical transport also minimised such
geographical limitations as mountains, climate and the
absence of water communications.

The general results were revolutionary. There was a new
mobility of goods and a new mobility of persons and a revolution in the commercial and industrial importance of States.
In fact, as far as three of the powers—Germany, Russia and
the United States are concerned, they may be said to have
been created as great empires by their railway systems while
the new British Empire of the nineteenth century was equally
a product of railway and steamer combined.

The new mobility of goods revolutionized the chief staples
of commerce bringing the bulky and perishable articles to
the forefront. The ease of communications caused new
forms of commercial organization to arise, a new international
finance emerged and new international rivalries were created.

The new mobility of persons created a revolution in social
life leading to the growth of towns, the rise of new classes of
workers concerned with trade and transport, an alteration
in the position of women, seasonal migrations of persons, and a
new emigration of Europeans and Asiatics. Changes of such
magnitude were bound to affect national policies. No great
power could afford to be without either railways or steamships for military as well as economic reasons. The ease and
cheapness of transport affected all tariff policies. The

European governments dyked up with higher tariffs against the flood of cheap food imports. The facility with which goods were transferred led them to protect their industries against those produced under more favourable conditions in other countries. Their railways were used either to prevent foreign goods coming in, or to assist their own goods out, *i.e.*, the rates from the ports inland were made higher and the rates from the interior to the ports were lowered in the one case hindering import, in the other giving a premium on export. Railways thereby became an integral part of a protectionist policy. Even free trade Britain found that " free trade " was of no effect to secure free international competition if the great international combines, the existence of which depended on rapid communications, chose to make treaties among themselves as to where they would or would not sell.

Railways could also be used to develop or assist certain branches of industry in which the State was interested. For instance, in Germany the timber and iron for ship-building were carried at specially low rates on the Prussian State railways.

The railways revolutionized the problems of government by making communications between all parts so easy that nations could become empires, with the result that separatist tendencies became modified and centralizing influences had much greater play. It increased the growing power of the State as it made direct government much easier and minimized the importance of local government. It not only added in this way to the growing power of the State which, as we have seen, has been one of the most marked features of recent years even in laissez-faire Britain, but it brought up new questions of State control. The railways of Russia, Prussia, and the various other German States belong to the government. The State in France owns a large part of the railway lines and has subsidized the building and working of the rest. The railways have become in those countries great departments of State thus making the State a kind of partner, through transport, in all business enterprises. In the United Kingdom and the United States the problems had not been those of State ownership, prior to 1914, but State control had become necessary and in both countries the railways stimulated the increase of State power to combat the evils of transport monopolies or railway amalgamations.

I.—THE REVOLUTION IN THE COMMERCIAL IMPORTANCE OF CONTINENTAL AREAS.

The revolution in the commercial importance of States was due first of all to the ability of railways to penetrate interiors. Up to the middle of the eighteenth century, countries having for the most part only unmetalled roads could not move heavy goods except along the sea coast or on unimproved rivers which often froze or were in spate or went dry or had shallows and shifting sand banks and other obstacles to boats. In any case, though it is easy to come down a river it is not easy to go up against a strong stream unless there is a tide. The alternative was a train of pack mules as carts stuck fast on the earthen highways. As new countries grow raw materials which they exchange for manufactured goods it was impossible for them to be settled up far inland except in proximity to rivers, hence over-seas settlement was confined to islands or a fringe of coast-line as in the case of the American colonies of England and Spain or the African and Indian possessions of Holland, Portugal, England and France. The interior of the continents of Africa and North and South America and Australia were undeveloped. Islands were the most important features of the old colonial system and a violent controversy took place in England in 1763 as to whether she should retain Canada or Guadeloupe and Martinique after the Seven Years' War.*

The railways by penetrating interiors enabled people to go inland and move away from rivers. In North and South America they opened up the central region and enabled the settlers who followed the railways to grow wheat, maize and cattle for export. From the coast this bulky agricultural produce was transported by steamers at cheap rates to Europe.

The developments hinged also on the fact that when interiors were penetrated the new methods of transport were cheap. The food stuffs could actually have been moved had transport been dear but they would not have been so moved because the people to whom they were to be sold would not have been able to pay the high price charged to cover the

*W. L. Grant : " Canada *versus* Guadeloupe " in American Historical Review, Vol. 17 (1911-1912), *p.* 735.

cost of high freights to Europe. Grain was, as a matter of fact, only transported from one European country to another in times of scarcity or dearth prior to 1850, the selling price at other times simply would not cover the cost of transport and to bring grain from far distant countries like America was almost unthinkable. Cobden calculated in 1845 that the English producer of corn would always have a protective tariff on corn of 10s. a quarter in distance even under free trade—that was the cost of moving wheat from Dantzig to London. He did not foresee the way the new transport developments would abolish distance and render bulky goods marketable at cheap rates.

A still further reduction in rates of carriage took place when the steel rail was substituted for the iron one. Bessemer had invented cheap steel in 1856. This gradually transformed the metallurgical trades by substituting steel for malleable or cast iron. As iron rails wore out they were replaced by steel. The result was not merely that the steel rail lasted longer than its iron predecessor* and therefore reduced the cost of maintaining the permanent way, but steel became the chief material for the rolling stock on railways, being lighter as well as more durable than iron. The result was heavier train loads, larger waggons and more powerful locomotives. Traffic was accordingly handled with greater economy and there was a downward trend of railway rates, all of which tended to create a new value in continental areas with a long haul to the coast. A characteristic feature of the nineteenth century has therefore been the development of continents instead of coast lines.

In the same way the possibility of penetrating interiors brought Africa prominently into the world economy. The interior of the continent being a great plateau, the rivers flow out to the sea over rapids which formerly barred the interior of the continent to the settler. The only method of transport over a large part of Africa before the railways was man, owing to the fact that the tsetse-fly kills horses. The railways enabled the rapids to be avoided, penetration of the interior took place and Africa became one of the great economic areas with a future The result was that " the scramble

*Depression of Trade Commission, 1886. *See* evidence of Sir Lowthian Bell on *pp.* 143-4 of this book.

for Africa " took place among the powers in the eighties which ushered in a new era of colonial rivalry. Meanwhile with the coming of the railways much of the labour which had been absorbed in sheer porterage was set free for agricultural development, such as cocoa in West Africa or cotton in Uganda.

In similar fashion Asia was opened up by the new transport developments. The Siberian railway penetrated the Northern and the Transcaucasian Railway the Central regions of Asia. India began to be transformed by its railway net, famines lessened, caste tended to break down, the volume of exports and imports increased. The Bagdad railway came into the forefront of politics as the great instrument to develop the Middle East and its oilfields. The steamers after the opening of the Suez Canal in 1869 linked India, China and Japan with Western Europe so effectually that " a fortnight in lovely Japan " could be advertised by English tourist agencies as an attractive and easy " trip."

While America, Asia and Africa were to an increasing extent focussed on Europe, that continent itself was powerfully affected by the new means of communication. Perhaps the most striking feature was the economic development of Germany after 1870—due largely to the creation of the European railway system. Hindered hitherto by a short coast line, by the Northern flow of her rivers and by the freezing of her canals in winter, she gained new outlets East, West and South at all times of the year. She became a Mediterranean power by the completion of the railway over the St. Gotthard in 1882. She obtained great economic influence in Northern Italy and Genoa became an important German outlet. In the same way the railway to Constantinople made her a power in the Balkans with commercial interests in the Levant. She was connected by railway with France on the West and Russia on the East and became the centre of the continental system of distribution, thereby affecting the hitherto unrivalled sea distributing position of England.*
Owing to the ease of her sea communications, this country had, before 1870, almost a monopoly of the carriage of goods

*The same railway gauge obtains all over Europe except Spain and Russia and goods can be transferred in a railway truck without breaking bulk all over Central, Western and Southern Europe.

from Northern Europe to the Mediterranean; after that date much of it went by land *via* Germany.*

The power to move bulky things has also led to the bringing of iron and coal together for purposes of manufacture. Previously the sheer cost of transport prevented their being developed to any large extent in any country but Great Britain, where they were found in proximity. We find Harkort complaining in 1830 that a German iron industry cannot develop in Westphalia as the coal and iron lie ten German miles apart.† The iron of Lorraine was brought after 1870 to the coal fields of Westphalia by railway and the German output of pig-iron which in the years 1870-1874 only averaged 1,800,000 tons per annum as against Great Britain's 6,400,000 had increased to 11,800,000 tons as against Great Britain's 9,800,000 in the period 1905-1908. In steel the 300,000 tons of 1870-1874 had risen to 10,900,000 tons annual average production in 1905-1908.‡ This output was second only to that of the United States.

Russia, hindered by frozen ports and rivers in winter and the absence of roads, was able when railways were developed, to communicate at all seasons with all parts of her vast Empire in Europe and Asia. She was enabled to bring the corn of the black mould zone to the Northern forest regions where there is a deficiency of food; she was able to provide the corn area with the wood it needed from the North for fuel in return. She was in a position to get an outlet to the sea all the year round through Germany or through Odessa, using ice-breakers in winter, or she could reach open water *via* Siberia and Port Arthur. Her iron and coal were developed by the possibilities of transferring the steel products which were made in the Donetz region in the South of Russia to the place of user. The cotton factories found a market not merely all over Russia but in Northern China and Central Asia, while drawing a large part of their raw cotton supplies from Turkestan or America.

The United States was another of the new powers that owed her development to mechanical transport. The natural trend of her trade was to follow the rivers North and South. The

---

*Report on British Shipping after the War, Cd. 9092, *p.* 87.
†Der alte Harkort, by Berger (1890) *p.* 170.
‡Fiscal Blue Book, Cd. 4954, 1909, *pp.* 3 and 4.

Mississippi and its tributaries form a magnificent network of waterways draining a million miles of territory right in the very heart of the Central region of the continent. The current was, however, so strong that boats took months to work up against stream. Goods came down the river on rafts or flat boats. There was no question of return cargoes. When steamers were started they soon became powerful enough to steam against this current and a revolution was created in the trade of the Central region, when the steamers became numerous.* There were 200 in 1829 and 450 in 1842 in the United States. The West could then feed the South as the South expanded its cotton plantations. There were three regions in the United States prior to 1860: the Northern and Eastern States formed one region separated from the West by the Alleghany mountains; on the other side of the mountains was to be found the second region, a grain-growing area becoming increasingly dependent on the South for its market; and thirdly there was the cotton belt in the South. The Civil War must have had a different issue if the Eastern region had not been able to divert the traffic East and West by building railways instead of its going North and South by the rivers. The Western States then (1861-1865) linked up with the East into one common system began to look to Europe as a compensatory market for the loss of the South. Instead of two distinct regions East and West were united. The Civil War gave an enormous stimulus to railway building in the United States and from that time onward " the history of the railroads was the history of the country." The railways became great emigration agents on their own account, they developed the interior of the continent, linked up both sides of it and enabled the grain to be transported.

" Over ordinary earth roads wheat will bear transportation for a distance of only 250 miles when its value is $1.50 per bushel at the market, Indian corn will bear transportation only 125 miles when its value is 75 cents. per bushel. When

---

* The Mississippi has so many curves in it that a voyage up or down stream is a long affair measured by time. Another great difficulty is that the bed and level of the river is always altering and sandbanks are a continual obstacle to transport. Hence the railway is far more effective as a means of transport than the river.

grown at greater distances from market these products have without railroads no commercial or exportable value."

Beyond a certain limit, consequently, " railways which transported at one-twentieth of the cost of transport over earth roads were the sole inducement to the production of these staples in an amount greater than that necessary for consumption by the producer."

" In point of importance the railroad interest now takes precedence of all other industries or enterprises. Its magnitude is greater than that of any other interest in the world and it has become so thoroughly a part of the economic system of the republic as to be second only to the government of the United States itself."*

It is interesting to notice that in the United States many railways were built for sheer speculation, they competed violently with one another and cut rates and the result was to stimulate the export of grain and meat to Europe by a sort of bounty in the shape of cheap railway charges. The result of this cut-throat competition in the eighties was amalgamation in the nineties and the amalgamations themselves proved so efficient in making up bigger train loads, and working more economically by adopting the most up-to-date equipment, that a further drop in rates took place.† With the beginning of the twentieth century there has been a rise in rates of transport.

The great metallurgical industries of the United States owe their importance to the same cause—transport facilities. Her iron and steel industries were developed by connecting the hæmatite ores of Lake Superior with the bituminous coal of Pittsburg and America became the greatest producer of crude steel in the world. The importance of railways in developing the steel industry may be gauged by the fact that the movement of the entire cotton crop of the United States does not equal the tonnage delivered to the railways

*Poor, " Manual of Railroads," 1889, p. xxiii.-xxiv.
†The decline in freight rates in the United States per ton per mile, Raper, " Railway Transport," p. 240 :

1867	-	1.92 cents	1895	-	0.839 cents.
1870	-	1.89	1900	-	0.729
1880	-	1.28	1905	-	0.748
1885	-	1.00	1908	-	0.754
1890		0.927			

by a single corporation, the Carnegie Co. of Pittsburg.*

It should be noticed that not merely have the railways created new Great Powers in the economic sense but they have exerted considerable influence in unifying those Powers. The great divergency in character, religion and history between Prussia and South Germany, which might have tended to separate North and South was largely counteracted by the railways after 1870. How were Bavaria and Wurtemberg, hemmed in by mountains on the South, to get access to the North or to the ᴊst of Germany except over the Prussian lines ? Prussia has had a factor of enormous economic leverage in her control of the whole of the Northern transport system of Germany.

It is difficult to see how such large areas as the United States and Russia could have been governed as a whole were it not for the uniting force of through communications. Another Civil War between North and South in the United States is extremely improbable. The railways have unified North and South as they did the East and the West. The Union of South Africa was the outcome of a railway conference while Canada acquired a new significance to the British Empire when the railways opened up the prairie belt, joined Vancouver to Quebec and diverted the traffic East and West instead of North and South, which had meant a growing deꞁeꞁdence on the United States. As soon as the railways cover the country they create an economic interdependence betweꞓn the various parts which makes for stability among the political units thus evolved by economic forces.

Mechanical and rapid transport may be said to have assisted in creating another great area, viz., the new British Empire of the nineteenth century. As far as Great Britain was concerned, railways were developed here into trunk lines in the forties, at least a decade before they became a coherent system anywhere else. It is not one or two lines of railway that revolutionize the traffic of a country. The railways must cover the country in something like a network proportionate to its area and this network must work together so that goods may be cheaply and rapidly forwarded over all the lines. Common working arrangements come only second to railway building in importance. The greater efficiency

*United States Industrial Commission, 1902, Vol. XIX., *p.* 266.

of railways and steamships over any other form of transport gave this country temporarily a bounty on production and distribution between 1850 and 1870. Great Britain could produce in masses, receive raw material in bulk and send away any quantity with despatch and punctuality in a manner that was not possible to any other country except France before 1870, whose railways had been developed into a working system by Napoleon III. between 1853-1857. The railway was especially valuable for the way in which it increased England's predominance as the forge of the world during that period. Manufactured iron articles such as machinery are particularly awkward articles to move but Great Britain was not only able to transport machines but the rolling stock for railways, gas pipes and drain pipes which were now in universal demand. As the builder of railways and steamships the provider of engineering tools, rails and locomotives England was unrivalled during the years 1850-1873. Of the 245 locomotives in Germany in 1840, 166 came from England.*

On the other hand, as England was the pioneer in railway construction " we could not in the earliest days of railways appreciate what the immense growth of railway requirements would be and what would be the dimensions of the loads we should be asked to carry or the weight per wheel which our bridges would have to bear. It is an extraordinary fact that on the British railways there are no fewer than 66 different loading gauges applicable to 150 sections of lines which have to be taken into account when considering the forwarding of rolling stock and the obstructions which exist on railways and dock properties which have sidings of their own."†

Thus England has had to suffer and reconstruct her railways just as she had to work her way through the chaos of the factory system to the Factory Acts, and the miseries of the early towns to the Health Authorities and sanitary engineers.

In the last quarter of the nineteenth century the railway itself created fresh rivalry for the United Kingdom but in its turn brought fresh compensations in that it developed

---

*Sombart, " Deutsche Volkswirtschaft," 1913 Edition, *p.* 243.

†Aspinall : Address to Institute of Civil Engineers 5th November, 1918.

the new British Empire and constructive imperialism instead of laissez-faire.

We have seen how the German railway system created a new industrial Germany and made her the great European land distributor while her steel and textile industries became serious rivals to this country. In the same way the railways created a new industrial and agricultural rival in the United States. Great Britain was forced to change her economic basis and relied after 1870 upon importing food and paying for it with high-class manufactures, coal, shipping and financial services. She left the manufacture of cheap standard articles to Germany or the United States. While she ceased to be the workshop of the world *par excellence*, owing to the industrial development of other countries, she maintained her position as the carrier of the world with the new steamships, and gained a new industry—iron and later steel ship-building.

American shipping in 1860 comprised 2,546,237 tons for foreign trade and 2,752,938 for the lakes, *i.e.*, 5,299,751 tons altogether. The United Kingdom had 4,658,687 tons. In 1858 Great Britain built 236,554 tons of shipping and the United States 244,713 tons. The United States did a considerable share of all the trade to and from British ports. The entrances and clearances at ports in the United Kingdom in 1860 consisted of 13,914,923 tons British and 2,981,697 American.*

When the iron steamer came in the situation changed completely ; the supplies of timber which had been so valuable an adjunct to the ship-building industry of the United States were no longer such an asset. The United Kingdom with her coal and iron fields situated right on the coast was admirably adapted for building the then most efficient type of ship, viz., the iron ship ; she had already nearly a hundred years' experience in dealing with iron in its newest forms, dating the revolution of the iron industry from the smelting of iron with coal which was adopted by the trade generally about 1750. As soon as a ship became a box of machinery, England with her supply of raw material for manufacture, her coal for fuel, her skilled engineers, able to make and work what was by far the most efficient type of boat, soon out-distanced all rivals.

*Tables showing the Progress of Merchant Shipping in the United Kingdom, 329 (1902).

The decline of the United States mercantile marine was rapid. She lacked the raw material in a convenient position for ship-building ; she lacked the technical skill in making marine engines. The revolution in ship-building technique occurred just at the time when she was absorbed in the Civil War. The next few years were occupied with reconstruction. Her capital was not put into shipping, it was invested in the seventies and the eighties in her great railway and industrial undertakings. Even had she bought ships she could not have repaired or manned them. No other country approached Great Britain during the whole nineteenth century either as a carrier or ship-builder. Invention followed invention. The compound engine economized coal and made it possible to carry cargo economically by steamers ; previous to its adoption between 1850 and 1860 the steamers had been used chiefly for passengers and mails. From 1860 onwards steamers were increasingly used for all purposes. The change from sail to steam was accelerated when the Suez Canal was opened as sailing vessels do not go through the Canal without being towed. Shipping differentiated into two types, the tramp that went everywhere and picked up cargo wherever it paid and the liner that kept to regular routes and regular sailings. Hence to supplement the punctuality and speed of the railways came a corresponding development in shipping.

Ship construction was revolutionized in the eighties by the utilization of steel instead of iron for ship-building just as railways were affected by the steel rail and steel rolling stock. All these changes occurred in England and tended to make her position in the shipping world almost monopolistic.

Up to 1912 the United Kingdom carried over half the sea-borne goods of the world,* and in the twenty-five years before

*" In 1912 the world's sea-borne trade as represented by imports into all countries probably amounted in value to about £3,400 millions, of which £510 millions (or fifteen per cent.) comprised the British Inter-Imperial trade. The trade between the Empire and foreign countries amounted to over £1,300 millions or thirty-nine per cent. Thus the trade of which one or both terminals were within the Empire aggregated to not less than fifty-four per cent. of the whole. The trade of which one terminal was in the United Kingdom was about forty per cent. of the world's sea-borne trade. We estimate that British shipping carried £1,800 millions, or about fifty-two per cent. of the total sea-borne trade of the world including ninety-two per cent. of the Inter-Imperial trade. sixty-three per cent. of the trade between

the war she built two-thirds of the new ships that were launched. This very development of shipping did much to create the new British Empire. With Russia and the United States it was the question of expanding into contiguous land territory, in the case of Great Britain it was her unrivalled command of swift steamer communication that made the British Empire possible, the steamer was as much a link for her as the railway was for Germany, Russia and the United States. All parts of the Empire were connected with one another by rapid sea communications supplemented by the cables and a new entity was created. Distance was abolished as a barrier; penny postage, the rapid transmission of letters and newspapers helped to bring the whole Empire into closer touch. When ministers of outlying parts could easily come " home " for a short period for conferences and responsible statesmen visit the Dominions, it was possible to evolve something approaching an imperial constitution. The importance of shipping to the Empire is seen from the fact that when the manufactures of England were taken in an English tramp steamer to South Africa, the ship picked up coal at Natal and took it on to India and in India got a cargo of raw materials and came home *via* the Suez Canal. The manufactures would have cost South Africa more if the ship would have returned empty instead of going on to India and the raw material from India would have cost the United Kingdom more if it had to bear the cost of the whole voyage. This again would have made manufactures dearer and possibly limited the sale. It is interesting to notice that the Dominions Commission seemed to consider that transport facilities outweighed in importance all other considerations in creating a link between the mother country and the dominions.* " So long as freights are cheaper and means of communication between the mother country and the Dominions overseas, and between the Dominions themselves than between foreign countries and the Dominions, so long will trade naturally follow Imperial channels. If, therefore, it is possible to devise some means of permanent betterment of sea routes within the

the Empire and foreign countries and thirty per cent. of the trade between foreign countries." Report on Shipping, Cd. 9092, 1918, *p.* 72.
  * *p.* 108, Cd. 8462 (1917).

Empire, a powerful impulse will have been given to Imperial trade while the strength and cohesion of the Empire will be notably increased."

In creating the new British Empire, however, the railway played its part. The British Empire of the eighteenth century had consisted mainly of islands and coastal towns, the Empire of the nineteenth century saw the colonisation under British rule of continents. The penetration of interiors which the railways facilitated enabled people to live inland, hence the development of such countries as Nigeria, Rhodesia, Uganda, the prairie belt of Canada, the Soudan and the interior of Australia. New commodities were produced in these areas which again made for the interdependence of the whole.

The wheat exports of Canada, as soon as the forest belt was penetrated by the railway and the prairie region opened out, grew yearly in importance, and outrivalled in the steadiness of its supply that of the United States. The wool of Australasia, the jute of India, the rubber of West Africa and the Malay peninsula, the oils of the Coast and the nickel of Canada are almost monopolies for the British Empire as is also the bulk of the gold production of the world. The cotton of Egypt supplemented with its fine staple the famous sea-island cotton of the United States of which there is not enough for British requirements, while India has proved itself to be the best customer for British goods since the eighties.*

*Annual Average Exports from the United Kingdom.

To	1880–84 Amount Millions. £	1880–84 % of Total Exports.	1885–89 £	1885–89 %	1890–99 £	1890–99 %
U.S.A.	28	12.2	28	12.2	26	11.
Germany	18	7.7	16	7.2	18	7.8
India	30	12.9	31	13.7	30	12.9

To	1895–99 Amount Millions. £	1895–99 % of Total Exports.	1900–04 £	1900–04 %	1905–08 £	1905–08 %
U.S.A.	21	8.6	21	7.4	26	7.
Germany	23	9.4	24	8.4	33	9.1
India	29	12.	35	12.2	47	12.8

Fiscal Blue Book, 1909, pp. 35-43.

The interesting thing is that many of these products such as rubber, jute and palm oil are new factors in international commerce and the volume in which all these products are transferred is also new. While valuable and non-perishable articles like wool and cotton could be transferred at any time, the quantities demanded would not have been so great except for production by machines, and machine production could not have increased in volume had there been no improvements in transport beyond that of the sailing ship, the barge and the cart. Thus while the development of raw material within the Empire was assisted by railways and steamships, it was the demand for raw material here that made it worth while to produce it in the first instance, and the corresponding market for the finished products afforded by the Crown Colonies, Dominions and India illustrate again the economic interdependence of the nineteenth century Empire as linked up by railway and steamship.

The railways and the steamships could not, however, create new industrial entities such as Germany, Russia and the United States without creating new industrial rivalries and therefore as a result of the railway era there is an increased struggle among nations to control the raw material producing areas and the markets of the world. There was a new stimulus to acquire colonies and spheres of influence. Mechanical transport has led very definitely to a return to protectionist tariffs to defend the home market against the ease with which goods can be sent from one country to another. In the same way labour has in some countries, notably America and Australia, supported the protectionist régime for fear that goods made by low-paid labour and now easily transferable might undersell goods made by well-paid labour, in which case the latter could not maintain its standard of comfort. Even a free trade government like the United Kingdom began to defend its people against unfair forms of competition which were created by the facilities for transferring goods. Prior to 1907 when patents were taken out in the United Kingdom, the patentees often made no attempt to start the industry but worked it in some other country and sent the goods here. Thus Englishmen were debarred from making the article in question while transport facilities enabled other countries to place it on the English market. This

o

was remedied under a Liberal government by the Patent Act of 1907 which made any patent taken out here by foreigners invalid unless the article were made here within four years.

This intensified national rivalry has led to the competition to secure new colonial areas and to the evolution of nations into empires. On the other hand, cutting across the new national protection and the greatly intensified industrial and commercial rivalry emerges the trust or combine, international in scope. There comes to be a point at which producers, manufacturers, merchants and transport agencies will not compete. They amalgamate or combine. In the same way there is a movement for a general approximation of labour conditions to prevent undercutting by sweated labour since import is now so easy. World-wide swift communications have enabled both labour and capital to organize on an international scale and it seems as if some form of economic international action would be necessary to counteract such of their activities as may be harmful since exchange has now outgrown the boundaries of nation or empire owing to mechanical transport. The solution probably lies in the national control of railways and ports and common agreement between States as to the extent to which they will allow combines to use the transport facilities of the State. In this way an effective method of bargaining with great international monopolies might be evolved.

II.—The Revolution in Commercial Staples and Commercial Organization.

Naturally with all this production at an intensified rate all over the world a revolution took place in the staples of commerce. Production previously designed for pack mules and carts and for canal barges drawn by horses and sailing ships, altered its character and was designed for world sale, distribution to all parts, world markets and world prices. A world economy supplanted national economy much as that had supplanted the local economy of towns and manors in the Middle Ages. Price is no longer fixed for most of the important articles by the national production for a national market but by international conditions and an article like wheat produced under such diverse conditions as those prevailing in England, Germany, Russia, Argentine, Australia and Canada fetched approximately the same price on the London or Mannheim Corn Exchange.* The same thing is true of cotton, wool, rubber, copper, oils, tea, coffee and sugar. There is not an English market and an American market but the world is one market.

With speed and regularity articles can now be transferred that were never transferred before and the value of world commodities alters. In the fifteenth and sixteenth centuries the spice trade was the source of wealth, later it was displaced as the chief objective by the growing demand for colonial products ; tea, coffee, sugar and tobacco held the field and if a person would write a story of the eighteenth century and wished to extricate his hero from pecuniary difficulties he ought to marry him to a sugar planter's daughter. In the nineteenth century the rich lines of trade are the bulky articles, iron, coal and food stuffs. While colonial products are still very important the movement of raw materials and minerals to the coal areas, the distribution of the manufactured articles and the movement of perishable commodities like meat occupy the centre of the economic stage.

In 1842, the economist, McCulloch, could write as follows :

" No country is likely to send any stock to England or indeed has any to send, with the exception of the countries round Hamburgh and the imports thence cannot be considerable. In the Ukraine and other countries in the South of Russia there is a remarkably fine breed of cattle but then it is impossible to import them alive into England or otherwise than salted and as already seen, our merchants have enjoyed the privilege of doing this for a series of years, under a moderate duty, without so much as a cwt. of beef having all the while been brought from Odessa or the sea of Azof.

" The same is the case with South America : no live cattle can be sent from it ; and no salt beef has ever come from it under the 12s., nor will an ounce ever come under the 8s. duty, or indeed though there were no duty at all. The South Americans rarely if ever send beef to the West Indies and how then is it to be supposed they should send it to England ? "†

*Fiscal Blue Book, *pp.* 194-202.
†Memorandum on the Proposed Importation of Foreign Beef and Live Stock, by J. R. McCulloch, 1842.

The wholesale import of meat, butter, eggs and fruit did not enter into commercial calculations before 1870 any more than fresh milk does now. With the steamer and rapid punctual voyages live cattle were moved easily in the sixties and seventies. Then cold storage developed at the beginning of the eighties, chilled meat, dairy produce and fruits came into commerce. New regions, as we have seen, were developed and wheat was transferred in bulk as it had never been transferred before. The trade in food stuffs caused the growth of great produce exchanges and commercial organizations with their speculation in " futures " gained in strength and changed in character. People generally enjoyed a more varied diet, the whole world was laid under contribution, and eggs from Central Asia or Morocco, fresh peaches from South Africa, butter from Siberia, pine-apples and bananas from the West Indies and Tasmanian apples came into ordinary household use. This was supplemented by the great trade in canned goods ranging from tinned apricots from California to tinned rabbits from Australia, tinned salmon from Alaska tinned muscat grapes from Spain, tinned tomatoes from Italy and tinned beef from Chicago.

Prices of food fell rapidly after 1873 and real wages therefore rose considerably. Coal, iron and machinery assumed a new importance because of their transportability. Three-quarters of the weight of English exports in 1913 consisted of coal, viz., 76 million tons out of 100 million tons, while 21 million tons of coal were supplied in addition as bunker coal to work the ships. The railways and the steamships themselves created an almost insatiable demand for iron and steel not merely for construction but for renewals. The life of a locomotive is at most thirty years and of a waggon twenty-five. A steel rail needs renewing on an average every twenty-five years according to the amount of use it gets. The English railways alone were calculated to consume as much as £30 million worth of material in the year 1913. A railway, like the steamship, does not stand still in matters of technique. Every period of ten or fifteen years calls for great changes in traffic conditions when the whole machine has to be overhauled with fresh demands on engineering skill and coal and iron to meet it. " A railway is not a museum for the retention of old machinery but a highly organized implement of commerce and to be efficient must progress."*

Some idea of the vast demands that the new developments in the engineering trades make on the mines and equally on transport may be seen from the following table of the amount of goods carried on the United States railways in the year 1900 :—

	Tons	% †
Products of agriculture -	53,468,496	10.35
„ „ animals -	14,844,837	2.87
„ „ mines -	271,602,072	52.59
„ „ forests -	59,956,421	11.61
Manufactures -	69,257,145	13.41
Merchandise -	21,974,201	4.26
Miscellaneous -	25,329,045	4.91
	516,432,217	

Not merely did the developments of transport stimulate mining and engineering to an extraordinary degree but the movement of food stuffs which had not seriously come into world economics before, created what the world had never previously enjoyed, viz., an insurance against famines. If the harvest fails in the United States, Australia will probably make it up ; if the locusts come in Argentina possibly India gets a good monsoon. There is the whole world to draw on.‡ While the sources of supply for an importing country like Great Britain vary extraordinarily from year to year, it is striking to see how the amount required was always forthcoming. So rapid and certain were the arrivals that the "Commission on our Food Supply in Time of War" reporting in 1905, could state that there were considerable periods when the first-hand stocks of wheat at the ports did not amount to two-and-a-half weeks' supply and in seven years

*Sir J. Aspinall : Address to the Institute of Civil Engineers, 5th November, 1918.

†United States Industrial Commission, 1900, Vol. XIX., *p.* 266.

‡The supplies from the United States in 1904 were 10,760,000 quarters less than in 1902 and yet the total imports of wheat and flour into the United Kingdom were 2,520,000 quarters in excess of those of 1902, the deficiency being more than made up by the large increase in the supplies from Russia, the Argentine Republic, the British East Indies and Australia.

out of the eleven between 1893 and 1904 the amount wa less than that for 102 weeks.* That the stocks could be allowed to fall so low as this shows the absolute certainty of the arrivals. In addition the minimum amount held in second-hand stocks in the millers' hands was at least three weeks' supply† while the bakers would hold at least one week's supply. There was, however, always some harvest of the world *en route* for England, the arrival of which was a practical certainty.

In normal times the supply of wheat for the United Kingdom was practically continuous throughout the year and shows clearly how the world had become one great market. In other words, owing to the developments in transport there is no longer an English market for wheat or a European market but a world supply and a world price, fixed by conditions of production, varying from wheat produced under primitive conditions by Russian and Indian peasants to that of the farmers on the prairies of the Middle West in the United States or Canada working with elaborate labour-saving machinery, and these compete again with wheat produced by highly intensive methods in the Lothians or Lincolnshire.

The arrivals of wheat in 1905 were as follows :‡

January cargos arrived from the Pacific Coast of America.

February			
March	⎰ ,,	,,	,,	Argentina.
April	,,	,,	,,	Australia.
May	,,	,,	,,	Calcutta and Bombay ⎱ Indian
June	,,	,,	,,	Delhi ⎰ Wheats.
July	,,	,,	,,	Karachi
July	,,	,,	,,	America (winter wheat).
August	,,	,,	,,	America (winter wheat).
September	,,	,,	,,	America (spring wheat).
October	,,	,,	,,	America (spring wheat) and Russia.
November	,,	,,	,,	Canada.

This is in striking contrast with the conditions of supply up to about 1850. Prior to that time if Great Britain required wheat in excess of her home supply she could only obtain it from Odessa or Poland and Prussia *via* Dantzig and then only at a high price as freights added considerably to the cost. Often, however, the continent suffered from the same type of weather

*Cd. 2643, *p.* 11.
†*Ib. p.* 13.
‡*Ib., p.* 9.

as the United Kingdom and a scarcity in England generally meant a scarcity in Europe when wheat would be unobtainable at any price. It was not for nothing that the English prayer-book has two special prayers against famines, two against scarcity arising from excessive rain and a petition in the Litany to be delivered from plague, pestilence and famine. A harvest thanksgiving must have had a very real meaning to the people of pre-railway days. Mechanical transport has meant security of life to an extent that was never previously realized.

Equally the steamship made for the safety of commerce by abolishing pirates. In the sixteenth and seventeenth centuries a ship was " fair game " for anyone strong enough to take her and piracy was the curse of commerce all through the eighteenth century. Now-a-days a pirate would need too much capital to equip a vessel capable of overtaking a modern liner. It would probably not be worth his while to overhaul a modern tramp as there would be a difficulty in disposing of a cargo of grain or wool. Coaling, too, would not be easy for a pirate. Yet early in the nineteenth century pirates sacked the port of Hamburg. Nor is it as easy for highwaymen to hold up a train as a coach. Hence there is a commercial revolution in the security of sea and land transport to say nothing of the greater freedom from accidents for the persons who travel.

Further, new bulky commodities came into commerce that had never seriously entered into the world's calculations till the last half of the nineteenth century because their weight or bulk in proportion to their value made it impossible to transfer them at any price that would pay. The evidence given before the Royal Commission on Canals by Mr. Royse, of Manchester,* shows how important these new bulky products have become.

" In 1850 we imported of wheat, barley, maize and other such food stuffs about 1,500,000 tons ; in 1905 we imported over 10,000,000 tons ; in 1850 of raw cotton we imported about 300,000 tons ; in 1905 nearly 1,000,000 tons ; in 1850 we imported no cotton seed—at that time it was not utilized ; in 1905 we imported nearly 570,000 tons. In 1850 we imported no petroleum oil—at that time it was not known ·

*Cd. 4979 (1909) p. 6.

in 1905 we imported about 1,250,000 tons—something like 300 million gallons; in 1850 we imported no wood pulp— science at that time had not produced it; in 1905 we imported nearly 608,000 tons; in 1850 we imported no iron ore; in 1905 we imported 7,250,000 tons, and I may say here that in my judgment this importation has largely assisted to save to this country its iron and steel trade, for without it it is a question if we could have produced from our own ores sufficient of the high-class quality of steel that is now required.   In 1850 we imported no pyrites; in 1905 nearly 700,000 tons. It is known that this cheap source of supply of sulphur has enabled us largely to keep the sulphuric acid and cognate industries in this country.   In 1850 we imported no phos- phate of lime; in 1905 over 400,000 tons.   I think I have given a sufficient number of instances to show how the trade of this country has changed during the last fifty years. . . . You see the enormous tonnage that now comes into the country where fifty years ago little or none came.   From those deductions I draw the conclusion that it is very impor- tant that we should have cheap inland transportation into the interior whether by rail or canal."

This movement of masses of raw materials that would not bear the cost of transport at an earlier period, and for which there would have been little demand since it could not have been utilized without machinery, the every day movement of food products like wheat, which previously were only transported at famine prices since freight added so enormously to its cost, these constitute one aspect of the commercial revolution of the nineteenth century.

Mechanical and rapid transport not merely altered the relative value of nations and commodities but it promoted a commercial revolution in business organization.   As traders could get goods swiftly and with absolute certainty, they no longer kept such large stocks.   They therefore needed less warehousing space and less credit from their bankers and were able to carry on business more economically.   A striking instance of this is to be seen in the revolution which took place in the London coal trade.   Up to the middle of the century the coal was brought to London by ship and the coal merchant owned a wharf and was a substantial man. Certain railways began to be anxious about 1850 to develop their coal traffic and gave facilities for storing coal in trucks at the stations. The result was that all sorts of small men could start as coal merchants since they had not got to provide expensive warehousing accommodation and could obtain the coal in truck-loads instead of ship-loads.

Sir Sam Fay explained this change to the Royal Commission on Canals as follows : " We are all living from hand to mouth : the consumer, the retailer, the middleman and the manufacturer all expect, and it gets worse I think every day, to telegraph or telephone for a thing to-day and get it delivered to-morrow . . . and you can get as a matter of fact pretty well everywhere, with the exception of extreme Scotland, your traffic that is sent away to-day delivered to-morrow."*

This did not merely apply to traders within a country but to nations. In the days when it took about a year to go and return from India, England kept the stores of goods for speedier distribution in Europe. She was the great *entrepot*. With the development of rapid transport it was easy to telegraph, say for tea, to India and have it sent on by the next liner or the ubiquitous tramp. It need not pass over England at all but could be dropped at Odessa for Russia, or Genoa for Central Europe, or Marseilles for France, the railways would carry it inland and distribute it.† With the development of the railways across the United States goods that used to go from China to London and be forwarded from there to the Eastern States, could actually be shipped to San Francisco and sent across the American continent.

As railways extended so local markets and fairs tended to disappear, the shops kept the stores, obtaining them daily if necessary from the manufacturers by rail. There was no need for an annual or semi-annual renewal of stocks at fairs by the householder. There was, therefore, a great increase in the number of retail shops.

While the railway thus altered the methods of inland and retail trading, the steamship, combined ..ith the telegraph revolutionized the methods of foreign trade. Up to the second half of the nineteenth century every ship carrying cargo was more or less " a venture." She was stocked with goods, a

*4979 of 1909, *p*. 81.
†Report on the Port Of London, 1902, Vol. XLIII-IV.

supercargo was put on board her and he sold the goods to the best advantage he could on arrival. In many of these cases the exporter owned the vessel in which the goods were sent. Ships belonged to the mercantile houses and were not, as now, common carriers for mixed cargoes or open to hire by any one. The exporter did not and could not know the state of the market when the goods should arrive; all trading was consequently something of a gamble. When goods could be telegraphed for and sent off at a few hours' notice the importer became the deciding factor not the exporter, as heretofore. The importing house decided on its wants, placed its orders and obtained them as it required them.

The effect of mechanical transport was to lead to the creation of large businesses and combinations which in their turn enormously stimulated the spread of machinery since capital was mobilized in these large businesses in millions and equipment and research were not restricted within narrow limits by the difficulty of obtaining capital for developments. The scale on which manufacturing is done in post-railway days is as different from the little factories of the days of turnpike roads and canals as those were from the domestic system of the reign of Queen Elizabeth.

After the development of railways, factories and workshops were not limited in size by the difficulty of getting either coal or raw material in sufficient quantities or hampered in dispatching masses of finished goods. Blast furnaces could treble and quadruple in size because there need be no physical limit to the amount of iron ore and fuel obtainable by rail. Everywhere the use of mechanism spread more rapidly; machinery became larger and more efficient; ships increased in size and became more economical in operation, railway equipment followed suit, train loads grew bigger and locomotives more powerful. The new factories had to be built larger to stand the strain of heavy machinery on the floors. The railways did not merely tend to create larger businesses owing to the facilities they gave for handling masses of goods but they often determined the transfer of businesses from one part of the country to another or even from one country to another. The wheat exports of the United States largely took place in the form of flour ground in Minneapolis or some other great milling centre. The railway and shipping companies found it easy to pour wheat into the truck or hold of the vessels and quoted lower rates for wheat than flour. The result was that wheat took the place of flour in the ships. To grind it great steam mills were erected at Liverpool and Hull. The development of the English milling industry at the ports is a direct result of cheap railway and steamer freights on wheat and higher ones on flour. This country gained at the same time the offals for cattle feeding which are a bye-product of flour milling. The localisation of the flour mills at the ports or the import of flour from America both meant the destruction of the country flour mills with which England had been dotted at intervals of ten or twelve miles.

The railways when developed into a network covering a country caused a new and fierce competition among businesses. Before the days of the railway most firms could only deal locally, the amount that could be distributed on a national scale was small. The railways broke down these barriers and made not merely for national but for international competition, and this became especially severe in the period after 1870.

The excessive competition cut profits so fine that firms began to combine to avoid ruin. The railways themselves led the way by amalgamating and continued to grow into larger and larger transport monopolies in both Great Britain and the United States. Shipping followed suit with rings and conferences. The ship-owners claimed that the rate agreements steadied prices and enabled them to give a better and more regular service. Other businesses did the same. It is now possible, owing to the ease of communications, for firms to control the whole industry from the raw material to the finished article. This form of combination is known as vertical and is to be found chiefly among the metallurgical trades where the coal mines, supplies of iron ore, blast furnaces, rolling and steel mills, steel wire and other works are frequently under the same management.* The great example of this is to be found in the United States Steel Corporation with its £369 millions of capital. The German Stahlwerksverband is another of these great combinations but is of a different

*Report on Trusts, Cd. 9236, p. 2.

type known as the cartel. The trust works as one great business firm. The cartel is an association formed by contract for certain periods of time. The combined firms retain their independence, merely joining for regulating production and sales but not for merger and they can recover their independence.*

Combinations sometimes take another form termed horizontal combines. Firms doing the same type of business sometimes unite for certain purposes, generally to fix prices or to push sales abroad. They do not necessarily control either raw material or finished goods. They simply unite all the businesses at their own stage of the manufacture. These horizontal combines may be national or international in scope. Such a combine is to be seen in Coats' Sewing Cotton, the capital of which was £10 million in 1899, when Sir A. Coats said that by far the larger part of the company's profits was derived from shares in foreign manufacturing companies and not from mills in the United Kingdom.†

Another form of world combination which has arisen in consequence of the ease of communications is one which is particularly noticeable in the metal industries, viz., combines to purchase raw material. Before the war the control of the world's non-ferrous metals was in the hands of a group of German traders who were primarily engaged in buying metal or in acting as selling agents for producers. A comparatively small number of firms controlled this trade all over the world and most of them were closely interrelated and fixed the price and regulated the production of metals and metal products.‡ Of these the most famous was the Metallgesellschaft.

There is not merely combination in manufacturing, selling for export, and for purchasing raw materials, but also in shopkeeping. Multiple shops have sprung up which, though not monopolies, show the tendency to co-ordination and large scale organization. They deal in all kinds of wares and aim

*The growth of this form of combination in Germany is very striking. In 1879 there were fourteen cartels, in 1890 210, in 1902 more than 400. Report on American Export Industries, *p.* 103. In the United States there were over 200 consolidations in 1913. Report on Commercial and Industrial Policy after the War, Cd. 9035 (1918), *p.* 36.
†Macrosty, " The Trust Movement in British Industry," *p.* 128.
‡Report on the American Export Industries, I., *pp.* 357-358.

at being " universal providers." To control supplies they
have their own tea plantations, their own orange groves, their
own fruit farms and their own food preserving factories.

A big business, in order to utilise its products fully will
often become the centre of a group of manufactures differing
widely in character. A firm making soap will have its palm
oil crushing works in West Africa ; it will convey the oil on its
own steamers and make, not merely glycerine and soap, but
other toilet preparations, as well as candles, margarine and
hard fat for chocolate and biscuits. A British explosives'
combine also makes motor cars, bicycles, artificial leather and
rubber tyres. It utilizes part of its strip brass for lamps and
wickless stoves and has developed a sheep dip in order to make
profit out of its chemicals. A still more striking instance
of what might seem to be the combination of very dissimilar
businesses is to be seen in a West African and Eastern trading
combine. To use up their cocoa from West Africa they have
bought chocolate works in England and have become manufacturers of both cotton and silk goods to sell in Africa.
As a rule it is manufacturers who add other departments
to their activities but this is an instance of traders taking
to manufacturing to use up their raw material and to provide
their trade goods themselves.

The " Big Five " in the United States controlled in 1918 over
seventy per cent. of the cattle, sheep and lambs slaughtered
by all packers and butchers engaged in inter-State commerce
but they were rapidly extending their control over all possible
substitutes for meat, fish, poultry, milk, eggs, butter, cheese
and all kinds of vegetable oil products. To their original
trade of preserved meat they added canned fruit and vegetables and also dealt in rice, sugar, potatoes, beans and
coffee and were said to be " dominant factors " in certain
of these lines. Armour & Co., primarily meat packers, had
become, in 1918, " the greatest rice merchants of the world."
They had extended their activities to other cereals. To deal
with them they had their country elevators, and sold to
the farmers fertilizers, cattle food, coal, posts, wire fencing,
builders'- hardware, binding twine, lumber, cement, lime,
brick, sand, gravel and roofing. It is interesting to notice
that much of the power of this huge combine was derived
from the fact that they owned ninety-one per cent. of all

refrigerator cars properly equipped for the transfer of fresh meat. They also enjoyed preferential treatment for their cars from the railway companies, getting them returned promptly when other firms have had to wait months. By reason of their heavy shipments they had a great leverage in all their dealings with the railroad companies and obtained special privileges as against rivals showing how close is the connection between transport and trusts.*

With the completeness of modern world communications it is possible for businesses in different countries to make treaties with each other as to where they will or will not sell and limit the extent to which they will " invade " other firms' territories. It is possible to divide the world as a trading area between English and foreign shipping conferences or between tobacco companies. These great international agreements override all national tariffs. It is no use for the English Government to declare for free trade if the members of the steel rail trust agree that neither German nor American rails shall be sold in the United Kingdom in return for the abstention of British steel rail makers from competition in Germany or the United States.† It is another instance of world economy overriding national economy. " Industrial combination knows no frontiers."

In the same way it is possible for an English or foreign firm to get the retailers who sell that type of article to agree to sell only their make of goods and in this way certain articles would find no sale, tariff or no tariff.‡

It is now possible to control world-wide interests as one great business undertaking. The result is the formation of combinations, cartels, rings and trusts which are the most striking feature of the modern business world. So far has the development proceeded in the United Kingdom that the Ministry of Reconstruction, reporting in 1916, could say, " We find that there is at the present time in every important branch of industry in the United Kingdom an increasing tendency to the formation of Trade Associations and Combinations having

*U.S.A. Report on the Meat Packing Industry, 1918 (Summary) pp. 17-21. This meat trust agreed in 1920 to confine itself to meat products only.

†Cd. 9236, p. 4, where other instances of international combines are quoted.

‡Op. cit., p. 2

for their purpose restriction of competition and the control of prices."

On the whole these combinations make for efficiency in production and the elimination of waste. It is possible to specialize branch factories to a very high degree, raw material bought in large quantities is bought cheaper and is easier and less costly to handle. Large scale businesses can afford to try experiments and carry out research as small ones cannot. Above all, they can assemble and utilize bye-products on a commercial scale impossible to small businesses. One of the advantages of combines is to be found in their capacity to obtain lower railway rates for their large shipments. They are also able to distribute from the nearest place of business and so save the expense of long railway hauls. The ability to control their raw material is perhaps their greatest asset. For pushing foreign trade these great businesses are unrivalled, they can afford to open up trade even at a loss ; they can give longer credits and employ better agents and travellers than small businesses. It is claimed that they can steady production and avoid fluctuation and consequent unemployment.

The old ideal of free competition between individuals has ceased, and the problem of twentieth century governments is to devise a scheme which shall allow the advantageous side of these combinations to have full scope and yet prevent the harmful effect of monopolies. These are to be found in the ability to keep up prices, in unfair methods adopted to squeeze out rivals or prevent new ones setting up, and in the undue influence they are able to exert over finance, politics or the press.

Businesses of this magnitude, national and international in scope, could not be carried on without daily correspondence to keep the whole in touch. They are therefore dependent for their existence on telegraphs, telephones, railways and steamships to link up the various branches in a common system. It is the minimizing of distance by rapid transport that makes the continued existence of these big combines possible, and enables them to work as one firm or group of co-ordinated partners. It also follows that one of the methods suggested to control the trusts is the Government ownership of railways. The combines would not then obtain facilities for distributing their goods by rail unless they complied with the Government conditions. The Federal Trade Commission recommended, in 1918, that the Government of the United States should acquire all the refrigerator cars and other rolling stock used for the transportation of meat and animals. If to these were added a Government monopoly of the stock-yards and cold storage plants and warehouses, they considered that an "adequate and simple solution" of the problem of the great meat combine in the United States would be solved. International trusts could probably be tackled in like manner by agreement between governments many of which already own and operate their railways as State undertakings.

The railways and steamships have destroyed the economic self-sufficiency of the nations; the world is interdependent and it would seem that economic isolation is no longer possible. Even China is building railroads.

The general result is that, thanks to mechanical transport, trade has outgrown national control; businesses can easily change their seat of direction and can operate as smoothly from one continent as another* and a system of international control will have to be devised.

III.—The Creation of a New Financial Era.

The railways did not merely affect the commercial position of States, the commodities of commerce and the organization of businesses, they caused new developments of finance, and here again we encounter the same problem of national control of international conditions. Finance has in consequence of the new ease of communications outrun national boundaries and national regulation.

The new methods of transport affected public finance, investment finance, raised new problems of taxation, and created a new financial mechanism to facilitate the new mass movement of goods.

*An example of such a transfer was given before the Royal Commission on the Income Tax in 1919. Vestey Bros., with a capital of £20 million moved the seat and control of their meat business from London to the United States and then to the Argentine in 1915 to escape the English Income Tax and Estate Duties. Royal Commission on Income Tax, Cmd. 288-293, *p.* 451.

In the first place railways tended to increase the national debts of States, and reacted upon their revenue and expenditure by making further taxation necessary in some cases or by yielding such profit in others that they proved to be a source of revenue in themselves.

With the exception of the United Kingdom, European governments have had to finance in some form or other the building of their railway systems. They have had either to guarantee the interest or raise loans for railway building and equipment. In the case of Prussia, for instance, the railways proved a very profitable investment ; in the majority of cases the State either made no profit or incurred a loss which seriously embarrassed the national revenue. Russia, for instance, was anxious to keep away from the West, yet was linked up to it by the necessity of borrowing money for railway building, the interest on which had to be paid by the corn export, thus furnishing another example of the world interdependence created by the railway. Steamships have also been subsidized in various forms by the various governments.* In every European count_y State finance has been influenced, favourably or adversely, by the new methods of transport. On the other hand, railways usually created such additional prosperity and security that any additional taxation necessary could be raised. In Russia, however, this taxation fell on a poverty-stricken peasantry and caused great hardship.

The private investor found new outlets in railway building, either by investing in railways where the State guaranteed the interest, or by building them in the continental lands outside Europe. The second effect of the railway on finance was to onen a new and vast field for the investment of capital.

The effect of this borrowing and lending was to unite the world still further in a bond of financial indebtedness ; the new countries took up the loan in the form of rails and locomotives and paid the interest in the form of raw materials and food. Certain countries, notably Great Britain, lent enormous sums to Governments for railway building and equipment or formed companies for railway construction in the undeveloped countries of the world, such as the Argentine Republic, Canada or Mexico.

*Report on Steamship Subsidies—1901 (VIII.) ; 1902 (IX.).

P

" In providing capital for railway construction this country has performed a great work. Most of the loans to colonial governments have been for railway construction, the major part of India's indebtedness to us is for railways, and a portion of the loans we have made to foreign governments has been used for a similar purpose. But beyond the money for railway construction which we have supplied to governments, we have formed a great many companies to construct and work the railways in other lands.

" From the capital we have supplied to railway companies working in the colonies, notably in Canada, we receive an income of £7,600,000 a year, from those working in India we derive nearly £4,800,000 per annum. The railways of Argentine, Brazil, Uruguay, Mexico, Chili and other foreign countries yield over £13,000,000 a year to us in the aggregate, and from the railways of the United States our investors receive no less than £27 million a year. The aggregate of these totals, which I have compiled from the companies' reports, and as far as possible from independent investigators, amounts to £82,777,000."

Sir George Paish calculates that this is the revenue of no less than £1,700,000,000 expended upon railway construction. " The capital has been supplied in about equal portions to the countries beyond the seas within the British Empire and to foreign lands."*

The capital invested in railways in the United Kingdom itself was given as £1,334 million in 1912 ;† in the United States it amounted to $11,491 million in 1900 ;‡ and the capital of the Prussian-Hessian railways was estimated at £437 million, the Bavarian at £77 million, and the railways in European Russia at £331 millions and £48 millions in Asiatic Russia§ by the Board of Trade in 1907.

These stupendous capital investments created a new financial era and as these shares were marketable the stock exchange business of the world increased enormously in volume.

---

*" Great Britain's Capital Investments in Other Lands," by Sir George Paish, J. R. Stat. Soc., September, 1909, p. 470.
†Cd. 6954 (1913) Railway Returns Annual.
‡Industrial Commission, XIX., p. 400.
§Return to House of Commons, 331, 1907.

The railways have made it possible to invest capital all over the world either in railway building itself or in the production of raw materials or food stuffs. Many businesses are not carried on in one country alone, but are international. The Rio Tinto Mining Company, for instance, works the ore in Spain, smelts it chiefly in the United States, is controlled in England, but a large proportion of its shareholders is French, and taxation is levied in each country on the undertaking.*
There thus arises the hardship of double, triple and quadruple taxation. The problem of adjusting international invest-ment, which is the necessary accompaniment of a world economy to national finance, is a problem of the greatest difficulty. It can only be solved by international agreement as to the proportion of taxation which shall be taken by each country.

The fourth great reaction of the railways on finance was comprised in the development, extension and specialization of the mechanism of credit for facilitating the new world movement of goods. Banking, exchange businesses, discount and accepting houses, produce exchanges and speculative markets all expanded and altered in character. The develop-ment of credit in all its various forms became very elaborate to enable these intricate world operations to be carried out smoothly.

International business combines, international finance and taxation and international exchange of commodities and human beings are the inevitable outcome of the new transport developments and national economics will have to be modified and readjusted to suit the changed conditions which, mean-while, have given rise to fierce international rivalries.

IV.—Social Effects of the Commercial Revolution.

The railways introduced a new personal mobility and that produced a social revolution no less remarkable than the political and commercial and financial.†

*Evidence of Sir A. Steel Maitland, Royal Commission on the Income Tax, 1920, *paragraphs* 27,953-27,957.
†Some idea of the increased movement of persons may be judged from the fact that in 1831 in what is now Germany about a million persons travelled by the public post waggons. In 1910, 1,541,000,000 persons were carried by the railways. Sombart, '' Deutsche Volks-wirtschaft,'' *p.* 244.

The personal mobility which the railways and steamships created stimulated what has been one of the most characteristic features of nineteenth century development, namely, the growth of towns. In addition, the new transport developments created a new industrial class, the transport workers, and also gave an impetus to a rapid increase in the commercial and trading class. The small shop-keeper, domestic worker and the small peasant farmer were all affected, the first two adversely, the latter favourably ; mechanical transport vitally affected the position of women, making it far less necessary for them to stay at home and provide the food supply for the household, while the new personal mobility induced an emigration and transference of population from one country to another on a scale hitherto undreamed of which, while still further linking the world together as an economic whole, created a series of fresh problems. A country had to consider whether it would allow its people to leave and take up residence somewhere else, also under what conditions it would permit other peoples or races to come in and acquire a domicile. When Asiatics began to move in considerable numbers, the question became urgent for the countries bordering the Pacific as to what extent they would permit or restrict the settlement of a people within their borders of a lower standard of comfort which might threaten the standard to which the white man had laboriously attained.

The growth of towns was originally due to the industrial revolution. Manufacturers set up works on the coal and iron areas to obtain cheap power or raw material ; the " hands " followed to get work. People also massed in the ports to deal with the growing quantities of exports and imports. The growth of towns corresponded in every country to the development of coal mining, factories and engineering works.* This tendency was, however, considerably accelerated and stimulated by the development of railways, not merely because people had increased facilities for moving into towns, but because the railways enabled towns to be fed. There is a definite physical limit to the growth of a town in size when its food stuffs are brought in by road or canal and even the feeding of a sea-port town is conditioned by the dock accommodation for vessels. Railways have enabled town millions to be fed every day§ and London draws its daily milk supply from an area of a hundred and fifty miles round the city.

The excellent railway facilities to and from towns encourage factory owners to set up in urban rather than rural areas. They can make sure of getting fuel, the distribution of which is already organized for household purposes ; they are near the market for finished goods in the shape of merchants who will attend to the sales, and with the railway facilities they can draw on several lines for the conveyance of raw material or the despatch of finished goods.   Apart altogether from transport, manufacturers tend to set up in towns because they can get hands there readily without the necessity of providing for their housing as they would have to do in a rural area.

The growth of towns seems to have been specially rapid in the United States where competitive railway companies try to induce manufacturers to use their particular line. Manufacturers are naturally attracted to these competitive facilities when setting up new businesses.   In their turn the

The growth of the population in England and Wales in the mining and manufacturing area may be seen from the following table :
oooo's omitted.

Date	Total Population	† Mining Northern Counties	‡ Manufacturing Midland	Rest of England and Wales excluding the County of London
1851	27.37	451	276	829
1861	28.93	529	364	953
1881	34.88	757	427	1030
1891	37.73	862	489	1126
1901	41.46	976	568	1255

It will be noticed that the population doubled in fifty years in the mining and manufacturing regions while the rest of England only increased fifty per cent.

*Bowley, " Manual of Statistics," p. 89.

†The Northern Counties include Cheshire, Lancashire, Yorkshire (West Riding), Durham and Northumberland.

‡The Midland Counties considered are Derby, Leicester, Nottingham, Northampton, Stafford, Warwick, Worcester, Monmouth and Glamorgan.

§Address by Sir Sam Fay to the Railway Students' Union at the London School of Economics, 1911, p. 2.

operatives gather round the factories and towns increase in size.*

"The entire net increase of the population from 1870 to 1890 in Illinois, Wisconsin, Iowa and Minnesota . . . was in cities and towns possessing competitive rates, while those having non-competitive rates decreased in population and in Iowa it is the general belief that the absence of large cities is due to the earlier policy of the railways giving Chicago discriminating rates."†

A thoroughly up-to-date railway in England having a monopoly of one area, or a State railway in Germany will, however, lay itself out to offer as good facilities to traders as can be obtained by competition in other countries. It pays a railway to develop business generally, while it also pays a railway having a monopoly of an area to concentrate business in that area rather than allow it to frequent another area. It pays the North-Eastern Railway in England to increase traffic there rather than allow it to concentrate in South Wales or the Midlands. It pays the Prussian State railways to attract business to Berlin rather than let it go to Bavarian Munich. There is a geographical competition of areas as well as that of railways owned by different companies frequenting the same area. The general result is, however, the same. Manufacturers are attracted into towns by railway facilities and the phenomenal growth of towns is due to a combination of the new industry and the new transport.

The concentration in towns not merely took place more rapidly in England after the railways developed, especially between 1841-1851, but also in France and Germany. In Germany the railways stopped the increase of population in the smaller cities except those of an industrial character and hastened the growth of the large cities ‡

The following table, taken from Weber, shows the increasing proportion of the population dwelling in cities of over 10,000 and the growth of very large urban entities of over 100,000. Up to 1851, the growth may be ascribed to the industrial changes; after 1851 it is largely the result of

*Weber, " Growth of Cities," *pp.* 152, 199, 200.
†Weber, *op. cit.*, *p.* 201.
‡Weber, *p.* 201.

transport which, in its turn, as we have seen, increased the impetus to the spread of the industrial revolution.

PERCENTAGE OF POPULATION IN CITIES OF OVER 10,000. *

		1800	1850	1890
England and Wales	-	(1801) 21.30	(1851) 39.45	(1891) 61.73
Scotland	-	17.	(1851) 32.2	49.9
Prussia	-	(1816) 7.25	(1849) 10.63	(1890) 30.
U.S.A.	-	(1800) 3.8	(1850) 12.	27.6
France	-	(1801) 9.5	(1851) 14.4	25.9
Russia	-	(1820) 3.7	(1856) 5.3	9.3

PERCENTAGE OF TOTAL POPULATION DWELLING IN CITIES OF OVER 100,000.

		1851	1891
England and Wales	-	22.58	31.82
Scotland	-	16.9	29.8
Prussia	-	(1849) 3.1	12.9
U.S.A.	-	(1850) 6.	15.5
France	-	(1851) 4.6	12.
Russia	-	(1856) 1.6	(1885) 3.2

Railways and factories do not wholly account for the growth of towns in the nineteenth century. People were gathered into towns not merely by the increased opportunities for employment which towns afforded, but by the attraction of town life and the excitement of living in the mass. The new transport facilities enabled them to find out in many cases how much they preferred town to country life by the trips and excursions which familiarized the rural population with urban conditions.

As women were ousted from work on the land by agricultural machinery they went into domestic service and left the village. The young men followed them. A woman once used to town life does not care to marry and settle down in the country. She feels lonely. There are no shops to look at just round the corner, and she may have to walk miles for a quart of petroleum for her lamp or other necessaries, while water, instead of being constantly at hand as in towns, often has to be fetched from a considerable distance for culinary purposes and washing. Town life has a great attraction for women of the poorer classes who have not the change of going out to work and coming home again like the man. Hence they use their influence to keep their husbands in the towns.'

*Weber, op. cit., pp. 144-145.

The growth of cities was no doubt accelerated by the sanitary reforms which became effective after 1850. More people now tend to be kept alive in urban areas than in the old insanitary days. Towns therefore grow not merely by the migration of people into them from outside, but by the decrease of the death-rate of their own inhabitants.

Perhaps one of the most striking features of nineteenth century town development has been the growth of a fringe of towns round the coast devoted primarily to catering for people who go to the seaside. This again is due to the possibilities of railways facilitating movement for a short period.

In England there is also a tendency for towns to grow up along the sea-coast for industrial reasons.* A business at a seaport receives its raw materials without transferring them from the ship to the railway, and it thus avoids the expense of carriage inland. In the same way, if the goods are exported, the cost of railway haulage to the coast is avoided.†

It is worth noticing that the growth of towns is not now hampered by the question of a water supply. It is part of man's control over nature so typical of the nineteenth century, that he can bring his water supply from long distances in the new pipes provided by the engineering developments of the century. The water comes from Loch Katrine for Glasgow, for instance, and from Cumberland for Manchester; while the bringing of water from Wales right across England, for London has been seriously discussed. Previous to the nineteenth century the site of towns was limited by the water available in the immediate area.

The development of railways created an entirely new class of workers and greatly extended employment for others. Men were wanted to make the railways and for that purpose the class of navvies who had made the canals and inland navigations were utilized. This class was not new, but the plate layers, drivers, firemen, cleaners, guards, shunters and station-masters were new, while greatly extended opportunities of employment opened out for people who transported goods to and from railways or who unloaded ships at docks.

*Report of Royal Commission on Canals, Cd. 4979, *p.* 98.
†Report of Royal Commission on Canals, Cd. 4979, *p.* 88.

The personnel of the British mercantile marine expanded rapidly, but although the numbers of English sailors rose foreigners were also extensively employed, as the following table shows :—*

Annual Average	No. of British persons employed not including Lascars	Foreign Persons	Proportion of Foreigners to 100 British	Lascars
1860–64	163,676	17,808	10.88	—
1865–69	176,114	20,630	11.71	—
1870–74	181,628	19,425	10.69	—
1875–79	174,407	22,393	12.84	—
1880–84	170,399	26,040	15.28	—
1885–89	171,710	25,709	14.97	—
1890–94	185,524	29,799	16.06	24,628
1895–99	176,773	34,130	19.31	31,126
1900–04	175,095	38,915	22.22	39,267
1905–08	190,128	37,556	19.75	44,152

The general result was that transport workers began to form one of the largest class of workers in any country, and tended to augment its numbers rapidly.

There was a rapid increase also in the trading class in consequence of the commercial revolution wrought by the railways. More people were needed to deal with the buying and selling of the new products that became subjects of exchange, as well as for the enormously increased volume of the transactions.

For instance, articles like tea, coffee, cocoa, sugar, rice, tapioca, raisins, currants, oranges and lemons, which were, even in the fifties of the nineteenth century, articles which were consumed chiefly by well-to-do people, became, by the end of the century, the necessities of the poor as well as rich. There was an enormous increase in the consumption of these articles by all classes and a much larger number of persons were required for the distributive processes. These products would not have been grown in such quantities had transport facilities not existed which enabled them to find world markets. The freights of these goods fell and lowered the price to the consumer, and railways and steamships brought them to his very door. But while railways and steamships were responsible for much of the increased consumption there were changes in the methods of production of these articles

*Fiscal Blue Book, 1909, *pp.* 102-103.

which also contributed. Sugar was made from beet and not merely from cane ; tea was extensively grown in India, and not only in China, while the coffee of Ceylon supplemented that of Brazil. This again lowered the price and stimulated consumption and increased the trading class.

The growth of this class was still further encouraged by the development of large towns, which, as we have seen, were largely the outcome of railways. In a town of over 10,000 people, direct trading between producer and consumer becomes more and more difficult. The picturesque days are practically over when the farmer's wife or daughter drove in to market with her butter and poultry and exposed them for sale and dealt with the housewife who knew her personally. Butter merchants, egg merchants, poultry merchants, wholesale butchers and milk companies, either supplying smaller shops or trading themselves, have taken the place of the old " market day," Even in the market halls the sellers to-day are middle-men and not producers.

The growth of towns has made it unnecessary for the housewife to keep large stores by her. Space is too precious, and the tradesman has better facilities with his iced chambers for keeping perishable articles. Hence the trading and shop-keeping classes increase from this cause also. Big businesses concentrated in one spot also needed intermediaries to carry out their sales and agencies and travellers increased to facilitate distribution.

The following table will show the increase in the transport and trading classes in the United Kingdom, and the increase in miners and metal workers whose occupation is so closely bound up with the new transport methods :—

NUMBERS EMPLOYED PER THOUSAND OVER 10 YEARS OF AGE.*

		Males			Females		
		1881	1891	1901	1881	1891	1901
Commercial†	-	30	34	41	1	2	5
Transport‡	-	75	85	95	1	1	1
Mining -	-	49	54	60	—	—	—
Metals§	-	75	79	91	3	3	4
Total occupied -		827	827	834	335	330	316
Retired or unoccupied -		173	173	166	665	670	684
		1000	1000	1000	1000	1000	1000
Total Persons over 10 years	-	12,55 0000	13,89	15,54	13.50	15,80	16,80

The growing numbers engaged in trade and transportation
in the United States are still more remarkable :—

1880	•	•	•	1,871,503
1890				3,326,123
1900	•	•	•	4,766,964
1910				7,605,730‖

The effects of railways and steamships on the small shop-
keeper and independent artisan class was to make his position
more difficult.   The growth of great giant stores which could
despatch goods by post or rail after reaching its customers
by illustrated catalogue tended to concentrate business still
further in large urban areas to the detriment of local industry.
The local draper with his limited range of patterns or styles,
or the small grocer whose new stock is being " expected in
every day " and which fails to arrive for weeks or even
months, is seriously affected by the despatch, often carriage
free, of the great distributing stores.   It is possible for a
woman in an outlying country district of Cornwall to shop
by catalogue with ease in either Manchester, London or Paris.
Shopping by post is a feature of the distributing business
in Germany, the United States and Great Britain.

In the same way the local artisan has been affected.   Previously nearly everything used in an area had to be made
within that area, as the difficulties of transport were insuper-
able.   There were thus in each area a number of independent
craftsmen or workers making things for the immediate needs
of the neighbourhood.   With the facility of transferring such
articles as furniture, cooking utensils or clothes from a large
centre the local industry tended to dwindle.   This is one of
the causes of the decline of the domestic worker.¶

This destruction of local life is all the more serious as it
is on the local men that so much of the business of local
government depends.

On the other hand, countries or areas with an agricultural

*Bowley, " Manual of Statistics," *p.* 91.
†This includes merchants, dealers, travellers and clerks.
‡This includes railways (but not railway construction), roads, rivers,
docks and the telegraph and telephone service.
§Metals include all work in metals except mining and the manu-
facture of tools, machinery and engines.
‖Occupation Census, *p.* 53, published 1914.
¶Schmoller, " Zur Geschichte der deutschen Kleingewerbe," *p.*
174 ff.

surplus are often benefitted by access to wider markets than those of the immediate neighbourhood. The result would be to raise prices in the locality, because a better market is obtainable by railway elsewhere.* This would be true of agriculturalists in the United States or Denmark. This is also true of the type of business carried on by the peasant farmer. If the collection of eggs, chicken, milk, butter or vegetables is organized by a middle man or a co-operative society for sale in a big town, the small cultivator is in a position to command much better prices than those obtaining locally, and is, therefore enabled to raise himself above the sheer margin of subsistence which used to be the characteristic of the small man in agriculture. He does not feel the effect of the foreign competition which arises principally in the bulky products such as meat or wheat, i.e., the small man with his fruit, vegetables and milk, caters for a national market, while competition is felt in the goods that are dealt with in an international market. On the whole large-scale agriculturalists in the United States or the English colonies benefitted because they could get a market abroad, but they had to encounter competition from all parts of the world which tended to diminish their profits. On the other hand a severe agricultural depression was engendered in Western Europe by the ease of import of agricultural products grown under prairie conditions.

The general result was that the agriculturalist in a new country benefitted since it gave him his chance to develop ; the small man got the growing market of the big towns, and did not encounter severe foreign competition, but the large scale producers that had hitherto enjoyed a monopoly of the home market suffered.

The fishing industry was another trade that increased owing to transport facilities. The local market for fish is soon glutted. It is a highly perishable commodity for which rapid transport with cold storage provides a wide sale, which is even now international. English fish is sold fresh in Switzerland, and Canadian salmon finds a market in Europe. Hence the fishing industry, instead of being an affair of small men owning their own boats, has become capitalized and

*When the Erie Canal was opened in the United States in 1825, the price of grain rose considerably in the North-West.

trawling companies are taking the place of the old fisherman.

Mechanical transport has also affected the position of women. The ease of distribution has led to many of the products formerly made at home becoming industrialized. Factories for biscuits, jam, pickles, cakes, sweets, laundry work, baking, food curing and food preserved in tins are all of recent development. Butter is no longer salted as a matter of course in summer by every housewife for the winter scarcity, since regular supplies are available in normal times from all parts of the world, ranging from Siberia to Australia, the summer in the Southern hemisphere coinciding with the Northern winter and *vice versa*. The artistic pleasure of the early Victorian housewife in " putting away a pig " and planning the strategic disposal of each item for future consumption is unknown to a generation that imports its hams very largely from Chicago and buys its Danish or American bacon by the pound at the grocer's. The result is that there has been a release of female labour from the work of the preparation of the food supply for home consumption. This in its turn has led women to seek for other employments and has led to their transference to other fields of industry.*

This large food import, as we have seen above, has led to much greater security of life. Instead of the ever-present danger of famine, the world is comparatively safe and ordinary people are relieved from that constant preoccupation about food and the harvests which is so characteristic of other centuries.

The new personal mobility has, however, brought other problems to the forefront. There is first of all a great migration of people always taking place within their national boundaries. The changes in the election registers alone show how marked is the contrast between the fixity of people to one town or spot in the early part of the nineteenth century and the constant change of abode that went on in the twentieth. The growth of the business of the furniture remover is another proof. This is as true of Germany and the United States

*In countries where communication is undeveloped a large amount of female labour is still employed in simply preserving food for the scarce periods. The account of the feverish activity in a Russian household in food preserving in the country during the summer months is interestingly described in Palmer " Russian Life in Town and Country," *pp.* 16-25.

as of Great Britain. This continuous movement raises serious problems of local administration. How can classes that are always changing their place of residence be governed ? The problem is still further complicated by the fact that many people work in one place and sleep in another, and feel no real responsibilities for the local welfare of either place. All this tends to throw more and more work into the hands of the central government, and so increases that reaction to State intervention which is characteristic of the years after 1870.

Accompanying this constant change of abode within a country are the seasonal migrations of persons to other lands. The Russians migrated in their thousands to get in the German harvest, as did also the Galicians.* The Irish came to England and Scotland to lift the potatoes. Many Italians went to Germany and even to the United States for the building trades in the summer and returned to Italy for the winter. Others went to the Argentine Republic merely for the harvest. Thousands of English skilled mechanics, stone-cutters, stone-masons, glass-blowers, locomotive engineers, etc., regularly visited the New England States in the Spring and returned home when the slack period arrived. " In this way they escape American taxation, perform none of the duties of citizenship, and they spend the bulk of the money outside the country in which they earn it."†

There is also a large emigration and immigration movement in all countries due to the abolition of distance by rapid and safe transport.

" Out of the remote and little known regions of Northern, Eastern and Southern Europe for ever marches a vast and endless army. Nondescript and ever changing in personnel, without leaders or organization, this great force, moving at the rate of nearly 1,500,000 each year, is invading the civilized world.

*The number of foreign agricultural workers of the migrant type (Russian and Gallician) who came to Germany for the harvest is officially returned as

1911	-	-	-	387,902
1912	-	-	-	397,364
1913	-	-	-	411,706

Consular Report (Germany) Cd. 7620, 1914.

†Johnson, " Emigration from the United Kingdom to North America," *p*. 319.

" It is a march the like of which the world has never seen, and the moving columns are animated by but one idea— that of escaping from evils which have made existence intolerable, and of reaching the free air of countries where conditions are better shaped to the welfare of the masses of the people.

It is a vast procession of varied humanity. In tongue it is polyglot ; in dress, all climes from pole to equator are indicated, and all religions and beliefs enlist their followers. There is no age limit, for young and old travel side by side. There is no sex limitation, for the women are as keen as, if not more so, than the men ; and babes in arms are here in no mean numbers.

" The army carries its equipment on its back, but in no prescribed form. The allowance is meagre, it is true, but the household gods of a family sprung from the same soil as a hundred previous generations may possibly be contained in shapeless bags or bundles. For ever moving, always in the same direction, this marching army comes out of the shadow, converges to natural points of distribution, masses along the great international highways and its vanguard disappears, absorbed where it finds a resting-place.

" The traffic in ocean passages has reached a stage of fierce competition, unscrupulousness and even inhumanity inconceivable to those not familiar with its details. Men who pro't by the march of these millions of people have a drag net out over continental Europe so fine in its meshes as to let no man, woman or child escape who has the price and the desire or need to go. Three great countries, Italy, Austria-Hungary and Russia, where the masses of the people are low in the social scale . . . are being drained of their human dregs through channels made easy by those seeking cargo for their ships."*

In the passage just quoted the reasons assigned for this vast movement of people are the intolerable conditions at home and the efforts of steamship companies who by the activity of their agents stimulate people to remove themselves to a place where conditions are represented to be better.

The causes of the emigration movement varied, however,

*Whelpley : " Problem of the Immigrant," *pp.* 1-3 (1905).

in the different countries.* As far as Great Britain was concerned the industrial and agricultural revolutions at home were the cause of their migration. The weaver was ousted by machinery, the peasant farmer by large farms. The distressed classes were assisted by charitable associations and land companies in the matter of their passage before the steamships competed for the emigrant traffic.† The successful ones in their turn sent money home to assist relatives to come out.‡

The sheer overpopulation of Ireland, the numbers of which were 8,175,000 in 1841, culminated in the famine of 1846, and was the cause of the emigration from that country, with a result that the population of Ireland fell to 4,459,000 in 1901.

The year 1847 was equally a year of scarcity in Europe which may have had something to do with the almost universal revolutionary outbreaks of the 1848. The result was that many people who had been mixed up with the Liberal movements found it desirable to leave the country, and a large German middle-class emigration took place. This was re-enforced during the nineteenth century by Germans from the country districts. Many of the freed serfs could not adjust themselves to individual, instead of communal farming, after the emancipation ; others emigrated because of the excessive sub-division of small farms in the West where the Napoleonic Law of the equal division of a large part of the property among the children at death held good, thus resulting in the sub-division of small farms to a point where no one

*Of the 7,783,503 British and Irish people who left the United Kingdom during the years 1880-1911 for North America, Australia or New Zealand, 4,407,253 went to the United States. Johnson, *op. cit.*, *p.* 346, Table II.

†Johnson, *op. cit.*, *ch.* 3.

‡Considerable sums of money were sent annually by successful immigrants in North America to friends in the United Kingdom. It was calculated by the British authorities to be

1849	-	-	-	£540,000
1859	-	-	-	£575,378
1869	-	-	-	£639,335
1878	-	-	-	£784,067

Only a few typical years are given as samples. For the amounts annually sent between 1848 and 1878, *see* Johnson, *op. cit.*, Table X, appendix, *p.* 352.

person could live out of the farm produce. The Germans went chiefly to the United States. In the same way the changing conditions of agriculture in Austria-Hungary and Russia, combined with the ease of transport by railways and the activities of the steamship companies which facilitated their passage, led to the starting of the hordes of Eastern Europeans.

After 1900 the English began to be increasingly attracted to their own colonies and the main stream of English emigration set in towards Canada. Between 1901-1912, 63 per cent. of the British emigrants migrated to places within the Empire ; only 28 per cent. had done this between 1891-1900.*

As England was the great shipping country, *par excellence,* the European emigration traffic took place, at first, *via* England. As other countries developed their shipping after 1880, they made a determined effort to capture the traffic as a basis on which to build up their new steamship lines in order that the carriage of their exports and imports might not remain to such a large extent in English hands.

Germany obtained an increasing share of the emigrant traffic, after 1894, by setting up control stations nominally to prevent the passage of diseased and undesirable persons.† These control stations were placed under the German steamship lines, and they took measures to ensure that such persons as passed through the control stations went by German and not by British lines. This was all the more easy as the emigrant traffic gradually altered in character after 1880. Owing to the development of German industry the German emigration fell off‡ as the people were absorbed by the new factories

---

*Dominions Commission, *p.* 88, Cd. 8462 (1917).
†Cd. 9092 (1918) *pp.* 8-9.
‡German Emigration

1881	-	220,902	1892	-	116,339
1882	-	203,585	1893	-	87,677
1883	-	173,616	1894	-	40,964
1884	-	149,065	1895	-	37,498
1885	-	110,119	1901	-	20,874
1891	-	120,089	1912	-	18,545

" Statistiches Jahrbuch."

As to figures of emigration from foreign countries during the nineteenth century *see* Tables, Charts and Memoranda on Emigration from the chief European countries in Second Fiscal Blue Book (Board of Trade) Cd. 2337 (1904) *pp.* 159-175.

Q

and engineering works. In the place of the Germans, Russians, Austrians and Italians began to move in increasing swarms. While the Italians went largely to the Argentine, the bulk of them went to the United States* but the Russian and Austrian hordes passed *via* Germany into the new world, and the Germans saw to it that they should build up the German Atlantic shipping business.

" This stampede has now reached such proportions as to occupy all the energies of a score of steamship lines in handling the traffic, to warrant the establishment of new and more direct routes and the building of new ships specially designed for the carrying of this cargo."†

It is not to the best interests of a nation that people physically in the prime of life should leave in excessive numbers. Too large a proportion of young and old remain, the physically unfit tend to preponderate. It costs money to rear and educate a man or woman.‡ He represents so much capital and to export capital in that form is nationally unsound, unless a definite gain can be proved to ensue to the mother country. The nation that lets its citizens go in the pride of their youth and strength incurs a loss in productive power. Both English and German agriculture have suffered from a lack of agricultural labour in the last quarter of the nineteenth century, while the emigration of English skilled artisans and miners to the United States has done much towards developing a rivalry there for the English engineering trades.§

While most governments have taken precautions that the emigrant traffic shall be conducted under conditions of reasonable security and decency for the person making the passage, the general tendency is for continental governments to restrict the outflow of emigrants by prohibiting advertising and curbing the activities of the steamship companies. But

---

*Cd. 9092, *p.* 6. 266,000 Italians went to U.S.A. in 1912-13 ; 81,000 to Argentina ; 32,000 to Brazil.

†Whelpley, *op. cit.*, *p.* 15 (1905).

Mr. Whelpley states that many emigration authorities hold the steamship companies responsible for fifty per cent. of the departures for foreign lands. *Op. cit.*, *p.* 11.

‡Marshall put it at £200. " Principles of Economics," *p.* 647, Note.

§For a discussion of the value of emigration, *see* Johnson, *op. cit.*, Chapter CXIII.

the countries to which these emigrants go have an equally difficult problem. They often underbid the labour market ; they collect in groups of the same nationality where they are difficult to control ; their standard of sanitation is deficient, and it is difficult to make them realise their duty to their adopted country. They must get work at once ; they crowd into towns, increasing the congestion at the ports and the general housing difficulty.*

Nearly all countries enacted laws and regulations governing the admission of aliens. Some merely prohibited the entrance of the diseased and the criminal ; others, notably the United States and Canada, imposed severe restrictions on immigration. The difficulty of the absorption of the Eastern European immigrants was not the least of the problems that the United States had to face in 1914. As long as the new arrivals were English, German, Irish or Scandinavian, their standard of life and general outlook was not so radically different from that of their hosts as to make Americanisation difficult. It is, however, a very different problem with the masses of illiterate Galicians, Russians and Levantines with which the United States has been deluged of recent years.

Nor is the British Empire without serious problems of its own arising out of the facility for general movement. The developments of transport have caused movement even in the changeless East. Indians, Chinese and Japanese began to migrate after 1850 to the Australian gold fields and were to be found in New Zealand and in the South African colonies, Indians were deliberately encouraged by the governments of Mauritius, British Guiana, Trinidad and Jamaica to migrate there as negro labour was so inefficient after the freeing of the slaves. The friction has come in the self-governing dominions. The Indian and the Chinese coolie have a lower standard of comfort than the European, and will work for lower wages than the white man. The self-governing dominions do not wish their standard to be undermined by cheap Asiatic labour. Hence they have all adopted restrictions intended to keep such labour out, either by penalizing the steamship companies, or by an impossible educational test, or by other methods. The Indian subjects of the King-Emperor complain, however, that as they are citizens of

*See Report of the Immigration Commission, U.S.A.

the Empire, they should be allowed free movement within the Empire, and much capital has been made by agitators out of the restrictions imposed by the dominions. A settlement of the question was arrived at during the war.

The question of the Chinese immigrant is rather different. He is not a fellow citizen of the same Empire. But the British fought on more than one occasion during the nineteenth century to make the Chinese open their country and their ports to foreigners. It is a difficult diplomatic problem to know how far the Chinese might claim a return compliment of the " open door," especially as Australia is a continent peopled only by four million persons, whereas the Chinese Empire is supposed to contain about 315 million, and to be over-populated. It is possible, however, that the development of industry, both in India and China, in the near future will absorb so many of their people that the question of a surplus Asiatic population seeking an outlet and not finding it may be temporarily shelved.

It is quite obvious that no country can afford to neglect either the question of creating or controlling railways and steamships. Alongside of the revolution in the methods of manufacture caused by machinery and engineering, it is the event of the greatest importance in the whole of nineteenth century economic development after the question of personal freedom had been solved.

PART V

THE DEVELOPMENT OF MECHANICAL TRANSPORT IN GREAT BRITAIN AND THE PROBLEM OF STATE CONTROL OF TRANSPORT.

SYNOPSIS.

British railways were founded on the model of the existing roads and canals.

I.—ROADS.

Earthen tracks were in the eighteenth century converted into metalled roads for carts by turnpike trusts charging tolls for the use of the road.

Road improvement and construction carried out by private individuals. Striking contrast with the State road system of France.

The " calamity of the railways " ruined the turnpike trusts.

II.—CANALS.

The need for increased transport facilities to move coal and raw material.

1—The development of the Canal System, 1761-1830.

The canal from Chat Moss to Manchester, 1761, followed by the Manchester and Liverpool Canal and the Grand Trunk.

Canals were built by private individuals who charged tolls for their use and were a financial success.

They varied in gauge and structure, belonged to many companies and no through system of carriage or rates was devised.

EFFECTS.—Great stimulus to trade and industry.

Gave facilities for distribution of food and so assisted agriculture and helped Northern towns to be fed.

Redistributed population.

Stimulated port development.

Trained a new class of workmen—the navigators.

Created commercial travellers.

In France the Canal system was assisted by the State after 1799 and then purchased by the State. Traffic on canals is free.

2—The relative decline of the canals.

CAUSES :—

(a) The greater efficiency of the railway system : speed. punctuality, through rates.

(b) The coasting steamer.

(c) The purchase of strategic links by railways.

(d) The reconstruction of English agriculture and English trading methods.

(e) The difficulty of effecting improvement when the canals were owned by so many companies.

(f) In vicinity of canals, areas are built over by houses ; difficult to enlarge canals.

III.—RAILWAYS.

1—The peculiarities of the British railway system—the absence of State aid, the absence of military motives, the catering for a developed traffic, the high capitalization, the short hauls, the supersession of the canals by the railways, the system of charging based on the analogy of canals and roads.

2—Periods of Railway History.

(a) The period of experiment, 1821-1844.

The waggon ways for coal distribution ; iron rails ; trucks drawn by horses.

The haulage by steam engine. 1821—The Stockton and Darlington public railway, opened 1825. 1826—The Liverpool and Manchester ; challenge to the canals ; carried passengers ; successful utilisation of steam for haulage ; provided waggons ; opened 1830.

System one of short scattered lines and two main gauges.

The question of control.

Act of 1844 providing for State purchase of future lines.

Commission to control railway promotions sat one year.

(b) The consolidation of the lines, 1845-1872.

The amalgamation of small lines into large companies.

Influence of Hudson, " the Railway King."

The railway mania, 1845-1847.

The influence of the Clearing House as providing a mechanism for common agreement between the companies.

The further development of the railway systems as part of a defensive policy against competitors.

The development of goods traffic by the railways and the decline of the Canals.

The attempt to control the growing railway monopoly. 1846—Railway Commission created ; ceased, 1851. 1854—Cardwell's Act prohibited undue preferences and facilitated through traffic. 1867—Railways to keep accounts.

Amalgamations continued. Control a failure.

(c) The development of State Control, 1873-1893.

Railway and Canal Commission, 1873, to control amalgamations and preferences. Appointed for five years ; became permanent.

Railway and Canal Traffic Act, 1888, arranged for new classification and new maximum rates which should be effective.

The Board of Trade set up as conciliator in disputes between traders and railways.

Railway rates fixed, 1892-1893.

The Act of 1894 limited the raising of rates to the permitted maximum.

(*d*) 1894-1914. The decline in dividends; the amalgamations and the question of nationalization.

    1—The increase in expenditure and the decline of dividends.

    2—Amalgamation to stop cut-throat competition—result, quasi monopoly.

    3—Labour troubles.

    4—Proposals to control railways :

       (*a*) By reviving canals as competitors.

       (*b*) By State ownership.

IV.—THE STEAMSHIP AND SHIPPING PROBLEMS.

The absence of State control of shipping.

REASONS :—Shipping companies formed under the Companies' Acts.

Greater competition in shipping.

Difficult to enforce fixed rates of freights for continuous voyages.

A ship is mobile and can unload at any port.

Competition of tramps and liners.

1—The change of policy from protection of shipping under Navigation Acts to free Competition.

The period of minor relaxations, 1796-1822.

The change from monopoly to reciprocity, 1822-1840.

The abolition of the Navigation Acts, 1849-1854.

2—The coming of the steamship and the progressive change in technique.

    (*a*) Change in material from wood to iron and iron to steel.

    (*b*) Change in the marine engine.

    (*c*) Constant growth in the size of ships.

    (*d*) Specialization of ships : liners and tramps.

3—The supremacy of the United Kingdom in the ship-building and carrying trades.

4—The growth of foreign shipping.

Subsidies to foreign ships.

The development of German shipping—the control of the emigrant traffic—the close organization of German ship-owners.

5—Combination in the Shipping World.

Violent competition in shipping between 1870-1880 gave rise to combinations to steady rates.

International combines for division of territory.

6.—The Government and shipping.

IT is no accident that the new form of transport should have been evolved in the pioneer country of the industrial revolution. The three things that are necessary before some great mechanical innovation can take place are the existence of capital to try experiments ; a demand for the new goods or new services to be rendered ; and the technical ability to construct the article required. As far as the railways are concerned, the people of Great Britain had accumulated the capital and were willing to sink it in the new form of transport* ; the development of coal, iron and other heavy materials was creating a demand for a new and improved method of moving them, as industry had really outgrown the capacity of the canals to handle its products with sufficient rapidity ; and in her iron-workers and engineers England possessed an unrivalled skill and capacity to make the locomotives, rails, marine engines and iron ships.

The English railway system possesses certain marked peculiarities of its own which differentiates it from all other railway systems of the world. This is partly due to the fact that it was evolved out of the existing system of roads and canals and copied many of their features.

I.—ROADS.

The public highways of Great Britain had been until the eighteenth century mere earthen tracks or bridle paths for pack mules and riders.† These unmetalled roads were kept in order, according to a Statute of 1555, by the labour of the persons of each parish who had to give compulsory service for six days each year on the roads. Those parishioners whose income exceeded £50 a year were obliged to provide the services of a man, horse and cart for six days. Wheeled traffic, though beginning, was still uncommon at the end of the seventeenth century. It was, however, increasing with the expansion of trade and the growing necessity to move larger quantities of goods. These wheeled vehicles wore the earthen surface of the highways into great ruts and the roads became more and more of a scandal just at the time when it became more and more necessary to be able to move masses of raw material or manufactured goods. The whole industrial development of the eighteenth century would have been held up if the roads could not have been improved. The tradition of the English Government was, as we have seen, to leave everything to individuals and during the eighteenth century

*Tooke and Newmarch, History of Prices, Vol. V.
†On the whole subject, S. & B. Webb, " The King's Highway."

the practice developed of certain persons, landowners and others, obtaining a Private Act of Parliament and reconstructing and paving a stretch of road in such a fashion that wheeled vehicles could easily pass to and fro. These persons formed a turnpike trust and were empowered to charge tolls to the users of the roads to recoup themselves and provide a fund for keeping the road in repair. One must therefore picture a network of fairly good high roads in the hands of 1,100 different turnpike trusts who had re-made the roads in varying fashion and had kept them in repair with varying degrees of efficiency. Outside their area lay a great net-work of parish roads that were still unmetalled tracks. One therefore finds the most extraordinary differences in the accounts of the roads at the time. One writer will talk of the " winged expedition " of coaches going at five miles an hour ; another will chronicle ruts four feet deep and a long line of carts broken down. It all depends on whether they were speaking of the improved turnpike roads or the unimproved parish roads. But even the turnpikes varied and some stretches were very defective. In other parts land owners were not enterprising enough, or did not have the capital to re-make a road and the wheeled traffic made the earthen road impossible. Towns were and expected to be cut off in winter, owing to the state of the roads, and each winter they salted in their provisions and prepared as if for a siege, as the roads leading into towns with the amount of traffic passing over them were peculiarly liable to fall into disrepair. It is difficult to picture the isolation which fell upon country districts in winter. " You too well know that in winter when the cheerless season of the year invites and requires society and good fellowship, the intercourse of neighbours cannot be kept up without imminent danger to life and limb," was the verdict in 1792.*

Throughout the eighteenth century the trusts were confronted with the great difficulty of getting any satisfactory surface for the roads. As late as 1808 it could be said that : " On examining the turnpike roads in the vicinity of London I find the materials by which they are repaired seldom last longer than a month or six weeks in winter before they are ground to atoms and raked off the road as puddle. . . . In

*Webb, *op. cit., pp.* 195, 226.

some places the tolls have been doubled yet are the roads sometimes almost impassable."*

The old parish roads continued to be mended by the six days' Statute labour or by levying a Rate and employing the paupers on the roads. In 1832, no less than 52,800 paupers were thus employed at a cost of £264,000. Out of a total length of recognized public highway in 1820, amounting to 125,000 miles, only 20,875 miles were under the turnpike trusts ; the remainder were cared for in 1830 by the inefficient labour of the poor, or the equally unsatisfactory labour of those who had to render six days' compulsory service.†

It is obvious that travel would be attended with considerable danger to life and limb and it is not surprising that people who were adventurous enough to make " tours " in England wrote of their adventures as if they had been to Central Africa.‡

At the beginning of the nineteenth century, however, three great reforms were revolutionizing the traffic on the main high roads. Macadam had invented a durable surface : Telford was showing how roads might be engineered ; and the turnpike trusts were beginning to combine into larger areas for which salaried officials were appointed and a more uniform system of maintenance and improvement was instituted.

The Highways Act of 1835 abolished the compulsory Statute labour on the roads and empowered each parish to levy a rate and appoint a salaried official for road maintenance.

Just as the highways were really improving they were overwhelmed by " the calamity of the railways." The coaches that had paid such a large proportion of the tolls were taken off the roads ; the turnpikes became bankrupt and the Government was obliged to abandon the policy of laissez-faire and do something for road maintenance. The turnpike trusts were gradually wound up but there were still 854 existing in 1871 ; the last toll was, however, levied in 1895 on the Anglesea portion of the Shrewsbury and Holyhead

*Adam Walker, Report on Highways.

†Webb, *op. cit., p.* 193.

‡It is interesting to notice that Dumas can find no greater instance of the power of money than to make his hero, Monte Cristo arrive in Paris at a day and hour fixed three months beforehand and yet the French road system was famous.

Road. In 1888, the care of the main roads was transferred to the County Councils, the others being given over to the Rural or Urban District Councils.

No greater contrast with the English road system can be imagined than that of her great industrial rival, France. From 1743 the main roads were improved and engineered by the Central Government over the larger part of France,* *i.e.*, in the Pays d'Élection. A school of engineers was set up and they trained men in road maintenance and road engineering. In the Pays d'Etat which retained a good deal of local autonomy the main roads were not under the supervision of the Government, but the force of example was bound to tell and the high roads of Languedoc were praised by so experienced a traveller as Arthur Young who was positively vituperative over the roads of the North of England. The French peasant had to give thirty days' work on the roads of his district and the French roads were the best in Europe.

The French Revolution saw the collapse of road maintenance in France, but Napoleon, partly to restore order in France and suppress brigandage, and partly in order to be able to move troops quickly, re-made the roads of France. They were divided, in 1811, into main roads or *routes impériales* radiating out from Paris, and local roads or *routes départementales*.† The former were maintained by the central Government, the latter by the departments. They were all engineered and cared for on a uniform plan, and the French proved themselves to be the greatest road-makers since the days of the Romans. The roads, unlike the English turnpike roads, were free from tolls. After the fall of Napoleon, the Restoration Government continued his policy and spent no less than 302 million francs on the roads, while between 1830 and 1848, 978 millions were spent on the two classes of roads. Contrast this with the bankruptcy of the turnpike trusts in the same period and the endeavours of the English Government to fob off the burden of the roads on to the local sanitary authorities. This is typical of the history of transport development in the two countries, whether road, canal

*Letaconnoux. Les voies de communication en France au XVIII. siècle in Vierteljahrschrift für Sozial und Wirtschaftsgeschichte, Vol. VII., *p.* 94.

†This distinction is still preserved but the main roads are " routes ationales."

or railway.   In Great Britain transport improvements were regarded as the business of individuals ; if they found the money well and good, if they did not then matters remained as before.   Transport was a business like anything else, and a Government had no concern with business undertakings except to prevent abuses.   In France the unity and welfare of the State was held to depend on smooth and rapid communication and transport has been under the peculiar care of the State during the eighteenth and nineteenth centuries, no matter how laissez-faire France might be in other matters. In France, transport developments came from above and were planned on a uniform system ; in England, they came from below, and grew up in a patchy haphazard piecemeal fashion with a tendency to amalgamate into bigger areas, but always retaining traces of the want of uniformity with which they started.

It was obvious that Great Britain with her growing traffic and growing industry must improve her means of transport beyond that of the turnpike roads, or she could never move the quantities required for the development of large scale production.

The result is that after 1760 we get the development of canals and the stimulus to the starting of canals came from the demand for coal.

II.—CANALS.

During the eighteenth century Great Britain began to require coal in increasing quantities, and some better and cheaper method of moving coal than in a cart or in panniers on mules became imperatively necessary.

The blast furnaces had developed rapidly after 1750 when the secret of smelting iron with coke got out into the trade and they required large quantities of coal.   The pottery industry was being successfully developed at the same time by Wedgwood, and it required china clay from Cornwall, coal for firing, and above all, some cheap and safe method of transporting the fragile ware when made.   With the timber famine coal was also needed for household fuel and the growing textile industries required it for steam power to work machines. It was also essential to Lancashire to be able to receive raw cotton in bulk and to be able to ship millions of yards of

Manchester goods with certainty and despatch. It is no accident, therefore, that the first canal should arise in the North where the road system seems to have been peculiarly deficient, if we may judge from the description of so experienced a traveller as Arthur Young.*

It is a little difficult to say whether the industrial revolution created the improved methods of transport or *vice versa*. The truth probably lies in the fact that each in turn stimulated the other. The improvement of the roads was due to increased traffic,† while the canals certainly owed their origin to the fact that they would " pay " owing to the increasing demand for coal. On the other hand, the factories could not have grown beyond the stage of small workshops if their coal and raw material had to be conveyed in small quantities.

The history of the British canals may be divided into two periods. Between 1760 and 1830 we get the rise and development of the canals as the most important part of the transport system and one on which the industrial existence of England had come to depend.

From 1830-1914 we get the period of the relative decline of the canals owing to the coming of the railway and the steamship. The industrial existence of the country began to hinge on mechanical transport and the spread of the

---

*Some idea of the growing mass of raw material to be handled may be seen from the following figures:

COTTON WOOL IMPORTED
(quoted Baines, " History of the Cotton Manufacture.")

		lbs. 000 omitted.			lbs. 000 omitted.
1751	-	2,976	1800	-	56,010
1782	-	11,828	1810	-	132,488
1787	-	23,250	1820	-	144,818
1790	-	31,447	1830	-	259,856

FOREIGN AND COLONIAL WOOL IMPORTED
(quoted, Cunningham, " Growth," Vol. III., *p.* 929, from " An Account of the Woollen Trade of Yorkshire.")

		lbs. 000 omitted.			lbs. 000 omitted.
1766	-	1,926	1830	-	32,305
1790	-	2,582	1840	-	49,436
1800	-	8,609	1857	-	127,390
1810	-	10,914			

† Webb, *op. cit.*, Chapter V.

industrial revolution was much accelerated and extended by the possibility of distribution of masses of goods in hitherto inconceivable quantities.

The first canal was built by the Duke of Bridgwater, at his own cost, to link up his colliery at Chat Moss (Worsley) with Manchester, and it was opened in 1761. As Manchester needed some better communication with the sea than the road and unimproved river afforded, the Duke built a second canal connecting Manchester with Runcorn and so with Liverpool. The pottery, salt manufacturers and others who were peculiarly hampered by want of good transport then combined to finance several canals in the Midlands. They were the Trent and Mersey (otherwise the Grand Trunk), the Staffordshire and Worcestershire (authorized in 1766), the Birmingham and Coventry (1768) and the Oxford canals (1769). The Grand Junction Canal which connected London with the Midlands was authorized in 1793. Of this last it was said that " the advantages to the Metropolis and indeed to all places on the line and its branches are incalculable. The staple goods of Manchester, Stourbridge, Birmingham and Wolverhampton—cheese, salt, lime, stone, timber, corn, paper, bricks, etc.—are conveyed by it to town ; whilst in return groceries, tallow, cotton, tin, manure and raw materials for the manufacturing districts are constantly passing upon it."* This quotation shows how far the Midlands were from being industrialized when, in 1831, the staple goods were of the nature described above.

In the last decade of the eighteenth century a great canal mania set in (1793-1797) and England was rapidly covered with a system of inland water-ways, built by numerous private companies. These companies had to obtain an Act of Parliament in order to have compulsory powers to take land for the purpose of the canal. Before granting the power to take land Parliament laid down certain maximum rates of charge for the use of the water way.

Over the benefits of the change contemporaries waxed lyrical : " The prodigious additions made within a few years to the system of inland navigation, now extended to almost every corner of the kingdom, cannot but impress the mind

*Priestley, The Historical Account of the Navigable Rivers, Canals and Railways, 1831, *p*. 335.

with the magnificent ideas of the opulence, the spirit, and the enlarged views which characterize the commercial interest of this country. Nothing seems too bold for it to undertake, too difficult for it to achieve, and should no external changes produce a durable check to national prosperity its future progress is beyond the reach of calculation."*

As many of these çanals linked up rivers, the rivers also had to be improved and these improved rivers were known as " inland navigations."

The general result was that by 1830 there existed 1,927 miles of canals and 1,312 miles of navigations, and 812 miles of open rivers in England and Wales, 183 miles of improved waterways and canals in Scotland and 848 in Ireland.† The canal system was, therefore, primarily an English development. This covering of the country with a network of water communication is really a remarkable achievement when one realizes how little experience the English had to go upon. Brindley, the Duke's foreman, was trained as a millwright and got £1 1s. a week from the Duke as wages. He had to work out by himself all the problems connected with canal making, including the method of making the canal water-tight. He had to act as surveyor, contractor, engineer, foreman of the works and inventor of the appliances required. No one in England could ever have seen a canal barge or a lock, and so little faith had people in the North in the Duke's schemes that he had to come to London to borrow the £25,000 he needed to complete the Manchester-Liverpool Canal. He could not raise even £500 for the purpose in the North.‡

When once the Duke showed the value of transport by canal the matter was eagerly taken up, and although England was fighting a great war which strained every financial resource these thousands of miles of water-ways were completed by private individuals. The English water-ways received no financial assistance from the Government, but in Scotland two canals, the Caledonian and the Crinan Canal,

*Aiken's " Lancashire " (1793), quoted in Royal Commission on Canals and Inland Navigations (1909), Cd. 4979, *p.* 3.

†Cd. 4979 (1909), *pp.* 14 and 20.

‡Smiles, Lives of the Engineers.

were constructed and improved by Parliamentary grants, but these were exceptions to the general rule and were thus assisted to enable ships to avoid the perils of the Highland sea-coast. It was not a commercial matter but a question of the safety of shipping. These are the only two water-ways in Great Britain which belonged to the State, and the Caledonian Canal was an annual loss. The general rule was that private individuals must find the money for transport improvements and might recoup themselves by charging tolls. So large was the traffic on the canals that many of them were not merely an industrial but a great financial success.*

These canals and inland navigations were built to compete with roads and therefore were only designed for small barges. Only a twenty ton barge can navigate from end to end in England though a sixty ton barge can be used on considerable stretches. The canals were built before steamers were known and were not designed for haulage by mechanical power; the banks would not be able to stand the wash of the steamers.

As a canal was only another kind of road, the canal companies did not undertake to carry goods themselves. Anyone could put his boat or barge on the canal if he paid tolls. With the exception of the Aire and Calder Canal Company, who were also carriers before 1845, no canal company set out to be a carrier and only a few attempted to develop into carriers under the stimulus of railway competition after 1845, and then many of them gave it up again.† Had the canal companies developed into carriers before the days of railway competition they would probably have unified the canal system for their own convenience; as it was, being mere toll takers paid

---

*The following figures, taken from the quotations of canal shares in the *Gentleman's Magazine*, December, 1824, and quoted Cd. 4979, *p.* 4, give some idea of the large dividends and profit:

Canal.		Dividends.		Price.
Trent and Mersey	-	75%	-	£2,200
Loughborough	-	197	-	4,600
Coventry	-	44	-	1,300
Grand Junction	-	10	-	290
Oxford	-	32	-	850
Staffs. and Worcester	-	40	-	960
Leeds and L'pool (Aug. 1824)	-	15	-	600
Birmingham	-	12,10	-	350

†Cd. 4979, *p.* 57.

very well, and when things were going well, they saw no reason for making a change.

Owing to the fact that the canals were built like the roads by hundreds of private companies they varied in gauge, depth, tolls, finance and upkeep. The locks varied, the tunnels were of different sizes and the bridges of different heights, and all were built on too small a scale for modern requirements and steamer haulage. The companies might have arranged a system of through *tolls* on a uniform basis, but as the canals were free to any person who chose to carry, and as these persons necessarily varied in their charges, no general system of through *rates* for the carriage of goods was possible and to send goods by canals often meant several sets of bookings.* A boat from Birmingham to Liverpool, for instance, would traverse six canals, and from Birmingham to Hull would pass over ten separate canals, and this multiplicity of authorities necessarily meant impediments to traffic.†

It must, however, be remembered that the canals and inland navigations were an enormous improvement on anything that had existed before. The carriage of goods was not merely rendered much cheaper but they were more expeditiously conveyed. In a pamphlet of 1770, it is said that merchandise from Leeds to Liverpool which is often three weeks or more in being conveyed by land at the expense of £4 10s. a ton and subject to damage, would be carried by these boats in the utmost safety in three days at the expense of 16s. a ton.‡

The general result was that the rates for carriage were reduced to about a quarter. It was possible to convey raw materials in quantities ; bulky goods, like coal and building material, received a new mobility, and a further stimulus to the movement of bulky goods took place when iron tramways

*In the evidence given by Mr. Corbett, of the Worcester Chamber of Commerce, he cited the case of a local timber merchant that took props from Stroud to the South Staffordshire mines. These props had to pass over seven different canals which meant five bookings, five different declarations of cargo and the necessity for entrusting the money for the five tolls to the man on the boat. Another witness described the disastrous consequences of this : " One boatman went on the drink and never took his boat for the potatoes and we lost about £4 advanced him." *Op. cit., p.* 7.

†Cd., *op. cit., p.* 6.

‡Killick, History of the Leeds and Liverpool Canal.

or waggon ways were laid down after 1767 to connect up the coal mines or quarries with the canals.

The pottery manufacture received a new impetus from the increased facilities for getting china clay from Cornwall, as well as for distributing the breakable ware, and the pottery district developed rapidly, the population increasing from 7,000 partially employed and ill remunerated persons in 1760 to about 21,000 in 1785, " abundantly prosperous and comfortable."* Wesley reported very favourably on the great improvement in manners and morality that had resulted in twenty years in this district.

It was possible for people to move away from the vicinity of woods and bogs now that fuel could be brought by canals ; the inland navigations meant cheap warmth for households, cheaper power for factories, and blast furnaces. Priestley, speaking of the Oxford Canal in 1831, said : " It is the means of conveying an immense quantity of coal from the coal district in the neighbourhood of Birmingham to Oxford and other towns situate on the Banks of the Thames." There was a great reduction in the general cost of distribution ; the centre of England was opened out, towns grew partly because food could be obtained and partly because building material was now available in quantities ; agriculture got a better market and the development of canals stimulated the growth of large farms and the general agricultural revolution.

Of Manchester it was said that " Since that time, 1788, the demand for corn and flour has been increasing to a vast amount and new sources of supply have been opened from distant parts by the navigations, so that monopoly or scarcity cannot be apprehended though the price of these articles must always be high in a district which produces so little and consumes so much. . . . Potatoes, now a most important auxiliary to bread in the diet of all classes, are brought from various parts, especially from about Runcorn and Frodsham, by the Duke of Bridgewater's canal."†

This extract shows that the " great industry " could not have taken root in the North owing to the difficulties of the

*Smiles, Life of Brindley in Lives of the Engineers, Ch. VI.
†Aikin, " Description of the Country from Thirty to Forty Miles round Manchester " (1795), p. 203.

food supply and not merely the coal supply had it not been for improved transport by the canals.

The reaction on port development was also very marked. Liverpool, from being a little place, chiefly engaged in the slave trade, had a great hinterland opened out behind it and became the gate for the entry of raw materials and the port of egress for manufactures of that region.

The inland navigations superseded a good deal of the coasting trade. When vessels were often hindered for weeks from carrying goods by sea from Liverpool or Bristol to London or Hull, the advantages of regularity and punctuality were on the side of the canal barge and canals such as the Leeds and Liverpool became great arteries of traffic for goods passing from the Irish Sea to the Baltic and Germany. The Kennet and Avon Canal, by joining up the Thames and the Severn, was " the central line of communication between the Irish Sea and German Ocean."*

With the growth of ports fed by this traffic new docks were needed and great constructional works were undertaken to cope with the increased traffic. New classes of contractors arose to make the inland navigations, and the experience they gained was invaluable for railway building later on. A new class of surveyors was trained and a new migratory class of workmen, the navigators or navvies, emerged, who could undertake great constructional jobs and who were available to provide the skilled labour for railway excavations. Trade changed its character. Merchants used to take round their goods on pack-horses and sell them as they went along. They were superseded after 1760 by commercial travellers, and " it may now be asserted," Guest wrote, in 1823, " that the whole of the internal wholesale trade of England is carried on by Commercial Travellers—they pervade every town, village and hamlet in the kingdom, carrying their samples and patterns and taking orders from the retail tradesmen and afterwards forwarding the goods by waggons or canal barges to their destination—they form more than one half of the immense number of persons who are constantly travelling through the country in all directions and are the principal support of our Inns."†

*Priestley, *op. cit., p.* 386.
†" Compendious History," *p.* 11.

The canals gave an impetus to all trade and communications and were an indispensable preliminary to large scale production.

Just as there was a striking contrast between the English and the French road system so there was between the canal system of the two countries and both were characteristic of the general evolution of their country. The French canals had fallen into chaos during the ten years after the French Revolution. It was impossible for individuals to find the capital in France, so great had been the destruction of credit and confidence during the years 1789-1799. If the French had any money to invest they put it into land. The canals were re-started and added to by means of concessions granted to companies assisted by loans from the State. As the dues charged by the companies were considered to be too high, the concessions were re-purchased by the Government. Unlike the English, the French Government reconstructed their canals to compete with the railways, and the waterways are standardized for boats with a carrying capacity of 300 tons. The canals are not a commercial venture in France and traffic is as free on them as on the roads, i.e., there is no system of tolls. Of course the tax-payer pays for them indirectly and between 1879 and 1900 the State expended £11,209,600 upon improving the rivers, £14,607,611 upon improving the canals, in addition to £30,384,073 spent on maintenance and repair between 1814 and 1900.* The result has been a great increase in the traffic on the French canals since 1880.

	Canals. Tonnage. (000 omitted.)	Increase.	Railways. Tonnage. (000 omitted.)	Increase.
1880	18,000	—	80,774	—
1905	34,030	90%	139,000	72%

This is in striking contrast with the fate of the English canals, which reached the zenith of their prosperity about 1830.† After that they were rapidly superseded as the principal means of transport by the railways. The canals actually carried in 1909 more goods than they had ever handled but the increasing bulk of the traffic was in the hands

*Cd., 4979, 1909, p. 100.
†Cd. op. cit., p. 5.

of the new mechanical transport. There had been a great
decline in canal charges and the financial prosperity of canals
had suffered heavily. For instance, the Grand Junction
Canal carried :*

			Tons.		Earnings.
1838	.	.	948,481	.	£152,657
1888	.	.	1,172,463	.	£84,981
1898	.	.	1,620,552	.	£100,075

The canals, compared with the railways, although carrying
more traffic, had entered upon a period of relative stagnation.

	Canals. Tons.	Increase.	Railways. Tons.	Increase.
1888	36,300,000	—	281,747,439	—
†1898	39,350,000	8¼%	378,563,083	34¼%

These figures are sufficient to show how the growing trade
of the country had been taken over by the railways and also
what a relatively insignificant part the once all important
canal system had come to play in British commerce. The
tonnage carried by the Great Western Railway alone exceeded
the total tonnage carried by all the water-ways of the United
Kingdom in 1905. These figures are also a striking instance
of the growth of trade in the railway era. The canals were
actually carrying more traffic in 1898 than when they were
the principal mode of transport and yet in that year an additional 378 million tons was transferred, which shows how
the railways helped to accelerate the change to mass production and distribution known as the industrial revolution.

When the railways came with through traffic, speed,
capacity to handle large quantities of goods, through rates,
punctuality, cartage and delivery at the terminals and civility
on the part of their officials, the canals rapidly fell into a state
of relative stagnation. This was increased when the coasting
steamers began to carry goods in large quantities without
the delays of the old sailing vessels and much of the canal
traffic was diverted to the coasting trade. The Government
was nervously anxious to preserve competition between the
railways and the canals in order to keep down rates and in
1845 authorized the canals to become carriers, with but

*For other instances, Cd. 3184, 1906, Appendix I.

†Cd. 3184, Question 64—Evidence of Sir H. Jekyll, Board of Trade.

little result. The canal tolls were reduced in some cases to as much as a seventh, but the canals could only have made effective headway against the railways by combination and reconstruction, and no one would have invested money in a method of transport so obviously inferior to the railways. The Government, already burdened with one Canal—the Caledonian—that did not pay, would not have dreamt of undertaking what the French Government did, viz., the enlarging and rebuilding of the canals for large boats and steamer haulage. In England transport was the business of individuals, not the State. The canals themselves were quite ready in many cases to sell out to the railways, to save their shareholders,* and in this way the railways acquired about a third of the canal mileage of the kingdom. The railways were sometimes forced to purchase the canals to get rid of the opposition they offered to projected railway bills in Parliament. The general result is that of the total mileage of canals and navigations 3,310 miles are not railway owned or controlled and 1,360 miles are so owned or controlled.† This still further prevented any improvement in the canals. The railways were bound by the terms under which they took over the canals to maintain them in repair. Maintaining them in the state of repair of 1850 does not necessarily mean that they are efficient for 1900. It is to the interest of the railways that goods shall be carried by rail because then the goods will not merely pay a toll for the use of the road but will pay for haulage, trucks and use of stations. If the goods go by canal, the barge owner gets the payment for haulage and use of the barge and someone else will get payment for the use of the wharf. It paid the early railways to divert traffic from their own canals to the railroads because they earned under three of four heads instead of one. The result is that the tonnage on railway-owned canals declined while that on the independent canals rose. The figures are as follows :‡

	Independent Waterways.	Mileage.	Railway Waterways.	Mileage.
1888	19,789,668	—	15,512,189	
1905	20,434,411	1,923	13,702,356	1,225

*One canal even became a railway company to save its traffic.
†Cd., *op. cit.*, *p.* 14.
‡Cd., *p.* 63.—Many instances are given in the report (*p.* 75) of

Moreover, the fact that the railways control about a third of the strategic links is an obstacle to the canals being brought up to a uniform system. It would not be in the railway interest to create a rival to itself.

The real reason for the decline of the English canals is, however, to be sought in the fact that English internal commerce had largely reconstructed itself and that the railway transport had come to suit it far better than water transport. English agriculturalists, for instance, had changed from selling wheat to selling dairy produce, and the water-ways were too slow for the transport of milk and butter, whatever they had been for cereals. The coal merchant was unwilling to provide large warehouses for coal ; he preferred to have it in railway trucks and get it as he wanted it ; he could then work with smaller capital. In the case of coal, the railways have the great advantage that their trucks can be brought on sidings up to collieries and filled at the pits' mouth. Coal can also be taken on sidings into different parts of the works or factory where it is to be used. For bunkering ships the coal can be shot from the truck into the ship at the quay side. Therefore, even in the case of coal which gave the original impetus to canals, the railway has proved itself the more efficient instrument and provides greater facilities. Builders wanted their material just when they were going to use it ; they did not like it lying about for months according to when it suited the barge owner to unload, and the same was true of road surveyors. The railway companies provided sheds where artificial food-stuffs or manures could be sheltered from the weather and the farmer preferred to get his stuff housed till he could spare his horse and cart to fetch it. Moreover, it is not every trader who can charter and fill a barge. Most of the traffic consists of small parcels and that is more conveniently taken by rail. Thus the greater efficiency of the railway service and the greater efficiency of the coasting steamer when large cargoes could also be transferred with punctuality, would in any case have adversely affected the canals. Barred, as they were, by their divided condition

Canals on which railway companies had spent large sums on improvements or had energetically promoted traffic, e.g., on the Birmingham Canal, by the London and North Western Railway ; the Trent and Mersey : the Forth and Clyde, but these were stated to be exceptions.

from making improvements and hampered by the fact that the railways controlled a large part of the system, it is no wonder that the English canal system became relatively unimportant.

The difficulty of effecting the revival of the canals is very great, partly owing to the multiplicity of authorities and partly owing to engineering difficulties, to say nothing of the great cost involved.

" A company whose canal forms only a section in a through route would be wasting its money if it improved its own section to carry larger through traffic while any other section of the route remained without corresponding improvements. In this way, the unwillingness or poverty or apathy of a company holding a few miles of a through route—the non-improvement of a single lock indeed—might block improvement all along the line."*

Another great obstacle to the improvement of the canal system lies in the fact that it was so valuable to be on a canal in pre-railway days that the canals are closely built upon in some areas, and to remove all the warehouses and buildings in the densely crowded district like the Birmingham region, in order to widen the canals, would mean a large initial outlay beyond the capacity of private individuals and the financial return for which would in any case be doubtful.

Probably if Parliament had not been so anxious to prevent the acquisition of the canals by the railways the latter might have gradually acquired control of the whole system of inland water-ways and would themselves have improved and unified the canal system over large areas and worked it as a feeder for the railways and subsidiary to their own service, but the Railway and Canal Commission was set up in 1873, one of the objects of which was to scrutinize and prevent the acquisition of canals by railways so as to preserve competition. Hence the patchwork nature of the English canal system was maintained. The survival of the existing traffic on canals may probably be ascribed to the fact that inland navigation was, before 1914, mainly worked by families of small means living day and night in barges, and this domestic system of transport, which avoided house rent, kept the rates of carriage lower than they otherwise would have been, and was an

* Cd., *op. cit., p.* 70.

índucement to certain traders to patronize the canals on account of sheer cheapness.

In France, Belgium and Germany the railway has by no means superseded the system of inland water-ways, but English commerce became so organized that speed, punctuality warehousing facilities and ease of handling became the important factors and in this respect the railways far out-distanced the system of inland navigations that seemed so wonderful to the men who remembered the old pack-horse and the earthen, unmetalled road. But it must be remembered that the canals served their turn and contributed materially to the predominant position occupied by this country during the first half of the nineteenth century.

III.—RAILWAYS.

With this tradition of private unco-ordinated enterprise in both roads and canals it was only natural that the railways when they came should follow a similar line of development.

The British differs from all other railway systems in several important particulars. They were started like the roads and canals, by private capital and on no uniform national system and they owed nothing to state aid.* They were built to accommodate an existing traffic which had outgrown the canals ; they were started for purely commercial reasons and were intended to be financially profitable to their promoters.

Compare this for a moment with the circumstances on the continent. In France, the French expected the Government to undertake the making and maintenance of the roads and canals, and similar assistance was also forthcoming in the principal German States. Therefore on the continent the Government was expected to either build or work, or assist in building and working the railways. Transport was a national and not a private affair. The Governments in both Germany and France were obliged in any case to undertake railway building for military purposes and the continental railways were largely planned for strategic reasons, although commercial reasons were present. But a great many railways, such as those of Prussia to the Russian frontier, would never have been built if the motive had been merely financial.

*This is not true of Ireland. *Cf. pp.* 174-175.

On the continent capital was scarce in any case and very shy of the new mode of locomotion. Thus the State had generally to find the capital itself or guarantee the interest. True to tradition, France made the road bed of some of her railways as she had made the roads and many of the canals and handed these new railroads over to private individuals to work. In Germany, most of the existing railways came into the hands of the various State Governments after 1870. State aid and the strategic military motive were wholly absent in Great Britain.

Nor did other countries imitate the patchwork nature of the English railway system. The roads of France were planned to radiate out from Paris and each great line was given a monopoly of its own district. The English Parliament, worshipping the fetish of free competition to regulate facilities and prices, were horrified at the very suggestion of a monopoly and did all it could to promote the competition of railway with railway and railway with canal.

The United States, which followed English tradition so closely in its turnpikes, did not follow it in its railways at first. Improved transport was so vitally necessary to a new country that the State Governments subsidized and encouraged many railways between 1830 and 1838, raising the money by loans. No less than $42,871,084 were spent by the States on railways before 1838. A great financial collapse followed in 1837 ; some States repudiated their debts and sold their railways, and the new State constitutions nearly all inserted a prohibition of the use of State funds for internal improvements.* After that they adopted the English method of allowing private individuals to finance the railways. The peculiarity of the railways of the United States lay in the fact that they were built in advance of the existing traffic in order to open up the country. They created their own traffic and made the country they passed through. They therefore received large free grants of land† and were not worried by stringent regulations as to safety appliances or the methods of the construction or equipment. Any railway was considered

---

* Bogart, " Economic History of the United States," *p.* 214.

† Before 1861, 31,600,842 acres of public lands were given away for internal improvements and the railways obtained by far the largest share. Bogart, *op. cit*

better than none and the more railways the more competition and the lower the rates. Hence railways were eagerly welcomed in the United States.

In Russia, with her autocratic tradition and lack of capital, the State was forced to build and work the larger part of the railway system though a certain amount of foreign capital had been tempted in by guarantee of interest.

The English railways, on the other hand, were not regarded as boons but as dangerous innovations that had to justify their existence before an elaborate Parliamentary enquiry. In opposing Huskisson on the question of the Liverpool and Manchester railway bill the following arguments were used : " What was to be done with all those who have advanced money in making and repairing turnpike roads ? What of those who may still wish to travel in their own or hired carriages after the fashion of their forefathers ? What was to become of the coach-makers and harness-makers, coach-masters, coach-men, inn-keepers, horse-breakers and horse-dealers ? The beauty and comfort of country gentlemen's estates would be destroyed by it. Was the House aware of the smoke and the noise, the hiss and the whirl which locomotive engines passing at the rate of ten or twelve miles an hour would occasion ? Neither the cattle ploughing in the fields or grazing in the meadows could behold them without dismay. Lease-holders and tenants, agriculturists, graziers and dairymen would all be in arms. . . . Iron would be raised in price an hundred per cent. or more, probably it would be exhausted altogether. It would be the greatest nuisance, the most complete disturbance of quiet and comfort in all parts of the kingdom that the ingenuity of man could invent."*

It is well known that the first attempt to get the Liverpool and Manchester Bill through Parliament failed because a noble Duke said it would spoil his fox covers and it cost £70,000 to obtain parliamentary permission to build the line. That was the outlay before a single bit of land was bought or a single sod cut. Even then curious precautions were taken that the railway should not be a nuisance. One of the clauses of the Liverpool and Manchester railway provided that " no steam engine shall be set up in the township of

*Francis, " History of the English Railway " (1851), p. 119.

Burton Wood or Winwick and no locomotive shall be allowed
to pass by the line that shall be considered by Thomas Lord
Lilford or by the Rector of Winwick to be a nuisance or
annoyance to them from the noise and smoke thereof."*
Northampton congratulated itself on making the railway
avoid the town by five miles.

In summarizing the opposition to railways, Francis
says :† " The country gentleman was told that the smoke
would kill the birds as they passed over the locomotive.
The public were informed that the weight of the engine
would prevent its moving ; and the manufacturer was told
that the sparks from its chimney would burn his goods.
The passenger was frightened by the assertion that life and
limb would be endangered. Elderly gentlemen were tortured
with the notion that they would be run over. Ladies were
alarmed at the thought that their horses would take fright.
Foxes and pheasants were to cease in the neighbourhood of
a railway. The race of horses was to be extinguished.
Farmers were possessed with the idea that oats and hay
would be no more marketable produce ; horses would start
and throw their riders ; cows even, it was said, would cease
to yield their milk in the neighbourhood of one of these
infernal machines.

" Vegetation, it was prophesied, would cease wherever the
locomotive passed. The value of land would be lowered by
it ; the market gardener would be ruined by it. The canal
could carry goods cheaper. Steam would vanish before
storm and frost ; property would be deteriorated near a
station. It was called the greatest draught upon human
credulity ever heard of. It was erroneous, impracticable and
unjust. It was a great and scandalous attack on private
property, upon public grounds . . . one class was
informed that the locomotive would travel so fast that life
and limb would be endangered, another was told that it
would be too heavy to travel at all."

Tunnels were an object of the greatest horror and supposed
to be injurious to the health.

Comic as these fears appear to us now they had a great
economic significance. Railways were things to be resisted ;

---

*Quoted Pratt, " Inland Transport and Communications," *p.* 249.
†*Op. cit., p.* 101.

to overcome that resistance the railways had to pay heavily : they were clogged from the outset by the necessity of fighting for their existence.

This difficulty of getting railways started partly accounts for another feature which is peculiar to the English railway system, viz., the high capitalization of the system. It is the most highly capitalized railway system in the world. The English railway capital was £54,152 per route mile in the United Kingdom, £64,453 in England and Wales. The Prussian railways cost £21,000 per mile and the American less than £13,000. One expensive item in this high capitalization was the cost of obtaining the railway Bills. It was not merely bringing up the surveyors and hiring barristers, but the expense lay in the struggle between the canals or other interested parties to prevent the railways being authorized or extended. The proposed railway company itself would marshal one great mass of expert evidence in favour of its proposals, its opponents would present an equally formidable array against the proposed concession, both would be anxious to have " a good team " of lawyers and it is not surprising that the cost of the preliminary expenses including surveying and legal costs have been put at £4,000 per mile.*

After the railway was authorized land had to be bought and as each land-owner in the early days thought his estate would be ruined by this terrible engine of destruction he charged as high a price as he could for his land. Francis, writing in 1851, quoted the following prices given for land by the railways :†

London and South Western	...	£4,000	per mile.
London and Birmingham	-	£6,300	,,
Great Western -	-	£6,696	,,
London and Brighton -	-	£8,000	,,

Land was valued for the London and Birmingham railway at £250,000, but cost three times that amount.‡

To allay the deep suspicion of the dangers of railway travelling the railways were built with great solidarity of

---

*Acworth, " Elements of Railway Economics," p. 11.

†p. 203.

‡The Duke of Bedford returned £150,000 and Lord Taunton £15,000 when they found that their land was not ruined.—Pratt, " History of Inland Transport," p. 254.

construction and that added to the initial cost. Parliament has increasingly insisted also on the railway managements providing the most up-to-date safety appliances such as the vacuum brake and inter-locking signals.

In addition the geographical situation has not made the English railways economical to construct. The North of Germany is a great flat plain where construction is cheap and engineering difficulties small, and this is also true of the Middle West of the United States and of Russia. As one approaches the West of England and Scotland the engineering presents greater problems, and far more capital is needed for making the track, the gradients and the viaducts.

Nor is the traffic of the same nature as the continental traffic. The area of Great Britain is small compared with that of the other Great Powers.

				Sq. Miles.
England, area	.	.	.	50,874
Wales, ,,	.	.	.	7,466
Scotland, ,,	.	.	.	29,797
				88,137 sq. miles.

France	.	.	.	207,054 sq. miles.
Germany (1910)	.	.	.	208,780 sq. miles.
United States (continental area, excluding				
water) -	.	.	.	2,973,890 sq. miles.
Russia (without internal waters)		.		8,417,118 sq. miles.*

The railway traffic of the continent necessarily consists of long hauls. In England there is no place more than ninety miles distant from a port. The ports of Great Britain are well distributed round the coast and the hauls for domestic use or export are for short distances only. It is cheaper per mile to move goods for long distances than for short ones. The charges of the English railways are calculated for short hauls, great speed, and small consignments, and these features make the whole range of English railway rates higher than those of the continent, to say nothing of the warehousing, cartage and delivery facilities which are again peculiar to the English railway system.

" The trader has grown so accustomed to ordering what he wants in very small parcels and having them sent in the

---

*These figures are taken from Statesman's Year Book, 1919.

afternoon to a railway station with the expectation of seeing them delivered with the regularity of a postal packet at his place of business the next day, that he will not tender anything like a reasonable waggon load. It has become a question with him of credit from his bankers and he will not hold a pound of stock more than is necessary to carry on his business, especially in the case of high priced merchandize."*

Moreover, it must not be forgotten that England had to make the experiments in mechanical transport by which, other people profited and that a good deal of capital was sunk in ways that would now be avoided. To be a pioneer is honourable but often hard on the pioneer. A good deal of the English railway system had to be remade before through traffic could take place and even now the English system is hampered by the platforms at the stations and the structure gauge which makes it difficult to use larger trucks and heavier engines.

We have already seen how the railways superseded the canal system and that this is peculiar to the United Kingdom.† Germany, France and Belgium enlarged and developed their canals alongside of their railways and made the water-ways supplement their railways. One must, however, realize that the railways of the United Kingdom were formed on the canal model, *i.e.*, they were thought to be like canals, another kind of road, of which the railway shareholders were only to be toll-takers. Like the canals they were started in life with a system of maximum tolls which they might charge as owners of the road and which were imposed by Parliament when they granted the railway bill. The original idea was that anyone might put their own truck on the road and pay a toll. The result is that a large part of the English railway equipment consists of trucks belonging to private persons, *The Earl of X's Collieries, The Y. Pantechnicon Furniture Removers*, and so on. At the end of 1913 the railways of Great Britain owned 786,516 waggons and it was estimated that the private traders owned 780,200.‡

*Aspinall, " Address to the Institute of Civil Engineers," 1918.

†This also happened in the United States where the development of canals was, however, relatively insignificant when compared with the English canal system.

‡Aspinall, *op. cit., p.* 17.—Sir John Aspinall points out the great hindrance these private waggons are to the efficient working of railways.

This idea that the railway was only a specialized road gave rise to the English system of rate charges which again is quite peculiar.

An English railway rate consists of ·

(a)   A toll for the use of the road ;
(b)   A charge for haulage ;
(c)   A charge for the railway waggon ;
(d)   A charge for collection and delivery ;
(e)   A charge for loading and unloading, covering and uncovering.
(f)   A charge for the use of the Stations.

The consignor or consignee may, however, provide his own waggons, load and unload himself, collect and deliver himself and provide his own stations and in this case the railway cannot charge him for these services. The only thing he is bound to pay for is the use of the road and the haulage. In addition, the railway charges extra for such services as it does perform or for the accommodation at the terminals over and above the road and the haulage.

Although the original idea was that the railway was a public highway it was soon proved that the railway must be a carrier as well as a road owner, that it must at least be responsible for the locomotive to ensure the safety of the trains. It was impossible to allow people to run about over

It is impossible to standardize the parts of all these privately owned waggons though something has been done in this respect. " In addition, if the tremendous labour and time involved in shunting out and sorting this vast array of traders' waggons which have each to be sent back empty to their own particular owners after each loaded journey can be got rid of by railway ownership, then without counting the cost of the return empty journey or without making any estimate of what the real cost of shunting out these traders' waggons comes to in a year, it is obvious that the trivial amount of one penny saved per day for one shunt per day for 300 working days on 700,000 waggons would amount to little short of a million sterling."

In addition, the abolition of private ownership would enable all waggon stock to be fitted with continuous brakes. " Until such brakes are fitted we shall not reap all the advantages of longer and heavier trains as far too much time will be occupied in bringing these trains to a stand. It seems almost a contradiction to say that efficient continuous brakes will enable a train to make its throughout journey in less time, but it is essentially true."

This illustrates the difficulties that have arisen from the fact that no one really understood in the beginning what a railway was going to be, and how Great Britain suffered from making the experiments.

the railway lines with their own little locomotives just as
freely as barges moved about on a canal or carts or omnibuses
on a road. By the time other countries started they could
see what railways really meant to the development of a country
and many of the early mistakes were avoided.

So little was known or realized as to the possibilities of
railways that the prospectus of the Liverpool and Manchester
Act said that the new railway held out " a fair prospect " of
being " a cheap and expeditious means of conveyance for
travellers," the receipts from which were estimated at £10,000
a year, an estimate which proved itself ten times too small.*
It was to facilitate the movement of goods and not passengers
that the railways were promoted. The great human mobility
that was to follow was practically undreamed of nor could
anyone have foreseen the commercial revolution that the
railways would cause.

British railway history may be divided into four periods :

> 1—There is the period of experiment from 1825-1844.
> 2—From 1844 to 1872 may be termed the period of the
> consolidation of the railway system, the formation of
> the great trunk lines and the elimination of canal
> competition.
> 3—The supposed railway " monopoly " so alarmed Parliament
> that it had to devise a system of control in spite of its
> laissez-faire principles. From 1873-1893 Parliament
> occupied itself, as far as railways were concerned, in
> setting up the Railway and Canal Commission to con-
> trol abuses and in 1888 and 1893 it went further and
> fixed maximum rates of charge on a new and extended
> scale.
> 4—From 1894-1915 an active period of competition in facilities
> set in between the railways ; dividends were reduced,
> and the result was to usher in a period of railway
> amalgamations after 1900, giving rise to protests
> from traders and from labour. The question then
> became one of devising fresh control or taking over
> the railways by the State.

(a)—*The Period of Experiment*, 1821-1844.

Coal started the canal system and coal produced the
railways. We have already seen that even in the seventeenth
century wooden rails had been laid down from the collieries

---

*On the London and Birmingham the passenger traffic was calculated
to yield £331,272, and produced £500,000 ; while goods calculated to
produce £339,830, produced £90,000 only.—Francis, *op. cit., p.* 203.

to the rivers and that iron rails began to be substituted after 1767 and were very much more efficacious. These rails were then laid down from collieries to canals and as the manufacturing towns grew we find instances of these lines being constructed to connect collieries with towns. These tramways or waggon-ways were, however, private tracks and could only be used by the colliery concerned. A line open to the public on which they might send any kind of goods was built between Croydon and Wandsworth, "the Surrey Iron Railway," in 1801, to carry lime and corn to the Metropolis. It was worked by horses. It was not a financial success, but the waggon-ways continued to extend, connecting up iron works with the canals and quarries with towns to furnish building material. Most of the traction on these tramways consisted of horses but experiments were being made at various places to see if the haulage could not be done by a steam locomotive. There were, however, all sorts of difficulties. It was held that no engine with smooth wheels could draw a weight, so that the early locomotives were furnished with toothed wheels to fit into a rack and were necessarily slow. In 1814, Hedley, at Wylam Colliery and George Stephenson at Killingworth both made locomotives with smooth wheels that would draw considerable weights. But then it was found that the weight of the locomotives split the iron rails; the speed was very slow and the locomotives clumsy and liable to failure. Stationary engines pulling trucks along a cable were found to be workable and were adopted.

The Stockton and Darlington Railway obtained an Act in 1821 to convey coal from the Darlington collieries to the port of Stockton-on-Tees, and it is famous as being the first public railway to use steam locomotion and carry passengers. Neither of these functions were contemplated at first, it was only in 1823 that the Company obtained an amending Act which gave it these powers. The railway was opened in 1825, and although the goods were conveyed by locomotives, horses were used for passengers.

Other small lines were projected, such as the Canterbury and Whitstable, in 1825, and the Bolton and Leigh, both of which obtained powers to work either by locomotive or fixed engine.

In 1826, however, the Liverpool and Manchester Railway obtained its Act and this opened a new era. It was promoted because of the inadequacy of the canals to deal with the growing traffic in the North. It is obvious that railways would have come in any case but the exasperation with the delays and high charges of the canals brought matters to a head more quickly than would otherwise have been the case and induced manufacturers and others to risk money in a venture, the success of which was doubtful. When success was obvious then the public followed and as eagerly found the money for railways as they had for canals.

The proceedings of the canal companies had become very arbitrary. They sent as much or as little as suited them and shipped it when and how they pleased. " Of 5,000 feet of pine timber required in Manchester by one house, 2,000 remained unshipped from November 1824, to March, 1825." Cotton took longer to go from Liverpool to Manchester than from Liverpool to New York. Huskisson stated in the House of Commons that " cotton was detained a fortnight in Liverpool while Manchester manufacturers were obliged to suspend their labours." Joseph Sandars wrote a letter in 1825, stated on the title page to be " An Exposure of the Exorbitant and Unjust Charges of the Water-carriers." He accused the Bridgwater Canal Trustees of charging double the amount of their authorized tolls and to have created a monopoly by securing all the available land and warehouses along the canal banks at Manchester. The Manchester and Liverpool traders also complained that in winter the canal was frozen and in summer there was often such a deficiency of water that the boats had to go half loaded.*

This was the first railway to throw down the challenge to canals. It was opened in 1830 and illustrated many new things in railway operating. It proved that the new mode of transport was above all suited for passengers, and up to about 1850 the bulk of railway revenue, quite contrary to previous expectation, came from passengers, i.e., started to carry coal, the railways mainly carried the most precious freight of all, viz., human beings. This was one of the first surprises of the railways. The Liverpool and Manchester,

*Francis, op. cit., pp. 78-80.

like the others, obtained powers to work by horse or locomotive. They offered a prize of £500 for the best locomotive; trials took place at Rainhill in 1829, George Stephenson's *Rocket* won easily and the superiority of the locomotive was demonstrated. Nevertheless, horse-drawn coaches, the property of private persons paying a toll to the company, continued to ply on the railways. The Liverpool and Manchester Company having successfully fought the canals, having successfully established locomotives and having successfully carried passengers next began to provide waggons and haul them, *i.e.*, unlike the canals it developed as a common carrier, and in order to have a monopoly it bought out the interests of the coaches in 1832. In other words it realised, after two years' working, that the railway transport must be under one common direction and that people could not be allowed to use the road just when and how they liked— a railway was not the same as a canal.

After the Liverpool and Manchester got its Act a railway was projected between Liverpool and Birmingham, but the Bill failed to pass till 1833. This same year saw the first of the long distance railways laid down, the Liverpool and Birmingham or Grand Junction was to be linked up with the London and Birmingham, and thus a railway connected Manchester with London in 1837. The Great Western was begun in 1835 by a line to connect London with Bath and Bristol, opened in 1838. The bulk of the railways were, however, little short lines scattered higgledy-piggledy here and there between two or three places so that Hadley could state that in 1844 the average length of the English railroad companies was only fifteen miles. In 1844-1847 there were chartered 637 separate roads with a total length of 9,400 miles.* For some time it was uncertain whether the steam locomotive was going to be the ultimate form of traction. Cable railways and atmospheric railways were proposed and tried, but the steam engine proved itself to be the most effective method of haulage.

In 1836 the financial success of the railways caused a small railway boom and many new railways were planned, the Midland Counties, the Eastern Counties (the longest railway yet projected, 126 miles, and the first in the Eastern counties).

* " Railroad Transportation," *p.* 167.

the South Eastern, the Great North of England, the Manchester and Leeds being a few of the most important schemes.*
By 1836 the nucleus of our present railway system was traced.
By 1838 there were 490 miles of railroad in England and Wales and fifty in Scotland, the construction of which had cost £13,300,000.† So efficient did the new method prove itself that by 1838 the railways were entrusted with the mails.

By 1840 Parliament began to realize the overwhelming importance of the new mode of transport and from this time onwards there are almost yearly committees or commissions to see how the new mode of transport might be controlled. There was a nervous anxiety not to interfere with private enterprise and initiative. On the other hand, Parliament realized that through their " superior accommodation and cheapness " the railways had " acquired command of the travelling in their district." In 1840 a Commission was surprised to find that the right secured to the public of running their own engines was practically a dead letter. They ascribed it to the fact that no provision was made for ensuring that the independent trains, although allowed to use the track should be able to use the stations and watering-places and so the railways could bar out private users of the line. It was difficult, in any case, for independent persons to work at a profit, and it had become necessary to place the control of the running of all trains under one head. Thus by 1840 it was obvious that the new railways were a success, especially for the conveyance of persons, that the railway must do the haulage, that there were signs that its efficiency was so great that it might prove a monopoly and that " something ought to be done." Meagre powers were, therefore, given to the Board of Trade which already had supervision of shipping. Meanwhile, the canals had undertaken drastic reforms and between 1820 and 1840 were carrying more goods than ever before, the railways, with their puny engines, being chiefly concerned with passengers who can always afford to pay more than goods for their transport. This feature is also noticeable in the case of steamships—the early steamships carried passengers first and later were adapted to mass traffic.

* For further details, " The British Railway System," H. C. Lewin, where there are excellent maps illustrating the growth of the lines.
† Report, 1867, p. IX.

The railways that Stephenson laid down were engineered on the 4-ft. 8½-in. gauge, and in this he followed the old tramway or waggon-way gauge. Brunel, however, constructed his lines on the broad or 7-ft. gauge, and thus England was covered with lines over which it was not possible to get through traffic without transhipment or the relaying of much of the track.

In 1842 the powers of the Board of Trade were enlarged (3 & 4 Vict. c. 97), Parliament announced that no new railways were to be opened without previous notice to the Board of Trade which might appoint officers to inspect all new railways. It was empowered to ask for returns of traffic and accidents. The powers entrusted to the Board were very limited. It could not interfere in the management of the railway but it could postpone the opening of the railway if it were not satisfied with the conditions for safety. The Act was, however, important as a declaration that pure unrestricted competition would need supervision. The Board of Trade really had very little power because it had no powers of coercion and public opinion in England at the time was always against any bureaucratic interference or control. It had neither law nor public opinion behind it.

By 1844, so nervous had Parliament become about the new power in the land that it took its courage in its hands and in the year 1844 it enacted that there could be a revision of the tolls, fares and charges if the dividend of any railway were more than ten per cent., and that the Treasury might purchase future railways, i.e., all railways before 1844 were not subject to compulsory purchase under this Act, but those made after 1844 were. This Act was again only a threat or declaration of the right to revise fares and to purchase. It was not acted on in any way. The Statute also prescribed one train daily along every passenger line stopping, if required, at every station and conveying third class passengers at 1d. per mile. This used to be known as "the Parliamentary."

Then another experiment in control was tried in 1844. A Commission was appointed to make preliminary reports to Parliament on railroad charters. After a year they ceased to exist. It was said that they died of too much work and too little pay. By 1844, however, the period of experiment was over as far as railway construction was concerned. Two things were quite obvious, viz., that the railways must be common carriers and that it would be to their interest as common carriers to work long as well as short distance traffic and that sooner or later consolidation of all these scraps of lines would take place. It was also obvious that the railways were going to be exceedingly important and that they must be controlled. But Parliament believed in laissez-faire and so we find nervous, tentative feelers put out to see what might be done and then these feelers are hastily withdrawn. It would set up a new department of the Board of Trade or a Commission and give them practically no powers. Parliament seems to have instinctively realized that the railways were going to be the great factor involving State control and they would not face it. Control must take place but it must be wrong if it is control was their attitude. Meanwhile it is interesting to see that an enquiry was held into the condition of the navvies who were building the lines. This is a remarkable departure from the rule that men could look after themselves and should do so under the idea of "free contract." The idea was that women and children—poor, weak, helpless things—ought to be looked after by the State as exceptions to the general rule, but men did not need this protection. It is a striking instance of the way in which railways were making for government interference that this enquiry should have been held at all. The Committee reported in 1846. It gave an appalling picture of the dangers of the navvies' existence. Their numbers were said to amount to 200,000. There were frequent accidents in making the tunnels and embankments. Blasting seems to have been responsible for the loss of many lives and that not merely because of the carelessness of the men, but because the employers did not furnish the proper appliances and safeguards. The Committee, therefore, recommended that the railways should pay compensation for all accidents. "By making the companies liable your Committee contemplates fixing that party with the liability who has the greatest power to prevent the injury and the greatest means to repair it." It considered that the burden of compensation should be put "on those whose works are the occasion of the mischief." It is interesting to see the doctrine of Workmen's Compensation set out so early. Only two years before there was the greatest difficulty in including women in the Factory Acts, and yet here is a Committee recommending drastic measures for the safety of the men themselves and that not even to secure the safety of the public.*

There was an appalling amount of truck brought to light by this enquiry. The wages were paid monthly ; the men got credit from a shop kept by the sub-contractors and were always in debt to the shop where prices were high and the quality bad. Every pay day gave rise to a riot as the men always thought that they ought to have more money and the contractors seemed to prove that it had already been spent at the shop. The Committee advocated weekly payments in money. The accommodation for the navvies was often shocking " with scarcely any provision for comfort or decency of living."

The Committee recommended that a Special Board should be created which should not sanction the beginning of a railway that did not provide proper accommodation for its workmen. They admitted that this was an " unusual kind of interference," but they considered that as " the State has an interest in the health and decency of its members " and " as it grants extraordinary and valuable powers to these companies " they should be forced to make suitable provision not merely for lodging the men but for their attendance at church on Sundays.

It is scarcely necessary to say that such drastic recommendations had little chance of being adopted in this or the ensuing period,† but it shows that Committees were prepared to deal far more thoroughly with railway companies than they were with manufacturers and that State aid, anathema though it might be, would be invoked more quickly to coerce railways than any other form of enterprise.

The next great experiment that lay before the railways themselves was to arrange for through traffic over hundreds of different lines with two main gauges, and the apportionment of the earnings of the traffic to the various lines over which the through traffic passed. As the railways had become

---

*Select Committee on Railway Labourers, 1846, Vol. XIII., *p.* 427*ff*
†Truck was forbidden in 1887, while the Employer's Liability and Workmen's Compensation Acts of 1880 and 1897 provided compensation for accidents.

carriers it was to their interest to promote an increase of traffic as well as take tolls and they realized that there must be through traffic for efficiency. How was it to be evolved and what would be the attitude of Parliament? Those were the problems for the next thirty years.

(b)—The Consolidation of the Lines.

By 1844 the period of experiments in railway *construction* was over, the next twenty-nine years were to witness the experiments in railway *operation*. The lines were there, a smooth working system had to be evolved which should both pay the companies and be beneficial to the public. The two principal events of the period, 1844 to 1873, are the consolidation of the lines into great trunk systems and the disappearance of the canals as important competitive factors.

Parliament saw the dreaded " monopoly " accomplished and tried by a new Railway Act (1854) by creating a new Board of Control, by giving increased powers to the Board of Trade, to institute some form by which the railway companies should be made to realize their duty to the public. On the other hand, it tried by the encouragement of canals and by sanctioning freely new and competing railway lines to keep " free competition " alive in the matter of transport rates.

In spite of frequent Parliamentary protest the consolidation of the lines was carried through in various ways because of the advantages of combination during the years 1844-1872, and competition in rates and fares ceased to exist because the great systems agreed to a common rate of charge between pairs of competitive stations.

The year 1844 saw the start of the consolidations and the process continued throughout the fifties and the sixties.*

*Cleveland Stevens, English Railways, *p.* 25.

	New Lines.	Acts for Amalgamations.	Acts for Purchases and Leases.
1844	37	3	7
1845	94	3	18
1846	219	20	19
1847	112	9	20
1848	37	5	7
1849	11	2	4
1850	5	1	5

Sometimes one railway would buy up another; sometimes it would lease another; sometimes they would arrange for running powers over each other's lines or they would simply amalgamate with a common merger of capital and management.

The individual who first gave the impetus to the consolidation of the lines into great systems was George Hudson, "the Railway King." He saw the necessity for through traffic and for combination in the interests of efficiency and financial success, and vigorously promoted such amalgamations between 1844 and 1847. Although he was proved to have been financially dishonest in some of his transactions, he did give the movement a powerful start. Combination was inevitable in any case, even the turnpikes had shown a tendency to combine, but it came quicker because of Hudson's operations and the boom that followed. He had the capacity to plan large railway schemes and make other people see them, and he had great skill as an administrator and introduced many improvements in the general working of the railways. The period, 1845-1847, was a period of railway mania. Lines were projected to all sorts of places, possible and impossible, vast sums of capital were to be spent on railways; a wild speculation in railway shares set in and then the boom collapsed. Many lines, however, survived and by 1850 Great Britain was covered with a good network of lines in proportion to its size, 6,621 miles being open on December 31st, 1850.*

While Hudson had given the consolidation of the lines a great advertisement, the most powerful factor working in the same direction was the Clearing House.

As the lines had become increasingly carriers of goods the old maximum rates of toll for the use of the road were no longer applicable. They had to evolve a list of *rates* of

---

*Cleveland Stevens, *op. cit., p.* 164.

Lines open December 31st,

	miles.			miles.
1842 -	1,857	1849	-	6,031
1843 -	1,952	1850	-	6,621
1844 -	2,148	1851	-	6,890
1845 -	2,441	1852	-	7,336
1846 -	3,036	1853	-	7,698
1847 -	3,945	1854	-	8,954
1848 -	5,127			

carriage for themselves as carriers. In the interest of their own business as carriers they wished to provide for through traffic without re-bookings which had been the bane of the canal system. The Clearing House was started in 1842, to facilitate the sending of trucks from one system to another and for adjusting the amounts to be paid to each company for the portion of the line used in through traffic. It was not compulsory for railway companies to join the Clearing House but gradually all the companies joined. In order to arrange through rates the companies had to agree on a system of classification before they could fix the through rate for each class of goods and apportion the amount due to each company for the use of their part of the line. This made it essential that there should be meetings at the Clearing House which were the only link between the whole. The arrangement of classification and the settlement of many questions relating to the through traffic necessitated constant meetings of the railway magnates and managers at the Clearing House. They began to realize their common interests and this was the foundation of the agreements as to rates and fares to avoid ruinous competition. The Clearing House has been described as "a sort of federal council for the English railway companies." Although founded in 1842, it was incorporated by an Act of Parliament in 1850, and it is interesting to notice that those railways that first joined it were the first to amalgamate.* "The Railway Clearing House, in fact, is an establishment conducted by the railway companies with the object of mitigating the evils of their independent constitutions."†

One underlying cause of these mergers, to which Hudson gave a start and the Clearing House provided a continuous opportunity for agreement, was to be found in the greater economy of working a large system than a small one. It was also more convenient for the users of the railway. Instead of passengers being kept waiting for hours for trains which did not coincide, or having to change stations, trains could be run to fit and use each other's stations. The Commission of 1872, which was anxious to control amalgamations, could not but report that the North Eastern gave better services and lower fares since it had become a monopoly.

* Cleveland Stevens, *op. cit.*, *p.* 177.   † *p.* 175.

While the greater efficiency and economy to be obtained by combination made the railways wish to combine, another strong motive was the desire to be large enough to be able to have powers of offence and defence against other companies. The larger lines began to be anxious to guard their territory from the intrusion of other companies. This was a defensive policy. To prevent rival lines setting up, the next step was to proceed to an offensive policy and occupy any neutral territory. Thus there were continuous Parliamentary contests between the companies to penetrate each other's districts. These contests usually meant that the big lines would build branch lines as feeders to themselves for fear some other line should build them and attach these feeders to the rival system. Thus the growth of railway mileage continued and as the new railways were attached to one or other of the older systems and as the larger lines tried to absorb the smaller ones that already existed either as part of the policy of offence or defence, the amalgamations went on side by side with the growth in mileage. But the railways could no longer afford to carry only high-class traffic like passengers; they had to lay themselves out for goods to make the new developments pay and the result was an enormous increase of the goods traffic and especially of such traffic as coal and bulky articles. New rates were made either to attract traffic from other lines or to attract goods that otherwise would not travel at all. It was at this point that the railway traffic began to surpass the canals. We have seen that they could offer through rates, punctuality and speed. It was now worth their while to lay themselves out to undercut the canals. It is scarcely necessary to say that Parliament viewed with horror the growing transport monopoly. It did not then see, what is now obvious, that the coasting steamer would prove a considerable competitor to the railway system. Only 447 British steamers existed in 1860 and only a few of these were cargo boats. The steamships of 1860 were chiefly liners and would not really compete seriously with railways, although it was stated in 1872 that three-quarters of the railway rates were subject to water competition.

Parliament, therefore, began to try to attempt some control in the interests of the public. It is almost pathetic to see a desperately laissez-faire government afraid to go back on its own principles of laissez-faire and yet forced by the logic of events to acquiesce in a monopoly it deplored because it was so much more efficient. The almost annual committees or commissions would point out the magnificent work the railways had done in traffic development and then they would wring their hands over the decline of the canals and the confusion of the Railway Acts out of which the railways were evolving a working system for themselves with which Parliament did not dare to interfere, and which would give the railways the monopoly of the area in which they operated which was contrary to the " sacred and glorious doctrine " of free trade and free competition held as a creed by the House of Commons. Therefore we find the Commission of 1867 reporting :*

"Whatever may have been the contradictions of the course of legislation which has thus been pursued it cannot be doubted that it has led to a very rapid development of the railway system and consequently of the national resources, probably far more rapid than could have taken place under any other conditions and has induced improvements in the construction and working of the railways which without the spirit of emulation engendered by it must have taken years to attain. Thus whilst in France there are less than $1\frac{1}{4}$ miles of railway in England and Scotland there are $2\frac{1}{2}$ to every 5,000 inhabitants."

The report then goes on to illustrate the consequences of the system under which the railways developed.

It explains that the special legislation is now (1866) contained in 1,800 Acts, while 1,300 Acts in addition modify the original Acts, " so that it has become an extremely difficult task to ascertain the precise law affecting any company or any particular portion of its lines of railway. In almost every Act sanctioning new lines there are special clauses conferring particular rights or benefits on individuals or other railway companies to carry into effect arrangements which have been entered into by promoters to avert opposition to their Bill."

This had produced extraordinary confusion in their charging powers. The Midland Company was empowered in one

*p. XXVII.

Act to charge 1d. per ton per mile for coals, in another 1½d., for grain in one Act it might charge 1½d., in two more 2d., and so on.

It was obvious that if the companies themselves could evolve order out of this chaos it would be better to leave them alone, and yet the railways owed a duty to the public. The amalgamations seemed to leave them masters of that public; how could the duty of the railways to the public be insisted upon and its rights safeguarded?

In 1846, Parliament set up a special tribunal of five highly paid Commissioners. Their function was to scrutinize proposed amalgamations, encourage competing schemes and generally take over any supervisory powers of the Board of Trade. It proved, however, to be a failure; amalgamations went on, and it was dissolved in 1851. Parliament "avoided all cause of offence by not giving them (the Commissioners) any powers" and they "died of too much pay and too little work."* Their powers were then vested in the Board of Trade on their dissolution.

In 1854, after another Commission, "Cardwell's Act" was passed, the gist of which was that all undue preferences by a railway to one person over another were forbidden and that the railways were ordered to provide facilities for through traffic. There had been a great danger that England would be covered with a network of lines having different gauges and that through traffic would be physically impossible, as railway trucks could not be shifted from a 4-ft. 8½-in. gauge to a 7-ft. gauge. By 1846, however, Parliament had definitely limited the broad gauge to certain districts and therefore through traffic over the larger part of England became physically possible. There were at that date 1901 miles of narrow and 274 of broad gauge. It was not till 1892 that the Great Western finally reconstructed its line from the 7-ft. gauge to the 4-ft. 8½-in. gauge, though it had begun to convert to the narrow gauge in 1868.

Commissions continued to sit to attempt to solve the problem of railway control and as the outcome of that of 1867, the railways were ordered to keep accounts in a specified form

By 1872, however, another great Committee reported that† "Committees and commissions carefully chosen have

*Hadley, *op. cit., pp.* 171-172.   †*p.* XVIII.

for the last thirty years clung to one form of competition after another, that it has nevertheless become more and more evident that competition must fail to do for railways what it has done for ordinary trade and that no means have yet been devised by which competition can be permanently maintained."

(c)—*The Development of State Control*, 1873-1893.

The twenty-three years between 1850 and 1873 were years of unexampled prosperity in Great Britain and the railways had been one of the principal factors in creating this prosperity. During the period the railways had come to an understanding that rates for competitive traffic should not be altered except by common consent.   It was obvious that free competition could no longer be trusted to regular rates and charges.   The chief characteristic of railway development between 1873 and 1894 is the progressive intensification of control of the railways by the State.   The development of Germany and the growing world competition were important influences in the general reaction from laissez-faire, but the tendency to abandon it as a maxim of State policy is, however, noticeable first of all in the railways and this is true, not merely of England, but of the United States.   A special expert body was set up to control the railways in 1873 and, unlike its predecessors, it really did control them to some extent.   It became permanent and was given additional powers in 1888.   The State further undertook the fixing of maximum rates as well as tolls between 1888 and 1894.

The new era of State intervention of which the railways became the touchstone was ushered in by the usual trouble about amalgamations and the consequent fear of monopoly. In 1871, the London and North Western proposed to almagamate with the Lancashire and Yorkshire Railway.   There was also another proposal for uniting the Midland and the Glasgow and South Western.   Nine other amalgamation Bills and seventy-one Bills for working arrangements were also submitted to Parliament and the result was the appointment of the Commission of 1872 to enquire into the whole subject once more.*   The feeling that the railways might abuse their monopolistic position was strengthened by an outcry against

* Cleveland Stevens, *op. cit., p.* 234.

preferences. These were alleged to be of two kinds. Of preferences given by the railways one was given to individuals either in the shape of lower rates or better facilities which enabled the favoured person to oust his competitors. The other type of preference was given to certain districts. Lower rates were charged from certain districts than from others where a shorter mileage existed, and these distant districts obtained an advantage over the near districts. If meat, for instance, were carried from Southampton to London at cheaper or even at the same rates as it was carried from Winchester to London, the former district would have a substantial preference in rates and prosper at the expense of the other region. Cardwell's Act had already prohibited " undue and unreasonable preferences," but no machinery existed for determining so technical a matter as what constituted an undue and unreasonable preference.

The Committee of 1872, which also investigated the question of the amalgamation of the railways and the canals, reported in favour of a technical Commission to control the railways and the result was the appointment of the Railway and Canal Commission in 1873. This body was only appointed for five years ; it consisted of three men, each paid £3,000 a year, one of whom was to be a judge and one a railway expert. Its functions were to hear complaints about preferences, and decide whether through rates were reasonable. They were entrusted with the duty of examining and if necessary preventing all proposed amalgamations or working agreements between railways and they had also to investigate and adjudicate upon all proposals of the railways to buy up the canals. They were empowered to decide between the railways themselves in case of dispute, they had to compel publicity of rates and decide on proper terminal charges. They took over some of the powers of the Board of Trade and were specially appointed to supervise and enforce the Act of 1854.

The Railway and Canal Commission was not, however, a very strong body. Parliament would never have delegated large powers of control to any strong body in 1873. " It has power enough to annoy the railroads but not power enough to help the public efficiently," was Hadley's verdict in 1886.*

The idea was that this new body should be a specialist

*Op. cit., p. 173.

tribunal for dealing with such technical matters as railways. It was thought that it would act promptly, it was to be easy of access and its procedure was intended to be cheap. In many of these respects, however, it proved disappointing. In the first place its powers were too limited for effectiveness. Its function was to enforce the Act of 1854 but, as we have seen, the greater part of the law relating to railways was contained in thousands of private Acts under which they were constituted, and the Railway and Canal Commission could not touch those. Nor could it enforce its decrees if the railways chose to disregard its findings. Nor did any standard of reasonable rates exist by which it could determine whether or no through rates were reasonable. Litigation before the Railway and Canal Commission was very expensive, and a complainant was said to be " a marked man," and the Commission was unable to protect him against the vengeance of the railroads.*

On the other hand, the very existence of such a court was a check upon arbitrary action in general. It probably made the railway managers more anxious to avoid giving occasion for an appeal to the Railway and Canal Commission. At any rate, the complaints about undue preference to individuals ceased. They also seem to have prevented the acquisition of any further canals by the railways. The result was that the jealousy of amalgamation perpetuated the chaos of the canal system. Had the railways been free to acquire all the canals they might have done so and have developed a unified system of waterways acting as feeders to the railways.

The creation of the Railway and Canal Commission does, however, mark a new era in that the State deliberately created a new body which for the first time had some real control over the railways—that new body became permanent and proved to be the forerunner of still more effective control.

At the end of the seventies a new cause of trouble began to arise over the question of unequal mileage rates. There were also complaints that goods were carried cheaper on the railways for foreigners than for Englishmen. A Committee was appointed in 1880 to enquire into the question of mileage rates, but the dispute soon merged into the question of an all-round reduction of rates. The great depression made the

*Hadley, p. 175.

high rates charged during the good years seem too heavy ; there was a great fall in prices and traders did not see why railway rates should not be reduced as well. The railways, on the other hand, had enlarged their stations and terminal facilities and were asking to be allowed to charge extra for these, *i.e.*, instead of reducing charges they were about to increase them. This was the last straw. British railway rates were not merely high, but they were unintelligible and the trader felt that he was being " done " somehow by a system he could not fathom. There was great confusion between the charging powers conferred by this mass of private Acts and the system evolved by the Clearing House. It was owing to this confusion that the Railway and Canal Commission could get no data for fixing reasonable through rates.

The result was the Railway and Canal Traffic Act of 1888 which prepared the way for the revision of the whole system of charging, and by 1893 maximum rates were fixed by Statute for all the railways.

This fixing of maximum rates was epoch-making. The State had hitherto limited its activities in the control of industry to fixing hours for women and children and seeing that the wages agreed on were properly paid. It now began to fix the prices of services. It had previously fixed maximum tolls for the use of the road because it had permitted railways or canals to appropriate property under compulsion and in many cases had fixed maximum rates, but it had made no attempt to see that these tolls or charging powers were adhered to until the Railway and Canal Commission was founded. As we have seen, its function was strictly limited owing to the inextricable confusion of the piecemeal and haphazard fashion in which the railways had grown up, which had caused its powers to be embodied in so many separate Acts. When, however, after 1888, the State fixed maximum rates with a view to being fair to the traders, the railways and the public, and also with a view to simplification and publicity, it repudiated laissez-faire and entered upon the great struggle to limit the profits of monopolies. Of recent years the feeling has grown that great monopolies ought not to be allowed to make more than a certain amount of profit out of the public even if they do it by fair means and greater efficiency. The fixing of maximum railway rates between 1888 and 1894 is the first intimation of the principle which found expression in the Excess Profits Duties.

Under the Act of 1888 every railway company was bound to submit to the Board of Trade, within six months after the passing of the Act, a revised classification and a revised schedule of maximum rates charged for each class. The rates were to be discussed in public when arguments and objections might be presented by the traders. Then the Board of Trade was to discuss the rates again with the railways and if they failed to come to an agreement, Parliament was to decide.

The Act further reconstituted the Railway and Canal Commission, strengthened its powers and made it permanent. An attempt was, however, made to ensure speedy redress and a cheaper procedure for those who felt themselves aggrieved by the railways. The complainant was given the right to lay his case before the Board of Trade. The Board of Trade had no compulsory powers to effect a settlement, but it acted as the candid friend of both parties and heard and helped to settle a large number of disputes.*

The Act further prohibited different rates for domestic and foreign produce of the same character carried under like conditions and a railway was not allowed to charge more for merchandize over a short haul being part of a greater distance than they were charging for that carried over the whole distance. All rates were to be posted in rate books and open to inspection by the public and any increase had to be advertised beforehand.

The settling of the railway rates was a task of some years. The rates were embodied in Provisional Orders and confirmed by Parliament in 1891-1892, and were to come into force on 1st January, 1893.

The task was a stupendous one. The Board of Trade in conference had to draw up a new classification of goods, had to fix the maximum price at which articles should be carried that fell within that class. It had then to determine what articles fell into those classes, whether the price should diminish with distance or with larger loads, and what charges the railways might make for terminals in each of the specified

---

*During the fifteen years ending 1903, some 3,126 cases had been dealt with by the Board of Trade under this Section. Cd. 2959 (1906).

classes. It had, therefore, to draw up a new classification and determine the charges for conveyance and terminals.

The Board of Trade did not attempt to be theoretical when it fixed the rates. Sometimes it took into account the cost of handling, sometimes the value of the goods, the damageability and the weight in proportion to bulk. Often it proceeded on the principle of " charging what the traffic will bear."

The general result was that there was a great simplification of railway rates. The railway companies gained the right to charge a higher rate in certain classes of goods and the power to charge definite terminals. The trader in heavy goods, such as coal and iron, got a substantial reduction ; the small trader was protected by the fact that the excess he might be charged for his small consignment was strictly limited, and he gained by the generally favourable treatment accorded to small parcels. The question of reduction for distance was settled by allowing the railways to charge so much per ton per mile for the first twenty miles, so much for the next thirty, and so much less for the next fifty, and so much for all mileage beyond. Thus the trader for long distances got a reduction on a cumulative scale.*

When the rates became law in 1893 the railways raised all rates to the maximum allowed by the Board of Trade to recoup themselves for the loss on reductions. There was a great outcry in consequence. A new Parliamentary enquiry was hastily held and the result was that, in 1894, an Act was passed saying that if the railways raised their rates above the level of 1892 they had to prove that such a rise was reasonable. The test that was taken by the Courts was that there must be some permanent increase in the cost of service. Therefore after having laboriously fixed rates, the railways were practically tied to the rates of 1892, even when their Provisional Orders sanctioned by Parliament allowed them to charge more. If they wished to raise their rates to the level permitted they had to risk an appeal by the trader to the Railway and Canal Commission and then justify the rise. Here again is another instance of an attempt to restrain profits within a certain limit. The maximum charges of

*On the whole question, see Mavor, " The English Railway Rate Question " in *The Quarterly Journal of Economics,*" 1894.

the railway companies were not merely fixed, but their power to increase the charges within the maximum limits was fixed in 1894. It was thought that the railways had abused their powers by charging up to the hilt and the Railway and Canal Commission was virtually put in charge of the rise of all rates over the 1892 level. This pegging of the railways to a quite arbitrary limit—1892—had several unforeseen results. In the first place the railways were afraid to try experiments in lowering rates for fear they should not be able to raise them again without an appeal to the Railway and Canal Commission. It therefore checked any downward tendency of rates. It has also been said that it discouraged economies since that would tell against the railways if they wished to raise rates. It finally killed any competition in rates that might have developed. Companies were afraid to lower rates to fight each other for fear they might not be allowed to raise them again and so nothing was left but competition in facilities. After 1894 the competition in granting facilities increased, and a new era in English railway history was begun.

(d)—The Approach to Nationalisation, 1894-1914.

The twenty years between 1894 and 1914 are remarkable for many developments. There was first of all a rise of expenditure with a decline in railway dividends. Competition which was acute during the years of depression continued in full force, and while the pressure on the railways became more onerous they could not raise their rates, being bound by the Act of 1894. The only hope seemed to be in amalgamations to stop competition.

The amalgamations coincided with a labour movement among railway servants that had wholly new features in the English trade union movement. The State was forced to intervene when a great railway strike took place in 1911 and again the question of State control became urgent. How could these big transport agencies be best made to serve the national interests in the matter of trade, and to what extent should their relations with their staffs be under some form of government supervision ? Meanwhile other countries were using their railways as instruments in the new world competition. Could Great Britain afford to leave her railways in private hands ?   When war broke out in 1914, the question of the nationalization of the railways had become a burning one.

After 1894 the railways had to meet a greatly increased expenditure without securing an increased revenue in proportion to the outlay.   The growth of traffic necessitated large increases of rolling stock, much more elaborate requirements were made by the Board of Trade for safety, such as continuous power brakes and interlocking of points and signals. The cost of labour rose—for instance, the rise was 43 per cent. in the running part of the locomotive department alone between 1891 and 1901.*   Rates and taxes almost doubled within the same period, rising from £2,246,000 in 1891 to £4,227,000 in 1902.   In addition to this the cost of coal doubled between 1896 and 1901, the general result was an increase in the cost of working and dividends declined.

A fierce competition in facilities began to develop after 1894 in which the passenger traffic was specially catered for. In 1872 the Midland had inaugurated the policy of treating third-class passengers well and the other companies had to follow suit.   The carriages were made more comfortable, corridors and dining cars were installed on the expresses in the nineties even for third-class passengers, with the result that 14-cwt. instead of 4-cwt. was hauled per passenger on the numerous fast through trains.

But not merely did the passenger traffic become more comfortable, the trains had to go faster, with long non-stop runs.   The bridges had to be strengthened, larger locomotives built, water troughs laid down and greater speed meant a larger consumption of coal.   Other trains had also to be put on to serve the intermediate stations.   Nor was there any great increase of receipts from passengers to cover the expense as there was a steady transference of persons from the first and second classes to the third and no extra charge was made for the expresses.   Excursions and week-end tickets also lowered the fares for passengers.

Meanwhile, in order to retrieve the situation, the railways began to compete not merely for passengers, but for goods traffic.   A large proportion of railway expenditure is for fixed charges. so that in many cases it pays better to take the goods

*Ross, *op. cit.*, *p.* 232.

and passengers at a low price rather than not take them at all. The goods carried will obviate some loss if they do not yield a profit. Hence it was worth while for railway companies to offer special facilities to traders where they could not make rate reductions because a loss would be incurred if the railways did not carry the goods in question. Thus competition between the railways continued to be acute, agents multiplied, booking offices increased, the railways were willing to fetch and carry small quantities and did not enquire too closely into the nature of the goods sent. Thus a dishonest trader might enter his goods as belonging to a class below that into which they would rightly fall and the railways shut their eyes to the fact that he might patronize another line next time. Almost any traffic was better than none. In addition to the economic motives there were personal ones. A great manager at the head of a railway likes to make his railway as successful as possible. " Impersonal as they may seem, great railway companies have not infrequently been made the fighting ground of strong individuals."*

In 1903, Mr. Grinling, in a lecture at Birmingham, said : " It is of the highest importance that Parliament and the public generally should grasp the fact that our railway companies so far from being bloated monopolists to be plundered on all hands have been reduced by recent legislation and a combination of adverse circumstances to the position of a threatened industry."†

Violent competition inevitably gives rise to amalgamations and the railways were no exception. A union between the South Eastern and Chatham was sanctioned by Parliament in 1899. They were worked as one railway but their capital accounts were kept separate. In 1909, the Great Northern, the Great Eastern and the Great Central sought power to combine in a similar fashion and the Midland and London and North Western were said to be tending in the same direction. Although the first combination did not obtain Parliamentary sanction the railways entered into working arrangements with each other which Parliament had no power to prevent.‡

*Cleveland Stevens, *op. cit., p.* 307.
†Ashley, W. J., British Industries.
‡Report on Railway Amalgamations 1911, Cd. 5631.

An official enquiry into amalgamations was held in 1911. It reported that " The present position is that the efforts which have been persistently made by railway companies to avoid active competition between themselves so far at any rate as regards the terms and conditions upon which the railway service shall be performed have now reached such a stage of completeness that they may, speaking broadly, be described as having prevailed. It must be accepted that the era of competition between railway companies is passing away and it was recognized by witnesses on behalf of the traders that this could not be prevented."*

This cessation of competition gave rise to new labour troubles, to proposals to develop the canals as competitors, and to the question of the acquisition of the railways by the State.

Hitherto the trader and the traveller had profitted by the competition and there was a great outcry for fear railways would offer fewer facilities. There was also a great outcry from railway employees. Amalgamations meant that fewer men were required, booking offices were shut up and even where dismissals did not take place, promotion was blocked as there were fewer posts. The railway men claimed that their work was so specialized that they could not easily get other jobs and that the companies got the men specially cheap because of the permanence of the employment, and that by being dismissed when there was a tacit understanding as to permanence they were being unfairly treated. The Labour movement, which culminated in the strike of 1911, had been growing for some years. The development of trade unionism among railway men had been late, only starting in 1871 and even by 1892 only one in seven of the railway workers was enrolled in a union. A strike took place on the Scotch railways in 1890 to secure a reduction of hours and the result was that in 1893 the Board of Trade was given power to enquire into the hours worked and fix reasonable limits if they were excessive.† Up to this time it had only been

* Report, *op. cit.*, *p.* 7.
† Any hours worked in excess of twelve had to be reported to the Board of Trade. The return of 1910 gives :
     108,562 railway men.   2,649,387 days worked.
     17,141 periods exceeding twelve hours, *i.e.*, .65% of the total days.

considered necessary to fix hours for women and children ; it was quite a new departure for a government department to be able to fix them for men in a trade not scheduled as dangerous, which again shows how railways contributed another s'one to the edifice of State control.

In 1897 a movement took place among railway men to place " all grades " under better conditions. They demanded among other things an eight-hours' day and a 2s. advance in wages.* The railway companies, with the exception of the North Eastern, steadily refused to meet the trade union representatives or even to acknowledge their existence. Discipline, so it was said, had to be maintained at a high standard, and it was held that trade union leaders might interfere or attempt to interfere in the management of the railways. A strike on the Taff Vale railway gave rise to the famous Taff Vale case by which trade union funds were made liable when the union or its officials had committed some unlawful act and the men were afraid to risk their funds in pushing the matter further by another strike. In 1906, the Trade Disputes Act made the funds of trade unions immune and the railway men then began to continue the fight for the improvement of the conditions of " all grades." A strike was imminent in 1907 when the Board of Trade intervened. The railway administrators still refused to recognize the Unions but agreed to Conciliation Boards being set up, which was another pioneer step as the Boards were to contain representatives of both the railway management and the men. Any application for a change in rates of wages or hours were to go first to the officers of the department concerned ; then to the sectional conciliation boards with representatives of the company and all grades concerned ; then to a central conciliation board formed of representatives from the sectional board, and finally the matter could be taken to arbitration, the arbitrator to be appointed either by agreement between the opposing parties at the central board or by the Speaker of the House of Commons or the Master of the Rolls. No one could be on these boards who was not in the employ

---

*On the whole question of Labour on the Railways, *cf.* Report of the Royal Commission on the Railway Conciliation and Arbitration Scheme of 1907, also Webb, " History of Trade Unionism." 1920 Edition, *pp.* 522-546.

of the company, *i.e.*, there could be no trade union official who was not also one of the company's servants. By 1911, however, there was so much dissatisfaction over the way in which the conciliation boards had worked and so much unrest caused by the economies the railway managements were introducing owing to the cessation of competition, that the strike broke out in August, 1911. It was also claimed that there had been vexatious delays under the conciliation boards and the awards had caused keen disappointment. The strike was for recognition of the unions which the railway management steadily refused to grant. The result was that the scheme of the conciliation boards was amended. The general effect of setting up conciliation boards was that the railway managements relinquished the position of being the sole persons to settle the wages and conditions of service. It had to be done by a process of bargaining with persons who had no financial responsibility for the result.

The joint action of the railway men on these boards led to the amalgamation of all the railway workers with the exception of the Associated Society of Locomotive Engineers and Firemen and the Railway Clerks' Association, in the National Union of Railwaymen, in 1913. It was open not merely to all those who were working on the railways, but to those who were employed by the railway companies in any capacity, " thus including not only the engineering and wood-working mechanics in the railway and engineering work-shops, but also the cooks, waiters and housemaids at the fifty-five railway hotels ; the sailors and firemen on board the railway companies' fleets of steamers."*

Thus the amalgamation of the railway companies was paralleled by an amalgamation of the workers on a scale hitherto undreamed of by any union. It has been termed " the New Model," since it was not merely confined to one particular branch of a trade, and to the men skilled in that branch, like the locomotive engine men, but aimed at including all grades though some preferred to be in separate societies. The Union also began to demand the nationalization of the railways. Thus while competition in facilities led to amalgamations, huge railway combines were met by huge combines of workmen and their proposed solution of the problem of control was

*Webb, *p.* 531-532.

State railways and a share in their management. All that was attempted by Parliament was to prescribe a much more elaborate and uniform system of keeping accounts in 1911.*

Meanwhile the railways explained that the improvements in the conditions of their labour and clerical staff were such as to necessitate a rise of rates to cover the expenditure and they demanded that they should be allowed to raise rates beyond the limit of 1892, as fixed in 1894, and this right was accorded in the Railway and Canal Traffic Act of 1913.

A strong agitation developed to revive the canals as competitors to the growing monopoly of the railways. A Commission began to sit in 1906 to enquire into the question, and the majority reported in 1909 in favour of the Government acquiring and re-building that part of the canal system which stretched from Hull to Bristol and Liverpool to London, crossing in the Birmingham area.† The proposal was that the canals of this " Cross " should be remade and re-built on a uniform gauge and deepened so as to take barges of 100 tons. The re-building was calculated to cost £17 million without reckoning the cost of the acquisition of the land necessary for the widening, which would have been very expensive, as so much of the canal frontage was built over, especially in the Birmingham area.

The arguments in favour of reviving the canals were that water carriage was cheaper than land carriage, and would provide cheap transport and be a great national asset. It was said that the waterways of Germany, Belgium and France gave a cheap and valuable alternative mode of transport for articles like coal, lime, bricks and manures and that the diversion of this low-grade traffic would really relieve the railways. It was suggested that whereas factories were leaving the Midlands and moving down to the coast to get raw material without the inland haul, improved water transport would check that movement, prevent the overcrowding of the seaports and the dislocation and loss that accompanies the migration of industries. It was urged, on the other hand, by the railway representative, that the scheme was costly, that it was unsuitable for the conditions of English trade which

*Report of Board of Trade Committee on Accounts and Statistical Returns by Railway Companies, 1909, Cd. 4697.
†Cd. 4679.

required swift delivery of small parcels and punctual delivery for steamers. It was said it was useless to compare English and continental conditions as regards canals. There were no long hauls in England as on the continent, nor were English canals capable of taking the 400 and 600 ton barges which made German canals so valuable an asset to a country with a poor sea-board. Indeed, it was doubtful if water enough could be obtained for barges of 100 tons. It was pointed out that the railways did in fact encounter a considerable amount of competition in the coasting steamers and in motors, and that these transport agencies had not got to pay for the upkeep of their roads as had the railways. Moreover, the districts not traversed by canals would have a right to complain of the preference given to districts through which the reconstructed canals were made. The whole country would have to pay towards the upkeep of certain favoured districts. Nothing was accordingly done towards carrying out the scheme recommended in the majority report. During the war so many of the bargees enlisted that the canals became almost derelict. The question as to their acquisition by the State has been made the subject of a new Government enquiry in 1920.

The difficulty of devising any scheme of railway control combined with the new labour pressure, brought the question of the acquisition of the railways by the State to the front. The arguments in favour of State railways were somewhat as follows :* It was said that the State would obtain a large revenue from the railways. It was also urged that an actual loss would not matter : the State could undertake improvements for the sake of the country as a whole, and could get its revenue back in the increased prosperity of the country. State management, so it was said, would be less costly since there would be an abolition of the various boards of directors and the wastage of competition would be prevented. It was even contended by some that State railways would be more efficient. Were the railways in the hands of the State, they could, so it was argued, make regulations for the social needs of the people. The traders thought they might get

*On the whole question, Gibb, Railway Nationalization, 1908 (Royal Economic Society), " The State in relation to Railways," 1912.

lower rates and the railway workers thought they would
be able to obtain better conditions from the State.

Against State management it was pointed out that only
two State railways in the world had a surplus after meeting
the charges incurred by the State, including interest on the
capital and the provision of reserves for renewals—Prussia
and South Africa. Prussia made a profit by forcing a
great deal of traffic on to the canals and starving the railways
of their proper complement of trucks. This became serious
when the canals froze. It was therefore doubtful if State
railways were more efficient.* They were said to be less
flexible and less adaptable. Private railways were said by
Mr. Acworth to have to their credit almost every important
invention and improvement.†

It was impossible to decide whether State railways were
cheaper or more expensive since the accounts of continental
and other State railways were kept in such a way as made
comparisons with English railways impossible. Moreover,
they did not offer the same kind of services.

The enormous cost of the necessary borrowings and the
dislocation of the money market was also used as argument
against nationalization. If the railways should not pay all
deficits would have to be made up out of taxation. Moreover,
the railways contributed in 1912 £5¼ millions towards taxation ; as State railways they would not contribute and the
traders would have to make up the deficit in higher rates or
increased taxation. It was said that there would be great
political corruption with State railways. Traders who wanted
lower rates and railway servants who wanted higher wages
would bring great pressure to bear. The wage bill of the rank
and file would rise and would swallow up any economies
arising from unified management and the abolition of a few
boards of directors. When railways are in private hands
the Government can step in as an arbiter in disputes between
masters and men. What if it were itself the master ? Moreover, it can be appealed to by the public as a third party,

*Cf. the account of the Western of France Railway, as given by
M. Leroy Beaulieu in a paper read before a Conference arranged by
the Royal Economic Society, published in The State in Relation to
Railways, 1912.

†The State in Relation to Railways.

but to whom could one appeal against the tyranny of the government-owned railway ?

The situation in 1914 was that the railway workers were demanding the reconsideration of the arrangements of 1911, that individual citizens felt that they were " the helpless subjects of huge monopolistic organizations." The *impasse* was that no effective system of control had been devised, that amalgamations had taken place to such an extent that competition could not be trusted to provide adequate safeguards and that the railways could not continue as they were with a growing wage bill and further demands for reductions of hours, which was creating an impossible financial situation. Meanwhile transport was becoming more and more an effective State instrument on the continent for the furtherance of national trade by cheap rates. The Great War staved off these questions and the railways passed temporarily under government control.

Further concessions in wages and an eight-hour day were granted with the result that the railways when released from government management were faced with financial insolvency. The government then permitted a considerable rise in rates and fares. During the war, when the railways were worked as one system, many economies had been effected, especially in the use of waggons. As it was undesirable to let the railways resolve into independent units again, the compulsory amalgamation of the English and Welsh railways into five great groups was effected.* Further State control was devised in 1921, when a tribunal of three experts, with salaries totalling £10,000 a year, was set up to fix and vary rates and fares without being bound by the old statutory maxima. The rates so fixed have to produce the same income for the railway as they obtained in 1913. †

Thus all through the nineteenth century the railways have been the great factor making for the extension of the sphere of State action and the abandonment of the idea of free competition, so dear to economists of the free trade era of the period before 1870 has been more forcibly illustrated in the case of the railways than in any other great industrial undertaking.

*Cmd. 787.   †See Appendix

IV.—The Steamship and Shipping Problems.

The commercial revolution, hinging as it did on the new methods of transport, was not merely brought about by railways. The steamship played a very important part in the transformation of economic conditions which tended to make the whole world practically one market. In this change Great Britain was again the pioneer. Although inventors in the United States developed a *wooden* vessel propelled by steam almost contemporaneously with the English, the United Kingdom made the engines to work the new form of transport in both countries and was the first to develop the *iron* steamer propelled by steam, and was also the first to develop it for ocean voyages and for cargo. In other words, the United Kingdom was the pioneer of the steamer on a commercial scale for world traffic.

Shipping is, however, fundamentally different from railway transport, in that while the latter was the great force making for State control, shipping was, between 1849 and 1914, almost free from control except in the matter of regulations to ensure the seaworthiness of the ship, the safety of the passengers and the proper treatment of the crew. There was no fixing of rates or fares by the State and no special Commission for shipping equivalent to the Railway and Canal Commission.

The reason for the absence of State control of steamships is due to many factors. In the first place the railways had to get land to make their roads and were obliged to appeal to Parliament for compulsory powers, and Parliament imposed conditions when they granted the powers asked for. In the case of shipping the road, *i.e.*, the sea, is free and shipping companies are formed under the Companies' Acts like any other Joint Stock undertaking. They do not require special powers for their creation. Moreover, a ship is relatively a cheap thing compared to a railway and she has not got to pay for making her road—the sea is free—nor does she pay for the making of her ports and docks. Therefore as there is a competition in shipping, because so much less capital is required, which does not exist in the case of railways, Parliament trusted to free competition to adjust matters. It would have been difficult in practice to enforce a limitation of

freights and fares because a proportion of English shipping
—about a fifth—did not come home for years together.
How would it be possible to fix rates for an English ship
trading between Buenos Ayres and Canton ? Further,
freights are not made up of one voyage there and back, but
of a continuous voyage in which cargo may be taken on
board and discharged many times over. Each item of that
cargo goes to make the profit of the voyage but it would be
impossible for a Commission in London to allocate the proportion of the charges of coal taken on in Natal and discharged in
India, of cotton taken on in India and discharged in Japan, and
of sugar picked up in Java, combined with wool in Australia
and discharged in the port of London along with tea loaded
in Ceylon for the homeward voyage. The freights charged
depend on the other goods available to make up a cargo and
they vary with every voyage. Moreover, a ship is not bound
to come to the port of London. She is mobile as a railway
train is not. A ship can discharge almost as well at Rotterdam
or Havre and the stuff for England could be sent across in
small coasting steamers. To regulate shipping to such a point
that it would leave the English ports would seriously injure the
British entrepôt trade. Finally ship-owners are faced by
foreign competition and railways are not. The ship-owner
has to be at liberty to quote such rates as will eliminate foreign
competition without being compelled to consider whether
he shall have to justify a rise of freights at some future date.
Moreover, there was still considerable competition in shipping
before 1914 between the tramps and the liners. If liners put
prices up too high the tramps, bound to no particular line,
would flock in and bring prices down.

All these factors have made for individual enterprise in
shipping and the history of railway and shipping transport
during the nineteenth century is in marked contrast, the one
standing for increasing regulation and the other for freedom.*

*Cd. 9092, *p.* 63 :—" Shipping . . . is everywhere exposed to
international competition and depends for success on elasticity and
freedom from restrictions. Private enterprise which has often saved
the country in the past, built up the splendid mercantile marine
without which this war could not have been carried on, and it is our
conviction that any departure from a principle which has been of such
material value to us, would be a dangerous experiment and a blunder
of the worst kind."

During the century, 1815-1914, there has been (1) a great change in policy with regard to shipping. The British mercantile marine had been developed under a series of strict regulations, known as Navigation Acts. These were repealed between 1822 and 1854, and British ship-owners were neither encouraged by bounties nor restricted in the way in which they carried on their business. There was (2) during the century a series of changes in technique which revolutionized the whole carrying trade. (3) Great Britain's predominance became more and more marked with the developments of the iron and steel ships and the elimination of her great rival, the United States, during the Civil War. (4) Competition between British steamship lines led to the growth of shipping rings ; in other words, the tendency to amalgamation, so marked in railways, becomes clear also in shipping. (5) This was followed by the growth of a new foreign competition started by States that wished to have their own mercantile marines, and therefore they gave subsidies and other assistance to their national ships. (6) This raised the question as to whether Great Britain should respond with similar subsidies and the matter became more complicated when the question of stimulating closer relationship with the colonies became prominent after 1897. To a sea Empire the development of shipping communications was all important. How could these best be assisted or could they be left entirely to private enterprise ? The colonial factor became an instrument making for the growth of State intervention even in shipping.

(1)—*Free Trade in Shipping.*

The change of policy known as the repeal of the Navigation Acts was carried out between 1822 and 1854.* The main object of the series of Navigation Acts, which dated back to 1381, had been to encourage the development of English shipping for the sake of defence. There was no specialized navy till the reign of Charles II., and all the ships of the realm were the defence of the realm, the King contributing his quota of royal ships with the rest of his subjects. Thus every ship added to the safety of the country. The prime object of the Navigation Acts was to create a reserve or preserve in which English shipping might be developed and

*Clapham, " Last Years of the Navigation Acts," *English Historical Review,* 1910. *pp.* 48 and 687.

U

they had a political object, national independence. There were also economic objects and these became clear as colonies developed in the seventeenth century. English trading ships rarely ventured outside Europe in the sixteenth century and one of the aims in founding colonies was to give English ships an opportunity of going long distance voyages. A larger type of ship would have to be built and it was hoped that England would produce ships both "tall and fair" for the colonial trade. The Dutch were, however, the great shipping masters of the seventeenth century and it looked as if they would absorb the profit of the English colonial trade. Therefore we find, under Charles I., a long series of Orders and Proclamations prohibiting the use of foreign ships in the trade between the mother country and the colonies or *vice versa*. In 1651 this took the shape of a Statute, which was re-enacted with additional pre-cautions in 1660, as no Act passed during the Interregnum was legal. The aims of the Navigation Act of 1660 and the various Acts of Trade, Customs regulations and Statute of Frauds which supplemented it, were to reserve an area for English shipping, to make England an entrepôt or distributing country, to make English ships go long distance voyages, to make the colonial trade centre in England, to out-distance the Dutch and employ Englishmen as seamen. To carry this out, the Acts prohibited or restricted foreign ships in certain lines of trade. The trade between the mother country and the colonies was reserved for English or colonial ships and the coasting trade was reserved for English ships. In the European trade certain "enumerated commodities," such as timber, corn and wine, might only come in English ships ; certain other commodities might come in foreign ships, but additional duties were levied by the Customs on goods permitted to be brought in foreign ships. To make England a great distributing country some colonial goods were also enumerated, of which sugar and tobacco were the chief, and these had to be brought to England for distribution. The enumerations aimed at securing part of the Dutch distribut-ing trade. To make English ships go long distance voyages, goods of non-European origin might not be brought from Europe, *i.e.*, coffee might not be fetched from Amsterdam, nor cotton from Marseilles. For foreign ships the port and pilotage dues in English waters were heavier than those for British ships but British ships had, however, to submit to regulations as to the nationality of the master and crew. The ships had to be English built, the captain had to be English as well as three-fourths of the crew. The Scotch did not count as English for purposes of a crew till after the Union of 1707. This elaborate code, designed to secure the development of English shipping against the formidable rivalry of the Dutch, was approved of by the free trader, Adam Smith, who said that as defence was more than opulence, the Navigation Acts were perhaps the wisest of all the mercantile regulations of the country. It was clearly understood by contemporaries that the colonial system and the Navigation Acts were part and parcel of the same thing. Colonies were regarded as estates to be managed for the benefit of the mother country, and the economic instrument of their control was transport, and to regulate shipping to and from the colonies was to regulate the colonial trade. The general effect of the Navigation Acts was to project English shipping into the ocean trades as distinct from the coasting trade round Europe and the Mediterranean. Whether it was because of these regulations, or in spite of them, the British mercantile marine steadily increased during the eighteenth century, as did also the shipping of the New England colonies, which were under the same ægis of protection and control. The growth of shipping may have been due to the large expansion of English industry and the corn export, which caused an increase of the imports and exports, which gave additional employment to shipping. As new colonial areas were added and the original colonies developed, the carriage of their produce would afford further scope for shipping. On the other hand, if it had not been for the Navigation Acts the Dutch might have absorbed the growing sea-borne trade of Great Britain as the English did that of the United States after 1870. It is possible to have a great exporting nation whose trade is carried on in foreign bottoms. The fact, however, remains that British shipping survived the French wars successfully in which the shipping of both the French and the Dutch was destroyed. The shipping of the United States had also been stopped during the war of 1812.*

* " Very few American ships were taken because they did not dare

Great Britain emerged in 1815 without a rival, as the one power able to carry on the shipping of the world in spite of the fact that she had lost about forty per cent. of her ships during the years 1803-1814.* The losses had been more than made good by increased ship-building.

By the end of the eighteenth century the English position was so strong that she could afford to consider making some modifications in her system. The period, 1796-1822, is a period of minor relaxations. Between 1822 and 1825 there is a change from monopoly to reciprocity, and between 1849 and 1854 the Navigation Acts were abolished and the colonial trade thrown open.

The period of minor relaxations was started by the revolt of the American continental colonies. The rule was that no goods from Asia, Africa or America should come to England except in British or colonial ships, and this shut off trade between England and the United States except in English ships. As, however, the English wanted cotton and foodstuffs during the war with France, and as the United States was a neutral and her ships relatively safe, goods from the United States were allowed, in 1796, to come to Great Britain in American ships. The inter-colonial trade had been reserved for British or colonial ships. Again the rupture with the United States introduced new complications. They had been accustomed to trade with the West Indies and Canada when they were British. This was now illegal. The West Indies were, however, so badly off for food that the prohibition had to be relaxed and the United States was allowed to trade with the West Indian Islands in 1796, and with Canada in 1807. Similar relaxations were extended to Brazil in 1808, and the new Spanish American republics in 1822.

In the twenties foreign nations were beginning to resent the British dominance of the carrying trade and threatened to retaliate by Navigation Acts of their own. The result was that in 1824 the Crown was empowered by Parliament to

to go to sea, with the exception of the few to whom exceptional speed gave a chance of immunity. While the enemy were losing a certain small proportion of vessels the United States suffered practically an entire deprivation of external commerce and her coast trade was almost wholly suppressed." Mahan, Influence of the War of 1812 on English and American Shipping, *p.* 221.

*W. R. Scott, Peace after War, Vol. I., *p.* 46

negotiate treaties on the basis of reciprocity in matters of shipping. As British shipping was much larger than that of foreign nations if concessions were made to get concessions, Great Britain stood to gain by this arrangement. Treaties were accordingly negotiated between 1825 and 1843 with Prussia, Denmark, Sweden, the Hanse towns, Mecklinburgh, Hanover, the United States, France, Austria, Frankfort, Venezuela, Holland (1837), the Zollverein and Russia. Each of these meant the abandonment of some portion of the Navigation regulations.

The Navigation Acts were still further modified in the direction of the freedom of the colonial trade between 1822 and 1825. The "enumeration" of colonial goods was abolished and the colonies were allowed to trade with foreign countries direct. Certain restrictions, however, remained. Goods of non-European origin could not be brought from Europe, the "enumeration" of goods from Europe was still maintained and the inter-imperial trade was still reserved for British or colonial ships. Nor could goods be brought from Asia or Africa in any but British ships.

By 1840, the free trade movement was in full swing and British merchants who honestly believed at that period that all restrictions were wrong, were anxious to be able to charter American ships freely. The American mercantile marine had developed rapidly. They had plenty of soft lumber for ship-building and American crews, trained on the fishing grounds of New England, were good seamen. It was said that the ships of the United States sailed faster and carried cheaper than the English. British merchants wished to lower freights by competition. There were certain inconveniences in the existing system of which a great deal was made. The colonies, too, saw their preferences dwindling as the free trade party became more and more victorious, and they wished for the abolition of restrictions not off-set by preferences. Canada, for instance, whose preference in corn had gone with the repeal of the Corn Laws in 1846, was anxious to charter the cheaper American ships for her trade with the West Indies. The Zollverein was threatening to penalize English shipping and free traders hoped that once the English had abolished all navigation laws that Europe would follow the English example. A large party was, however, opposed to the repeal on the ground of the growing competition of America. It was feared, if the ports were thrown open to all ships, that the British would never hold their own with the greater expense of ship-building in this country, owing to the dearness of timber. It was said that the merchant service was the training ground for the navy, and that the repeal of the Navigation Acts meant the fall of the English sea-power, which rested so largely on the mercantile marine. Nevertheless, the free traders gained the day, the Navigation Acts were repealed in 1849, the coasting trade was thrown open in 1854, and after 1853 there was no longer any obligation to man ships with British seamen. It is worth noticing that shipping has a long tradition of control behind it—473 years, from 1381-1854.

(2)—*The Coming of the Steamship and the Continuous Change in Technique.*

The repeal of the shipping restrictions came at a time of a revolution in sea transport, and all the forebodings were falsified. Between 1850 and 1860, the iron steamship began to prove a success both for passengers and cargo, and England's great capacity for iron working, with her abundant supplies of coal and iron, gave her the lead in ship-building and the carrying trade. The evolution of the iron ship was, however, gradual. Wilkinson, in 1787, had built an iron canal barge, but it was considered to be " against nature," and only gradually was it discovered that iron ships were not merely stronger but lighter than wooden ships. The new motive power, the marine engine, was developed separately from the new material for ships, *i.e.*, the marine engine was worked originally in a wooden ship. The *Charlotte Dundas* was the first vessel successfully worked by steam and she plied, in 1802, on the Forth and Clyde Canal. She was fitted with an engine constructed at Carron. In 1820, an iron ship to be worked by steam was built at the Horseley Iron Works at Tipton, in Staffordshire.* The new material and the new

*She was brought in sections to the Thames and reconstructed in the Surrey docks. She then steamed to Paris. Her builder, Manby, set up works at Charenton, on the Marne, and built steamers there, another instance of English technique reacting on the continent. The first iron steamer built on the Clyde was built in 1832, the *Aglaia*, of

engine were thus combined for the first time. Steamers in general were, however, only small craft that sailed most of the way and only turned on steam when the wind failed.* Even by 1860 the sailing ship was still the preponderant type of ship. There were at that date, 6,876 in the United Kingdom, as compared with 447 steamers, and of these only 91 were between 1,000-2,000 tons and only four were over that size. The steamers were, therefore, small vessels of less than 1,000 tons and that meant that they could not carry coal enough for long voyages if they wished to carry any considerable amount of cargo. The idea was also prevalent that steamers would spoil the flavour of comestibles carried as cargo. The first steamships, therefore, developed as liners to carry passengers and mails.

By 1850, however, coaling stations were established and vessels no longer needed to carry such large quantities of coal. The compound engine greatly economized the use of coal in working a ship, and was recognized as a commercial success at the end of the fifties. Other improvements followed, leading gradually to the almost complete displacement of sail by steam.† The opening of the Suez Canal gave a great

thirty tons. Ten years later an iron steamer, the *Prince Albert*, was built on the Tyne. Of the four vessels that crossed the Atlantic in 1838 under steam, the *Great Western* was built of iron. Kirkaldy, British Shipping, *ch.* 4 and 5.

*Some of the principal dates in the history of steamships are as follows : In England the first successful passenger steamer was the *Comet*, built in 1812, but Fulton, in America, had designed a steamer which was engined by Boulton & Watt, and ran regularly between New York and Albany from 1807. In 1814 a steamer built on the Clyde ran regularly on the Thames. In 1813 no less than four steamers were built on the Clyde ; in 1816 the number rose to eight, and in 1822, forty-eight. These were wooden vessels. In 1819 the *Savannah*, an American ship, crossed the Atlantic and used steam as an auxiliary to sails. By 1838 four ships crossed the Atlantic using steam all the way. They took from fourteen to seventeen days and proved the practicability of steam for long ocean voyages. In 1825 a vessel, the *Enterprise*, went to India, using steam as auxiliary to sails. After half a century of experiment it was demonstrated between 1850 and 1860 that the steamship was not only practicable but would pay well financially for both cargo and passengers. For the history of the liners, *see* " *Quarterly Review*," 1900, Volume 381. Ocean Steamships.

†In 1890 the United Kingdom possessed three million tons of sailing vessels ; by 1900 the amount had declined to a little over two million tons, and by 1913 to 850,000 tons. Cd. 9092, *p.* 54.

impetus to the adoption of steam, as sailing vessels cannot navigate the Canal.

The four things aimed at in building and working a ship are economy of fuel, economy of labour, space for cargo and cheapness of construction.

Economy of fuel was obtained by the compound engine and this was improved upon by the triple and quadruple expansion engines of the eighties and the nineties, until 6-lbs. of coal per horse power per hour was reduced to 1¼-lb. of coal per horse power per hour.*

Economy of labour in working was obtained partly by the adoption of mechanical appliances and partly by building larger and larger ships as fewer men in proportion are required to work large ships.† " Such vessels are more economical and in peace the country that can run its vessels most economically will necessarily be the predominant carrying power."‡

The cargo carrying capacity of a ship was increased by two things. In the seventies the material of which a ship was built began to be changed and steel began to be adopted in the place of iron. Steel was lighter than iron, and this meant that more cargo could be stowed before the load line was reached, *i.e.*, the displacement of a steel ship was less. Moreover, it was cheaper to run steel ships than iron, as the life of a steel ship was longer than an iron one.

The cargo carrying capacity was also increased by developments in the marine engine. In the first place, with economical engines, less coal was carried and more space left for cargo. Then the engines were improved upon so as to take up less space. The turbine and the geared turbine for cargo boats have been pioneers in this respect. Their weight is also said to be less and therefore the loading capacity of the ship is

*Kirkaldy, British Shipping, *p.* 131. The first steamer fitted with triple expansion engines was launched in 1881 and the quadruple expansion in 1894. Kirkcaldy, *op. cit.*, *pp.* 131-132.

†An article on Merchant Shipping in the *Quarterly Review*, 1876. Volume 141, *p.* 263, says that the proportion of men to each 100 tons was, in 1852, for sailing ships, 4.55, and for steamers, 8.04. In 1874, the proportion for sailing ships, 3.19, and for steamers, 4.10. . . . " A great deal of the heaviest work formerly done by men is now done by machinery, especially in steamers. The steam winch is the best man in the ship."

‡Cd. 9092, *p.* 54.

increased. The great example of economy in space in connection with the motive power of the ship is the motor or oil engine using oil fuel. The engine takes up less space, less space is needed for storing oil fuel than for coal and oil needs fewer men as stokers.

In addition to these changes in technique by which the small iron steamer became the huge steel ship worked by turbine or motor, there has taken place a considerable specialization of shipping, such as oil-tankers and refrigerating ships.

In the days of the old sailing ships there were two types of ships, the great East Indiaman and the West Indian free trader, a smaller, handier and less specialized type of ship. With the coming of the steamship two types again appeared, the liner and the tramp. The liner had regular routes, scheduled sailings and carried passengers and mails as well as that type of cargo that required speed, such as things liable to a change in fashion. The liners were therefore swift as well as regular. The tramp was a cargo vessel, often chartered to third parties, and was free to go anywhere and carry anything that would pay. In the tramp were moved the great seasonal cargoes of rice, cotton, wheat and wool. The tramps would often break ground for the liners. They would pick up cargo in likely or unlikely places and make the breach and as the trade developed the liner would come in with its regular sailings. Tramps thus often acted as the scouts of the regular service. Before the war, about sixty per cent. of British shipping consisted of tramps and forty per cent. of liners. The tramp dealt with the larger part of the bulky cargoes. A further specialization developed in the vessels built for the frozen meat trade and the tank steamers for carrying oil.

(3)—*The Supremacy of the United Kingdom in Ship-Building and in the Carrying Trades.*

In iron and steel ship-building, as is only natural with her start in engineering, the United Kingdom has been the world's ship-builder. Before the war the twin industries of ship-building and marine engineering employed together well over 200,000 work-people ; the capital invested was not less than £35 millions, and the annual output exceeded a gross selling value of £50 millions. The normal production before the war was greater than that of all foreign ship yards put together.* At the outbreak of the war the British mercantile marine was the largest, the most up-to-date and the most efficient of all the merchant navies of the world. It comprised nearly one half of the world's steam tonnage (12·4 million tons out of about 26 million tons net) and was four times as large as its nearest and most formidable rival —the German mercantile marine.† The United Kingdom energetically sold her old ships to foreigners and equipped herself with the newest and largest type of vessel. The result was that " the merchant tonnage of foreign countries was as a whole older and less efficient than the tonnage of the British mercantile marine."‡ Of the tonnage on the register in 1913, eighty-five per cent. had been built since 1895, and forty-four per cent. since 1905. These vessels consisted of the large and most efficient type of ship.

1913.	No.	Tons.
Steam vessels of less than 1,000 tons - - -	8,855	1,100,000
Steam Vessels above 1,000 tons	3,747	10,173,000
	12,602	11,273,000

This was the strength of the United Kingdom in peace and her weakness in war, as these large ships made such a target for submarines and, when sunk, the loss of a single vessel was disproportionately great.

The world's ship-builder was also the world's carrier and the United Kingdom not merely built the ships but she used them. The reason for the predominant position of the United Kingdom as a carrier was due to the fact that she was the industrial centre of a world-wide Empire and drew on the whole world for food-stuffs and raw material. The great coal resources of the United Kingdom not only provided outward cargoes for a large amount of shipping that would otherwise have gone out empty but supplied bunker coal to much of the shipping engaged in the foreign trade.§ This

*Cd. 9092.
†Cd. 9092, *p.* 53.
‡*Op. cit., p.* 55.
§Twenty-one million tons was supplied for bunker in 1913. This is not included in the figures of the exports of coal. Cd. *op. cit.,* 75.

island possessed coaling stations specially suited to British ships distributed all over the world and had established by a long start historical lines of connection that would be difficult for new comers to sever in their favour.

The very ubiquity of the British demands and commercial connections makes the United Kingdom a country where nothing comes amiss and where most things can always be disposed of somewhere else if not required here. The result is that she became the pivot of the world's sea-borne trade.

In the period before the war the British were the ocean carriers of the world and forty per cent. of the world's sea-borne trade was with the United Kingdom while British ships carried about fifty-two per cent. of the total sea-borne trade of the world. The strength of British shipping lay in the ocean trades, *i.e.*, the trade with countries outside Europe and the Mediterranean. As regards *volume*, the greater part of the trade before the war was with countries in Europe and on the Mediterranean, but as regards *value* the greater part of the trade was with countries outside Europe.* "The possession by England of the bulk of the world's over-seas trade not only gave British ships ample cargoes but also made it possible for English ship-owners to lay out the trade routes so as to insure fullest possible cargoes for their ships at all stages of their voyages."† The Navigation Acts had deliberately projected this country into the long distance voyages or ocean trades and as the colonial trade was the connecting link in the past, so the Empire is to-day of great importance in maintaining the British hold on the ocean trades.

It is interesting to notice that this great revolution in ocean transport involved in the change from the sailing ship to the steamer was assisted in the first place by subventions given by the British Government for the carriage of the mails. These were paid to the Cunard line and to the Peninsular and Oriental Steam Navigation Company, the Royal Mail Steam Packet Co. and the Pacific Steam Navigation Company. Since that date subventions have been given for fast mail steamships that complied with Admiralty requirements in building and which could be used as cruisers

*Cd. *op. cit., p.* 75.
†Government Aid to Merchant Shipping, G. M. Jones, U.S.A. Department of Commerce, Special Agents' Series No. 119, 1916.

in time of war. Cargo ships have, however, received no aid from the British Government and about sixty per cent. of British tonnage consisted of tramps. In 1894 it was calculated that the proportion of British shipping receiving postal or construction subventions did not amount to more than three per cent. of the total.* It is, therefore, true to say that English shipping received no government aid except " for services rendered," and that by far the larger proportion of her ships were built and launched without any government aid whatsoever at a time when her ports were thrown open to all the world. In both English railways and English shipping the new transport developments were the work of individuals who were able to obtain financial backing either from the Banks or from the public who subscribed to the companies. It was, however, the mobilisation of capital in Joint Stock companies that enabled these great undertakings to be carried out on so large a scale.

During the nineteenth century Great Britain witnessed the eclipse of one great shipping rival and the rise of another. In the middle of the century the United Kingdom had a formidable competitor in the United States for both ship-building and carrying. Hawthorne writes of America " disputing the navigation of the world with England." Returning from his mission to England, Buchanan publicly declared that " our commerce now covers every ocean, our mercantile marine is the largest in the world." On the eve of secession, Alexander A. Stephens said, in a speech delivered before the Georgia Legislature, " We have now an amount of shipping, not only coast-wise but to foreign countries, which puts us in the front rank of the nations of the world. England can no longer be styled the ' Mistress of the seas.' "†

Within twenty years all was altered. The Civil War caused a good deal of destruction to shipping, but more disastrous still from the point of view of America was the English skill in iron working and ship-building and the ever progressive nature of the English ship-yards. The large quantities of vessels turned out by the yards of the United Kingdom gave all the advantage of mass production ; the

*G. M. Jones, " Government Aid to Merchant Shipping," *p.* 22.
†The British Mercantile Marine, *Quarterly Review*, 1904, *p.* 333. For figures, *see pp.* 160, 193.

cost of ship-building was much larger in the United States than in England and, accordingly, American capital was sunk in other directions—in railways and other constructional works, and the United Kingdom remained the mistress of the Atlantic.

(4)—*The Growth of Foreign Shipping.*

In the middle of the eighties a new form of competition began to develop. With the accentuated nationalism of the period after the Franco-Prussian War, nations began to wish to develop their own shipping and to be less dependent on Great Britain. They began to stimulate the creation of national shipping by means of subsidies in various forms. Sometimes they took the form of bounties on construction or bounties given in proportion to the number of miles worked or a direct government grant ostensibly for the mails but in excess of " services rendered." Indirect aid in the shape of special railway rates, special concessions for the carriage of ship-building material over the State railways, the free admission of goods required for railway construction, whatever the tariff on other things, the reservation of the coasting trade to national ships, the payment of the Suez Canal dues, exemption from taxation, loans to ship-owners and reimbursement of port dues are all forms of State aid to shipping practised since the eighties by the various States. France started this elaborate bounty system in 1881 and was followed in 1885 by Germany, Italy, Austria, Hungary, Japan, Russia, Denmark, Spain, Belgium and the United States, all of whom adopted some of these forms of State encouragement to national mercantile marines and they were still in force in 1914.[*]

Whether it was in consequence of these aids, or in spite of them, the most noticeable phenomenon was the rapid development of the German mercantile marine which soon became a formidable competitor to Great Britain. In any case, with or without government aid, German shipping would probably have developed rapidly to carry the growing export and import trade of Germany. The fact that the foreign

[*]For particulars, Committee on Steamship Subsidies, Cd. 1902. Royal Meeker, History of Steamship Subsidies   G. M. Jones, Government Aid to Merchant Shipping.

trade of Germany was concentrated in a very few ports,
mainly Hamburg and Bremen, which assured a maximum
of cargo for ships calling at these ports, and also that the
imports and exports of Germany were well balanced in tonnage
giving freight both ways, were of considerable assistance to
the German mercantile marine. The development of Germany's great iron and steel industries in the eighties provided
the raw material for ship-building which the State railways
carried at specially low rates to the ports. While specially
low railway rates were granted on goods for export going
out by *any* ship, a special tariff was drawn up for goods
going out by *German* ships to the Levant and East Africa.*
Postal subventions, undistinguishable from bounties, were
given to the North German Lloyd for services to the Far
East and Australia, and to the German East African Line and
these were only given to German built ships.†

The most successful development of German shipping was,
however, in the Atlantic trade and here no subsidies were
given. The success of the liner traffic was, however, based
on the emigrant traffic and the emigrant traffic was so worked
through the control stations that it was the foundation of
the German Atlantic trade. These control stations were first
established in 1894, after the outbreak of a cholera epidemic in
Russia, and all emigrants from Russia and Austria coming
out *via* Germany, had to pass through them. Their erection
and management were given by the Government to the
Hamburg Amerika and North German Lloyd companies and
no steamship was allowed to carry emigrants from Germany
without a license. Such a license was only granted to English
steamship companies under special restrictions‡ which hampered their business. Furthermore, emigrants in the control

---

*The Report on Shipping of 1918 seems to consider that these special
export rates were given when there was thought to be a sufficient
reason for so doing to goods shipped by German lines to South America,
the Middle and Far East, and even Australia. *Op. cit.*, *p.* 99.   §8250.

†The North German Lloyd received £30,450 for services to China,
the East Indies and Australia ; the Hamburg Amerika, £10,000 for a
service to Heligoland and Borkum and the German East Africa line
£67,500 for the service to East Africa. The Belgian Government
also paid the North German Lloyd a subsidy of 80,000 fr. a year to
call at Antwerp *en route* for Australia and the East, and refunded
light and pilotage dues. Cd. 9092, 8228-8234.

‡Cd. 9092, *p.* 8.

stations who intended to ship by British lines were
"subjected to every kind of inconvenience to make them
travel by the German lines. Their tickets were often forcibly
taken from them and not returned to them for days, whilst
in the interval these poor people were left to incur expenses
at the control stations which they could ill afford and were
often forced to return home. Meanwhile, the agents of the
German lines sought to cajole them into buying tickets
from the German lines, generally by threatening to have them
sent back to their homes if they did not comply."* These
control stations became the great German weapon in the
struggle for the Atlantic. It must be remembered that liners
carrying passengers and relying on them to pay the bulk of
the cost are enabled to carry goods at cheaper rates in consequence and thus the development of the liner traffic based
on emigrants assisted the whole German export trade to
North and South America.

A great deal of the success of the German mercantile marine
was due to the organization of German shipping. Over sixty
per cent. of Germany's shipping was held by a group of ten
lines working with one another, and those lines outside the
ring were in close relation with the original ten constituting
the "Reederei-Vereinigung." A contract with one of these
lines was a contract with the group, each line was prepared
to support every other against foreign flags, and they brought
their united weight to bear upon their rivals. German
competition was felt in all the ocean trades and it affected
the British entrepôt trade and transhipment business.† The
general effect of the development of the national mercantile
marines was that although the volume of trade carried in
British ships was increasing, the proportion of the world's
trade carried in British ships tended to decline.‡

(5)—*Combination in the Shipping World.*

A new feature became noticeable in British shipping in
the eighties, viz., the growth of rings or conferences fixing

---

*In a letter to the Foreign Office, 5th November, 1913, certain of
the British lines wrote as follows: " The arbitrary action of the agents
of the German lines has taken the form in the past of compulsorily
separating friends and relations at the control stations." Cd. 9092, *p*. 9.
†Cd. *p*. 88.
‡*See* Table, *p*. 75, *op. cit.*

rates by agreement. After 1873, there ensued a period of great depression in shipping. Steam tonnage had become more effective than sailing ships, steamers could make more voyages, and therefore added considerably to the tonnage available. The Suez Canal shortened the route to India and again made more tonnage available. Meanwhile ships were constantly being produced which made still more tonnage available and as the new ships were increasingly equipped with the latest type of engines and were built of steel they, too, added to the effective tonnage competing for freights. On top of this came the foreign subsidizing which called into existence a certain amount of tonnage which would otherwise not have existed, and supported by the Government, it could work at low rates. There was in consequence of all this tonnage a cut-throat competition and a great fall in freights.*

This great drop in freights, due to the cutting of rates through competition, made shipping unremunerative and highly speculative. Shipping became a sheer gamble. The result was that shipping rings or conferences were formed to stabilize rates and to introduce some sort of sound basis for the industry.† There was a ten per cent. charge called *primage*, made to merchants for the use of the ship's gear in loading and unloading. The custom grew up of giving this

---

*The following will serve as instances. " Report on Agricultural Depression, 1894," C. 7400, II, *p.* 662.

	Jan., 1874	1880	1884	
New York to United Kingdom or Continent per qr. (grain)	10s. 6d.	6s. 0d.	4s. 0d.	
Philadelphia to United Kingdom	10s. 6d.	—	4s. 0d.	
		Jan., 1881	1885	*Jan.,* 1892
San Francisco to United Kingdom, per ton (wheat)	67s. 0d.	70s. 0d.	40s. 0d.	22s. 6d.
	Jan., 1872			*Feb.,* 1893
Odessa to United Kingdom	45s. 0d.			7s. 3d.
Coal from Wales to Aden (ton)	27s. 6d.			11s. 0d.
Coal from Wales to Bombay (ton)	24s. 0d.			11s. 0d.

†On the whole question, Royal Commission on Shipping Rings, Cd. 4668.

ten per cent. back to merchants who shipped regularly with the same company. Out of this arose the deferred rebate by which the conference lines worked their monopoly. When a number of ships entered into a conference or pool they fixed the rates of charge and in order to prevent their customers shipping by other lines outside the pool they gave a rebate to those shipping regularly by them for six months, but this was not paid for another six months, so that there was always six months' rebate in hand which was forfeited if the person to whom it was owed shipped by a line outside the conference. They had thus a certain amount in hand to guarantee "loyalty." The first conference was formed in 1875 to regulate the trade to Calcutta and it started the deferred rebate system in 1877. In 1879 the China conference was formed, in 1884 the Australian, in 1886 the South African, in 1895 the West African and the North Brazilian and in 1896 the River Plate and South Brazilian and the West Coast of South America in 1904. The British coasting trade was not in a combine because they had to meet the competition of the railways. In the North Atlantic trade a conference existed for the passenger service only. These conferences not merely fixed freight rates but they entered into understandings or agreements to respect each other's spheres of influence. " Thus an understanding is said to exist between the Rangoon and Calcutta Conferences under which the lines of each abstain from trespassing on the domains of the other."* The conferences existed mainly in the export trade, there were few in the home import trade. The outward cargoes, being chiefly manufactured goods, they were despatched in small quantities and it would not pay a merchant to charter a whole ship ; he therefore took the terms he could get for the portion of a ship. On the homeward journey the cargoes were wool, corn, rice, ore and timber for which a whole ship could be chartered, and here it was the rates at which tramps could be hired that to a large extent determined the price, and the rates at which tramps could be chartered varied and it was impossible to keep up a regular rate of charge when faced with tramp competition. Yet in some cases conferences did work even in the homeward trades. The advantages claimed for the shipping rings were the provision

*Cd. 4668, *p.* 12.

X

of regular sailings and stable rates of freight. They were not high at one time and low at another. With greater steadiness of rates a better type of steamship was said to be provided and there was greater economy of management, hence freights charged were on the average lower. Charter parties tended to become standarized for each route and with the growing regularity of sailing and arrival it was possible to reorganize the dock labour at the ports so as to de-casualize much of it. The whole object of the conferences was to abolish rate wars with the inevitable wastes they entail. On the other hand, the great objection urged against the conferences was that they sometimes charged lower rates from the continent or from the United States than they did from British ports and thus affected unfavourably the export of British goods by giving a sort of bounty to foreign goods in the shape of low rates. In some cases it was held that by lowering rates on foreign goods the shipping rings nullified the effect of the colonial preference system.* The shipping companies urged that they gave these low rates to prevent foreign steamship lines entering the trade. So great was the objection to the rebate system which tended to limit the shipping competing to South Africa that the South African government, in 1912, refused to give the mails to any line working on this system.

With the growing competition of Germany and the great pressure her organised lines were able to bring to bear conference agreements were entered into with German ship-owners. Divisions of territory were arranged by which the United Kingdom trade was confined to British lines while the German trade, and sometimes that from Dutch and Belgian ports, was reserved to German lines. It was an arrangement " to keep off one another's territory."† The Germans did not, however, observe their agreements loyally, or rather they observed them "only so far as it paid them to do so." Where an important object could be served by disregarding an agreement an excuse was generally found. It would appear that the British lines " preferred to put up with a certain amount of evasion rather than engage in a ruinous rate war."‡ It is interesting to observe the same tendency to amalgamation becoming prominent in shipping as it had previously

*Dominions Commission, Cd. 7210 (1911) §95-98.
† Cd. 9092, *p.* 103.   ‡ Cd. *p.* 105.

become prominent in railways. There is the same objection to a monopoly and the same answer that combinations secure a more efficient service.

British shipping was vitally affected by the war, in which her losses were disproportionately heavy since she lost the finest and most efficient types of ships. Between August, 1914, and August, 1917, over three million tons net of British shipping and one million of Allied shipping was destroyed * There was a great loss of specialized vessels such as tank steamers and refrigerating ships. The withdrawal of the ships from the distant trades to economize shipping for the near-by trades had the result that other countries, notably Japan, filled the gap. British shipping was limited in its profits by the fixed rates under Government control and by the taxation of eighty per cent. of the Excess Profit Duties ; neutral ship-owners, such as the Dutch and Scandinavians, were not so limited and were able to realize enormous profits which it was feared would enable them to build and run ships in effective competition with British ships. The loss of shipping was practically all in the ocean trades and these long distance voyages† were England's specialité. Great Britain, therefore, came out of the war with her great position threatened by the rise of two very important rivals, the United States and Japan. While she has emerged from the war with less tonnage and much of the replacement of a less efficient nature she has also emerged with a productive capacity for ship-building‡ that is in excess of anything she possessed before the war. She has also gained by the elimination of a rival whose unscrupulousness and efficiency combined to make her a formidable menace to an industry

*Cd. 9092, *p.* 58.

†Cd. *op. cit., p.* 62.

‡Cd. 9092, §64. " With regard to the supply of Ship and Engine Forgings after the war representatives of the Forge Masters assured us that, taking into account the new plant which they had added during the war, their power of production after the war will far exceed any prospective home demand. They stated that they had no fear of fair competition from abroad, but that the " dumping " which had taken place before the war had severely injured their trade, and might have led, but for the orders of the Admiralty and others who called for British materials, to the final closing of their works. This dumping they held was largely a political move designed to destroy the British Forging Industry." *p.* 27.

which was a vital necessity to an island people. It is possible that the development of oil as a fuel for steamers may influence British shipping adversely since coal formed such a large proportion of the tonnage carried in British ships. In 1913 no less than 65.6 million tons of coal were shipped to Europe and the Mediterranean and 10.6 million to countries outside Europe. The coal sent to countries outside Europe was carried almost exclusively in British vessels and prevented ships which brought in food and raw materials going out in ballast. Since this country requires a great many bulky products, such as food and raw materials, she requires for her service more inward space than outward. Therefore if ships can fill up on the outward voyage with coal the price of the goods brought home does not have to cover the cost of both voyages. The United Kingdom, therefore, received her food and raw materials under specially advantageous conditions as regards freight charges.

(6)—*The Government and Shipping.*

Although the whole development of British steam shipping has been on free trade lines, signs are not wanting that if necessary the Government would intervene to protect its shipping against a concerted attack by foreign governments. When the *Kaiser Wilhelm* secured the record for the fastest Atlantic passage and when it seemed as if the blue ribbon of the Atlantic might pass to Germany, the United Kingdom gave, in 1903, a loan of £2,600,000 to the Cunard line at 2¾ per cent. to build two turbine vessels of twenty-five knots, which proved themselves the fastest things of their kind afloat, —the *Lusitania* and the *Mauretania.* Again, when there was a question of developing trade between the West Indies and England, a subsidy of £40,000 a year was given by Jamaica and England for the years 1900-1910 to the Elder Dempster line to provide the shipping facilities required. At the colonial conference of 1907 the British Government also agreed to subsidize ships for an " all-red route " but no further steps were taken. The English load line was raised in 1906 to the same height as the German to put English ships on the same footing as German ships as regards carrying capacity. It was suggested by the Committee on Shipping Subsidies, in 1902, and frequently at colonial conferences, that the inter-imperial trade should be again reserved for imperial shipping. The United Kingdom has never repealed the clauses of the Act 16 & 17 Vict. c. 107 §324-326 by which Her Majesty by Order in Council, was endowed with right of retaliation against any foreign country that penalized the shipping of the United Kingdom. Only for sixty years (1854-1914) has British shipping been completely free from government control except for the safety regulations imposed by the Board of Trade and the requirements laid down for the protection of seamen as to wages, accommodation and food.

The following figures show the relative superiority of the various shipping nations in 1914 and 1925:

WORLD'S SHIPPING.

Sea Going Steel and Iron Steamers and Motor Vessels.

Country.	Gross Tonnage.*		Percentage of World Tonnage.	
	July 1, 1914. Mill. Tons.	July 1, 1925. Mill. Tons.	July 1, 1914.	July 1, 1925.
The World ..	42.5	58.8	100.0	100.0‡
British Empire	20.3	21.5	47.7	36.6
U.S.A.† ..	1.8	11.6	4.3	19.7
Japan ..	1.6	3.7	3.9	6.3
France ..	1.9	3.3	4.5	5.6
Germany ..	5.1	3.0	12.0	5.1
Italy ..	1.4	2.9	3.4	4.9
Holland ..	1.5	2.6	3.5	4.4
Norway ..	1.9	2.6	4.5	4.4
Sweden ..	1.0	1.2	2.3	2.0
Spain ..	0.9	1.1	2.1	1.9
Denmark ..	0.8	1.0	1.8	1.7
Greece ..	0.8	0.9	1.8	1.5
Belgium ..	0.3	0.5	0.7	0.9
Other ..	3.2	2.9	7.5	5.0

The increase in motor vessels has been the most striking development since 1914.

						Nos.	Tonnage.
Great Britain and Ireland			..	..		305	754,495
Norway	..	..	..	..	..	233	345,965
Sweden	..	..	..	..	..	211	277,947
Germany	..	..	..	..	..	196	275,656
U.S.A.	..	..	..	..	..	197	267,119
Denmark	..	..	..	..	..	112	191,837
Italy	..	..	..	..	..	96	142,158
Holland	..	..	..	..	..	128	138,397
Other	..	..	..	..	..	667	320,499

"Lloyd's Register, 1925-6," Table 13.

* "Lloyd's Calendar, 1926." *p.* 378 : Wooden and composite ships not included. In 1925, they comprised 1.1 million tons.
† Steel and iron steamers on lakes not included amounting to 2.5 million tons.
‡ It must be remembered that in 1925 much tonnage was laid up. According to the Chamber of Shipping the tonnage laid up in the principal ports of the United Kingdom on July 1st, 1925, was 420 ships of 777,000 tons, i.e., an equivalent of 4 per cent. of the gross tonnage. At the same date the tonnage laid up abroad was six million tons or 10 per cent. of the whole. "Lloyd's Calendar, 1926," *p.* 386.

PART VI

THE INDUSTRIAL AND COMMERCIAL REVOLUTIONS AND THE NEW CONSTRUCTIVE IMPERIALISM.

SYNOPSIS

The coming of machinery and mechanical transport gave a new value to continental colonies.

I.—PERIODS OF COLONIAL HISTORY.

1—The first Empire and its disruption. 1603-1776, Causes of the revolt of the thirteen continental colonies.

2—The Period of Drift, 1783-1870. General dislike of colonies. England organized for world exchange had no use for the narrower limits of the colonial system.

3—The creation of new colonial values by the development of mechanical transport, 1870-1895. The scramble for colonies by European powers. The new chartered companies a link between the old policy and the new.

4—Reaction from world economics to imperial economics, 1895-1920. The period of constructive imperialism. Chamberlain as Colonial Secretary. Great Britain influenced by the growing rivalry of Germany, the colonies influenced by the presence of Russia and Germany in the Pacific and the rise of Japan.

II.—THE EMPIRE IN ALLIANCE

1—Conferences, 1887, 1897, 1902, 1907, 1911. Imperial War Conferences, 1917 and 1918. The War of 1914 accelerated Imperial Consolidation. India admitted to the Conferences.

2—Communications: Penny Postage. Cables.

3—The Preferential System.
    (a) Tariffs. Denunciation of the Belgian and German treaties by Great Britain in 1897. Preferences given to British goods by the self-governing colonies and the West Indies either by surtax or by rebate. Extension of the system to the inter-imperial trades. Subsidized mail services in the inter-imperial trade.
    (b) Preferences in finance. Through the Colonial Stock Act the Colonial Governments obtained from Great Britain cheap loans. Preferences in income-tax treatment.
    (c) Tariff Preference given by Great Britain in 1919.

4—The development of inter-imperial trade.   The appointment of
trade commissioners to act for the Empire, 1908,
1917.

5—The development of the resources of the Empire.
    (a) The Dominions Commission.
    (b) The importance of rapid sea communications.   Imperial
        Shipping Committee.
    (c) The appointment of the Bureau of Entomology, 1913, the
        Imperial Mineral Resources Bureau, 1918, and the
        Imperial Economic Committee, 1925.

III.—THE EMPIRE IN TRUST.

The development of the undeveloped estates.

1—The financing of railways, Uganda, West Africa, Sudan.   Colonial
Loans Act, 1899.

2—Encouragement of scientific agriculture and experimental stations.
Grant of £10,000 a year for cotton growing experiments.

3—The encouragement of investigation into tropical diseases.

4—Preferences on export of raw material.

5—The changing position of India from the Empire in Trust to the
Empire in Alliance.
    (a) Admission to Imperial Conferences.
    (b) Permitted to increase cotton duties.
    (c) Trade Commissioners for India.
    (d) Adjustment of Indian migration within the Empire.

6—The advantages of being a member of the British Empire.

THE industrial revolution had created a demand for new
commodities, increasing quantities of raw material were
required, markets were needed for the new mass production,
a new commerce was inaugurated which in its turn made
demands on new forms of transport.   Transport again
quickened the whole volume of transactions and stimulated
a new industrial and commercial development which proved
to be a veritable commercial revolution in that it altered
the relative value of the commodities which were the subject
of commercial dealings ; it brought new articles into commerce and created a further demand for raw materials,
food-stuffs and markets.   The result of the industrial, combined with the commercial revolution was to give a new value
to colonies and to inaugurate a new scramble for the
unoccupied territory of the world on the part of all the Great
Powers.   As far as the United Kingdom was concerned a
great change took place in her attitude towards the overseas
possessions.   At the beginning of the century she regarded
them as burdens, at the end as assets of value, and they
proved to be one of the great factors in producing that reaction

from laissez-faire which is so characteristic of the modern economic policy of States.

The function which the Empire played in the commercial revolution was to provide increasing quantities of raw material increasing quantities of food, increased employment for shipping in the long distance trades, a great field for the investment of British capital abroad and important markets for British manufactures.

I.—PERIODS OF COLONIAL HISTORY.

The history of the economic relations between England and her colonies may be divided as follows :

1—1603-1776.	The Old Colonial System.
2—1783-1870.	Colonial laissez-faire.
3—1870-1895.	Reaction due to foreign competition.
4—1895-1920.	Constructive Imperialism.

(1) 1603-1776. *The First Empire and its Disruption.*

Under the old colonial system colonies were regarded as estates to be worked for the benefit of the mother country. England had a small population of about 4-5 millions in the seventeenth century ; why should she allow her people to leave and be burdened with the cost of defending them from Spain, Holland or France unless they were going to develop her power to a greater extent by leaving the country than by staying ?   From the very outset it was understood that the colonies must help to form a self-sufficing Empire by supplying England with the tobacco she would otherwise have to get from Spain, or the sugar she would have to get from Holland.   With the great shortage of raw materials in the eighteenth century, colonies acquired a new value as producers of timber, naval stores, flax, cotton and silk, and bounties and other forms of encouragement were given on their production.   The colonies had to get their manufactures either from England or *via* England and the few competing industries, such as cloth or hats that were started in the colonies were discouraged.   The plantations were not to compete with Great Britain, only supplement. After all, England had the burden of their defence and they must not cripple her resources.   These restrictions were enforced through the Navigation Acts and Acts of Trade.   If foreign ships could not frequent the colonies and if colonial ships could not trade with foreign countries, then manufactured goods were bound to come from or through England and tobacco and sugar were bound to come to England. The colonies were favoured by a preferential tariff in the British market and in the case of tobacco, English tobacco growing, which was a considerable industry, was suppressed.

A great deal of independence had been allowed to the colonies in other ways ; they enjoyed a large measure of political independence and complete religious toleration. The British colonial system was far more liberal in every respect than that of Holland, France or Spain. This very liberality made the colonies develop a vigorous political life of their own and in the continental colonies a sense of nationality was created which resented any expression of overlordship on the part of England. The Northern colonies had been peopled with religious dissenters who had shaken the dust of a godless England from their feet ; they felt they owed her no loyalty since she had driven them out by her adherence to wrong courses, for which she would assuredly feel the vengeance of the Lord. These colonies were also recruited from political dissenters, adherents of the King, who did not find the Cromwellian régime to their liking, and were disappointed when Charles II. returned and it was found that he could not compensate everyone. Adherents of Cromwell found the England of the Restoration an undesirable place to live in. A stream of emigration to America set in from the North of Ireland and again these men, driven out, as they considered, by Test Acts and the commercial restrictions imposed upon Ireland by England, felt no loyalty to the land they quitted. Foreigners of many kinds formed a substantial element in the new population and Huguenots and Palatinates could not be expected to feel themselves part of England, or be willing to sacrifice anything for her sake. Nor were the first two Georges the type of men to inspire any deep personal loyalty. Labour was provided partly by free emigration but also largely by kidnapping. Convicts were sent out to the American colonies and many persons who could not pay their passage were sent out as indentured servants and auctioned at the quay-side for service for a term of years. As labour was scarce negro slaves were also imported.

It is impossible to understand the revolt of the continental colonies in the eighteenth century and the loyalty of the English colonies of the nineteenth unless one realizes the different elements which constituted the two Empires. The bulk of the people who went out in the nineteenth century went out merely to better themselves ; they were proud of being British, they felt a great loyalty to Queen Victoria. Steamships and railways abolished distance and the emigrants were able to keep in touch with " home " in a way that was impossible in the seventeenth century ; in the nineteenth century, emigration ceased to be exile. The bulk of the Irish emigration went to the United States and there was a homogeneity of population in the nineteenth century Empire which was lacking in the eighteenth century, when colonies were so largely composed of foreigners and dissentients.

The old colonial system envisaged two types of colonial possessions, a trading Empire based on the production of staple tropical products, and a colonial Empire which gave opportunity for the expansion of the race in new and un-occupied countries. Of the two the former was considered by far the more important. This Empire consisted in the seventeenth century of trading posts in India, of trading posts in West Africa and of some of the West Indian islands. In the colonial Empire resting on racial expansion, Virginia, which grew tobacco, was one of the regions most favoured in British eyes. Spices, sugar, tobacco and cotton, these were the great staples that England was anxious to secure. West Africa was valuable as furnishing the labour which grew the sugar and tobacco. The most favoured spots of the whole Colonial system were, however, the West Indian Islands ; they not merely supplied the goods England needed, but they did not compete or attempt to compete with English products as did New England ; they were big customers for her manufactures and gave great employment to shipping.

The continental colonies although developing a vigorous life of their own, remained loyal to England because they were afraid of the French and because they could not agree among themselves for any concerted action. The fear of the French was removed in 1763, when England acquired Canada. In reorganizing the defence of the colonists against the Indians, who had hitherto been kept in check by the French, further

expenditure was necessary on military grounds. Great Britain considered that the colonists should help to bear part of the burden and tried to get it by taxation. During the war with the French, the Northern colonists had consistently smuggled food to the French and prolonged their resistance against the English. The British Navy was deputed to prevent this illicit trade, and after the peace they continued to stop the evasions and smuggling under the Navigation Acts. Thus, just as the colonists were freed from the fear of the French they became acutely aware of England's overlordship, both by the action of the British navy and by the proposed taxation. However justifiable that taxation may have been, it was an expression of the right to tax a people who were becoming conscious of a separate nationality. When a people is in that mood, any demand, whether it be really harmful or not, is a burden that cannot be borne—everything except perfect freedom is a grievance. The English economic restrictions were few and light, but they were stigmatised as intolerable by a people, the majority of whom had from their very origin no inherent loyalty to the English tradition and no pride in English achievements. It is doubtful whether Englishmen at that time realized the importance of the loss of a large part of a continent and the South had not begun to grow cotton so that her value as provider of raw material for the new cotton manufacture was as yet undreamed of. To most people the Antilles were the really important part of the English system; had Jamaica revolted Great Britain would certainly have put forth more effort to retain her than she did to retain North America. But the continental colonies were only a strip of coast-line ; the great inland penetration was as yet unthought of, and the rebellion in America soon became only a side show of the great struggle with France. When France joined in to support the colonists, she was followed by Spain and Holland. Great Britain then devoted her efforts to sweeping her two great rivals off the seas. French finance became so involved that the King was obliged to call an Assembly, which precipitated the Revolution and put England's greatest industrial rival out of action. The commerce of the Dutch was destroyed and the Dutch East India Company was ruined. Great Britain emerged, with, it is true, the loss of a continental strip of coast-line in which Virginia had been the most important region, but she was able to offset to that the fact that she was left without a rival to carry on the overseas commerce of the world.

(2) 1783-1870. *The Period of Drift.*

The effect of the successful rebellion of the thirteen continental colonies was to produce a great change in English colonial policy ; colonies began to be regarded with dislike and distrust. It was considered to be inevitable that when strong enough they would " cut the painter." Why should Great Britain undertake the onerous duty of defending them, why not let them be free ? If the United States had revolted because of the economic restrictions, why not leave the colonies alone to carve out their own economic path ? There was a pessimistic feeling that colonies were " no use," and that this country would be better off without them. With Great Britain's monopolistic position in manufactures and shipping, they would still have to trade with this island as before.

The generally depressing outlook was accentuated by the agitation against slavery. If slavery were to cease, what would be the value of West Africa, and what would become of the West Indies whose labour supply and sugar growing was bound up with negro labour ? South Africa, too, rested on slavery, and was held to be an undesirable possession, although it was the great strategic post on the high way to India. The slave trade was abolished in 1807 and slavery in the existing British possessions in 1833, and this created further friction with the colonists in both South Africa and the West Indies. The compensation was inadequate and they lost control of their labour supply.

A preference system existed between the mother country and the colonies and to free traders this served to make them still more undesirable since they had involved this country in mistaken economic courses.

" England is sufficient to her own protection without the colonies and would be in a much stronger as well as more dignified position if separated from them than when reduced to be a single member of an American, African or Australian confederation. Over and above the commerce she might equally enjoy after separation, England derives little advantage except in prestige from her dependencies, and the little she does derive is quite outweighed by the expense they cost her and the dissemination they necessitate of her naval and military force, which in case of war or any real apprehension of it requires to be double or treble what would be needed for the defence of this country alone."

Such was John Stuart Mills' opinion. " Any party," it was said, " would rather lose a colony than a division." A royal commission in 1865 passed a resolution that it would be unwise to extend the British dominions in West Africa and said that the great thing was to give over the administration more and more to the natives prior to our withdrawal from the West African coast. Froude, writing in 1888, speaks of a conversation he had " seventeen years ago " with an official. He " informed me that a decision had been irrevocably taken. The troops were to be withdrawn from the islands and Jamaica, Trinidad and the English Antilles were to be masters of their own destiny, either to form into free communities like the Spanish American republics or to join the United States or to do what they pleased with the sole understanding that we were to have no more responsibility. . . . I was told . . . that it had been positively determined upon and that further discussion of a settled question would be fruitless and needlessly irritating."

It is true that the anti-colonial feeling was not unanimous. A little group of enthusiasts, consisting of Wakefield, Molesworth, Buller, and others, strenuously advocated the value of colonies, but the great bulk of the educated people who regarded the world as their sphere had no use for the narrower limitations of a colonial system resting on preferences. One after another these preferences disappeared as the free trade movement gained its victories in the years 1842 to 1860. Self-government was granted as a step on the road to the complete independence which *The Times* on February 11th, 1850, declared to be " an inevitable event. '

The period was one of colonial laissez-faire. Colonies were looked upon as an antiquated encumbrance from the past.

(3) *The Creation of new Colonial Values by the Development of Mechanical Transport.*

The years 1870-1895 witnessed a reaction. Gold had been discovered in Australia in 1851, and a considerable emigration had set in ; the wool exports from the Antipodes were the mainstay of the British woollen industry which was being rapidly converted to machinery in consequence. As markets the colonies became increasingly valuable after 1870, France began to try to obtain compensation for the loss of Alsace and Lorraine by an extension of her colonial possessions in Africa and Asia. Italy, Belgium, Portugal and Germany also began scrambling for Africa. The railways and steamships not merely brought all the British colonies nearer to the mother country and so abolished the great barrier of distance and exile, but they made all the world approachable for all countries. The possibility of penetrating interiors gave a new value to continents. People instead of staying on a fringe of malarious coast-line could go inland. The real value of the old colonial system had centred in islands ; the new colonial system was concerned mainly with continents. Germany, in a wave of nationalism and sentimentalism, set out to acquire territories in Africa under the impression that colonization was good business and had contributed to England's greatness. France was seeking in Africa a compensation for Alsace-Lorraine and a revival of her prestige. Belgium, a great manufacturing nation, needed an outlet for her surplus manufactures. It seemed to the British as if these protectionist nations would acquire territory and would not pursue the policy of the " open door." Great Britain would then be shut out, or her world interests would suffer with a series of protectionist tariffs against her. Therefore to preserve the " open door," she, too, abandoned her expressed determination not to acquire another yard of territory in Africa, and joined in the partition of the eighties.

While railways and steamships gave a new value to colonies in their new accessibility and expansive possibilities inland, they were soon to acquire a new value as markets and as providers of raw materials for the great increase of manufacturing consequent on the railway and shipping developments.

One effect of the great depression of the period 1873-1886, was an " over-production " in excess of any prices that would pay. Germany had become a manufacturing nation ; the United States had filled her own market and the continent had gone back to protection, making it more difficult for British goods to find an entrance.

It was suggested in the minority report of the Depression of Trade Commission that the United Kingdom might find some relief in the formation of an Imperial Zollverein or Customs Union, in which the Empire would be united by preferences against the protectionist world outside.

Imperial federation, economic or political, was a topic much discussed, and the new position the colonies were beginning to assume found expression in the fact that the first colonial conference was held in 1887, after the Jubilee of Queen Victoria. *The Times*, in 1887, said : " In these communities, as we are all beginning to feel, there is a great reserve of strength for the mother country." The Press woke up to the fact that we were owners of a considerable portion of the world and published new lists or explanatory accounts every morning of the British possessions. Mark Twain's comment, " And the meek shall inherit the earth," was not regarded as a joke, but as a just appreciation of the situation rather remarkable in a foreigner.

While this great change was going on in public opinion certain people had been acting in advance of the change. Great Britain would not have obtained the considerable share of Africa that became her portion had not claims been staked out previously by chartered companies. The East India Company, the Royal African Company, the Levant Company and the Hudson's Bay Company had, in the sixteenth and seventeenth centuries, undertaken the pioneer work of opening up new trades. The Crown was too poor to act and individual merchants and others joined together for the venture. They sought out the best trading areas they could find so as to recoup themselves. Their trading posts in the case of the East India Company and the West African Company expanded into territorial possessions. It has been said that the British Empire was acquired in a fit of absence of mind. There was not much absence of mind about the worthy merchants who embarked their money in chartered companies—they went for the paying areas and disputed them with the French and the Dutch. They lost money and they made money, but they were not absent-minded. Chartered companies were also started in the seventeenth century to take out colonists who alone could make the territory in a new country valuable, and although colonization companies did not survive the seventeenth century, the trading companies did, and the tradition of expansion by chartered company without government aid and in advance of the government was never lost. Even in the nineteenth century period of laissez-faire, colonial companies were formed to colonize new areas such as South Australia, New Zealand and Canada* by taking out emigrants and regulating the disposal of the land.

When railways and steamships opened up new possibilities for continental development, new companies were formed. The British North Borneo Company received its charter in 1881; the Royal Niger started as the National African Company in 1882, and obtained its charter and its new name in 1886; the British East Africa Company followed suit in 1888 and the British South Africa Company in 1889, and all formed a bridge between the anti-colonial laissez-faire period and the new constructive period. They pegged out a claim for Great Britain in these regions at a time when the home government had no decided colonial policy. They were a compromise between taking over the territories in question by the British Government and letting foreign nations acquire them. The object of the new companies was to establish British control by private initiative over regions that would otherwise have been annexed by a foreign power. They had no monopoly of commerce as had the old chartered companies and they differed from them in yet one other important particular—they dealt with continental interiors whereas the sixteenth and seventeenth century companies had been concerned either with islands or with trading posts on a sea front or river.

The general result has been that large areas were acquired, opened up rapidly and new markets and sources of raw material exploited at very little cost to the Imperial Government. In taking over the territory the Government paid the Royal Niger Company £300,000 in 1900 and the East African

*1—Canada Company, 1825. 2—South Australia Co., 1834. 3— New Zealand Co., 1837-1850.

Company £50,000 in 1895 for their territory with £200,000 in addition for their claims against Zanzibar, and the shareholders took out their dividends in philanthropy.

It was the good fortune of this country to possess in the eighties a group of able men, Taubman Goldie (Niger), Cecil Rhodes (Rhodesia), Mackinnon (East Africa), and Dent (North Borneo),* who were able to combine financial and imperial interests and so secured the predominance of England on the African Continent at practically no cost to the nation.† The chartered companies were skirmishers in front of the main body of organized British possessions and were the link between the period of disinclination to acquire more territory and the idea of the development of the undeveloped estates formulated by Chamberlain. Promoters of the chartered companies devised a way by which the government could annex territory without seeming to do so by reviving methods considered antiquated and monopolistic in the free trade days of the seventies. Parliament would vote money to maintain rights whereas it would not grant money to acquire new areas. The question of reconciling the necessity for a new expansion without laying a heavy burden on the tax-payer was solved by the chartered company stage.

*Of this last company it has been said that " it is remarkable in that it acquired its territory from an American citizen, its charter from a Liberal Government and that it marks the re-establishment of conditions which everyone had believed had disappeared as the last of the old monopolies."

†Mr. Chamberlain's verdict in a speech in Parliament, 13th February, 1896, shows the importance of the work done by the South African Company. *Hansard*, 4th ser., Vol. 37, *p.* 223:
" I cannot conceive that such a Department could do the work that had been done by the existing chartered company or by any of its successors. I am perfectly sure that if the persons responsible for the development of these territories had to go as I have had to go, over and over again, to the Treasury to ask their assent to an expenditure of £5, it would have been perfectly impossible for them, or for anybody in my position to have done what the Chartered Company have already done or another Chartered Company might have done in their place, to make railways, to make hundreds of miles of roads, to do everything to bring into rapid occupation the territories which have been submitted to their rule. Therefore let the House understand, as to this question between the Chartered Company and the Government that you may have if you like, a system which may be more controlled by the House of Commons but you will not have in it a system that will be productive in the long run of the success or speedy development of these untried countries."

(4) *Reaction from World Economics to Imperial Economics.*
1895-1920.

In the nineties a new era began—the era of constructive imperialism. It meant a change from a sea psychology to a land psychology.

The United Kingdom was during this period at the parting of the ways. Her world position was challenged. Foreign governments were throwing all their weight on the side of their own people ; they were subsidizing steamships, using the railway as an effective weapon in the commercial struggle, increasing their tariffs and striving for colonies and markets. With the question of Africa temporarily settled in the eighties, there began a scramble for Asia in the nineties and with that the question of the dominance of the Pacific was raised. The Siberian Railway had brought Russia to the Far East. Japan was developing as a great maritime power, the swiftness of modern sea communications put Germany within striking distance of Australasia once she had obtained a footing in the East Indian archipelago. The Germans, with their possessions in South-West Africa and East Africa, were an ever-present menace to South Africa, holding her, as it were, between two half-closed nippers. The British were no longer an unchallenged world people, and the British colonies were no longer safe from foreign aggression. The day of small economic units was over. Railway transport had created three great land empires : the United States, Russia and Germany, while sea transport had developed a fourth, Japan. Could Great Britain reorganize her vast sea and land possessions on new lines and evolve a new colonial system which should be her bulwark and defence in the growing economic struggle ? Could she so organize her quarter of the earth that each part, though scattered, would supplement the other and prove a unit big enough to weigh in the scales against the other giants ? The colonies, too, realized that their great unoccupied territories were a standing invitation to aggressive or overcrowded neighbours and that their immunity was gone. This was even more marked when it became obvious between 1904 and 1905 that Japan could beat a first-class power like Russia. The dominions began to be willing on their part to fit in to some common scheme for defence

or trade.   Their readiness to do this was accelerated by the
pride in a great race tradition common to all.

The United Kingdom had to choose between abandoning
laissez-faire in colonial matters in order to establish some
closer tie with her overseas possessions and trusting to being
able to hold her own in the new rivalry as she had done in
the past.   Should she fail to do this her fate would be that
of Holland or Greece, an economic football for the Great
Powers.   Would Great Britain subordinate her world position
to the imperial idea with the possible retaliation and loss
involved?   Would she try to strengthen her economic
defences by alliances with the dominions and the development
of dependencies or would she still trust to the policy of letting
things take their course?   That was the great problem of
the nineties.   In 1895 Chamberlain came to the Colonial
Office with the view of not letting things drift, colonially
speaking, and he remained in office eight years.   When he
left the government in 1903 the new lines of constructive
imperialism were laid down and England had begun to think
in terms of land development and not in terms of the sea
approach.

The difficulty about the co-ordination of the British Empire
is that it is really two Empires and not one.   There is no
such homogeneity as is to be found in the United States,
Russia and Germany.   One part consisted of the regions
where white men could settle and rear children and make a
home, and these had become self-governing dominions.
They were, however, sparsely populated.   They reproduced
the institutions and language of the mother country but had
developed their own tariff system on protective lines while
the mother country remained free trade.   They constituted
with the mother country a community of loosely allied nations
the tie being largely one of sentiment with a growing interest
in the necessity for a common system of defence.   These
self-governing dominions constituted an Empire of autono-
mous States and may be termed " the Empire in alliance."
It has also been called the " Empire of Settlement," and the
" British Commonwealth of Nations."

On the other hand, the United Kingdom ruled over a
densely populated tropical and semi-tropical area where no
self-government had been evolved and where the coloured
inhabitants were still in tutelage.   They were governed by

the mother of Parliaments autocratically for their own good and were compulsorily free trade.*

This is " the Empire in Trust," or " the Empire of Rule."

Its great economic value lay in furnishing such valuable products as tea, coffee, cocoa, sugar, rubber, fibres of various kinds, such as cotton, hemp and jute, edible oils and nuts and spices of all kinds.

Here were two Empires governed on entirely different principles, differing in colour, in racial origin, at different stages of economic development, comprising in one whole at one time the peoples in the stage of development of all centuries from the fourth to the nineteenth, and whose inhabitants ranged from cannibals to Prime Ministers. Was there any possible ground of union between regions that lay centuries asunder in their economic civilization ? Was there, further, any possible connection between even the self-governing dominions themselves when one compares the problems of Canada with its long winters, and Australia with its heats and droughts ; Newfoundland with its fish and fogs, and South Africa with its gold and ostrich feathers and its coloured population. Its size and the magnitude of its trade may be judged from the following figures :

EXTENT OF THE BRITISH EMPIRE

(*Statistical Abstract*, 1915. *Cd.* 7827).

AREA : 11,273,000 square miles of which the
    United Kingdom is -   -   121,142 square miles
    Excluding
      Egypt -   -   •   • 350,000 ,, ,,
      Sudan -   -   •   • 984,520 ,, ,,
      Protected Malay States -   - 146,000 ,, •

POPULATION : 417,268,000 (1911 census) excluding—
    Egypt -   -   -   - 11,189,978
    Sudan -   -   -   • 3,380,531

POPULATION OF INDIA : 315 millions.

*Mr. Harcourt, when Colonial Secretary, in a speech in the House of Commons on June 12th, 1912, described himself as a " despot under democracy." " The position of the Colonial Secretary on the Crown colony side of his Department carried with it the powers, duties, responsibilities and anxieties of a practical and laborious despot controlled only by the forces of nature, by his own discretion and by the sporadic curiosity at question time of friends or opponents inspired either by imagination or information."

WHITE POPULATION OF THE EMPIRE (1911 census) :

United Kingdom	-	-	-	45,000,000
Australia -	-	-	-	4,455,005
New Zealand	-	-	-	1,008,468
Canada -	-	-	-	7,204,838
Newfoundland	-	-	-	238,670
South Africa (total population 5,973,394)			1,276,342

AREA OF SELF-GOVERNING COLONIES (approx.) : 7 million square miles.

TOTAL TRADE OF THE EMPIRE IN 1913 (Statistical Abstract, 1915) :—
    (a) with foreign countries   -   £1,557,159,000, *i.e.*, 73.8%
    (b) Inter-imperial -   -   £ 551,527,000, *i.e.*, 26.2%

    TOTAL   -   £2,108,686,000

The only possible connection between such varying entities was their connection with England as a sort of common meeting-ground on the basis of either common defence or common interests in trade. The tropical and self-governing Empires were both providers of raw material for the great manufacturing centre, England. They produced indispensable food products and afforded employment for shipping in the ocean trades. From Great Britain they received the capital which furthered their rapid development ; they were enabled to enjoy all the security and prestige which came from being a member of the greatest community that the world had ever known ; they obtained through England unrivalled opportunities for the distribution of their products. Although there were solid advantages on both sides there was a large group of people who considered that more ought to be done to create some closer tie, economic or constitutional or both, and that the resources of the Empire should be still more energetically developed for the benefit of the whole. In 1895 Chamberlain deliberately chose the office of Colonial Secretary, hitherto considered second rate and humdrum, to try to carry out a closer connection between the self-governing dominions and the mother country and to develop the resources of the Crown colonies and protectorates, a policy he regarded as self-preservation for the whole.

" It seems to me that the tendency of the time is to throw all power into the hands of the greater empires, and the minor kingdoms—those which are non-progressive—seem to be destined to fall into a secondary and subordinate place.

But if Greater Britain remains united no empire in the world can ever surpass it in area, in population, in wealth or in the diversity of its resources."*

" I have long believed that the future of the Colonies and the future of this country are interdependent."†

He therefore laid down the beginnings of a new policy of constructive imperialism which was continued and developed by his successors.

As the two Empires needed different measures the policy adopted towards the Empire in Alliance may be termed that of giving the loose alliance a definite economic tie ; with regard to the Empire in Trust, the policy pursued was one of development by science and railways.

II.—THE EMPIRE IN ALLIANCE.

The policy of closer alliance was pursued along both the political and economic path. A colonial conference had been held in 1887 on the occasion of Queen Victoria's Jubilee ; another was held at Ottawa in 1894. Chamberlain took advantage of the presence of the colonial representatives at the Diamond Jubilee of 1897 to hold another conference which discussed economic matters affecting the Empire and questions of defences and preferences. Yet another was held in 1902 at the coronation of King Edward, which coincided with the end of the Boer War.

In 1903, Chamberlain had left the Government and the Liberal party came into power in 1906. Although the parties had changed, the policy continued and another colonial conference was held in 1907. Hitherto the conferences had been between the Colonial Secretary and the Premiers, now it was attended by the Prime Minister and the Cabinet, and changed its name to Imperial Conference and at that meeting the conferences were made permanent institutions to be held every four years. The next was accordingly held in 1911.

Each conference resulted in some arrangement for closer economic union, either along the lines of preferences given by the colonies as an offset for defence, the main burden of which fell on the mother country ; or by a closer approxima-

*Speech, March 31st, 1897. " Mr. Chamberlain's Speeches," ed. C. W. Boyd, Vol. II., 1914, *p.* 5.
†Speech at Canada Club on Colonial Federation, 25th March, 1896.

tion to a common commercial law, a common patent law, a common shipping policy and a common emigration policy. Not merely were these meetings held every four years but a permanent secretariat was set up to preserve continuity of policy and to diffuse information in the interval. The United Kingdom thus became the clearing house of imperial policy. The outbreak of war in 1914 postponed the conference due in 1915, but an Imperial War Cabinet was summoned in 1917 which was followed by an Imperial War Conference. To this representatives from India were summoned. It was again affirmed that " each part of the Empire having due regard to the interests of our allies shall give specially favourable treatment and facilities to the produce and manufactures of other parts of the Empire." Representatives of India were present at this conference and a resolution in favour of their inclusion in all future conferences was passed.* The Empire in Alliance, therefore, ceased to be merely a racial alliance of white peoples.

The war, though in many cases it ended one epoch and started another, only accelerated the work of imperial economic consolidation. Another Imperial War Cabinet Meeting was summoned in 1918, followed by another Conference.† It was here agreed that it was necessary to secure for the British Empire the command of certain essential raw materials and a Committee of the Conference was appointed to see how this best could be carried out.

Thus, by the end of the war, the declared policy of the Premiers in Council and the representatives of India and the United Kingdom was in favour of an Empire serving itself first in the matter of raw material and developing preferences with all its parts.

Preferences were, however, only one item in the programme of closer inter-imperial relations. Postal and cable facilities and shipping communications are equally, if not more important. Regular and rapid communications abolish distance and make for unity and increased trade.

Imperial penny postage was introduced in 1898, giving a new postal unity to the Empire. Cable communications were subsidized by the mother country and the colonial governments in 1900 and the cable is the joint property of

*Proceedings, Cd. 8566.   †Cd. 9177.

the governments.   A Pacific Cable Board was set up in 1901 which contained representatives of the colonies concerned and the United Kingdom, forming another link of common co-operation.   The treaties which prevented the United Kingdom from giving or receiving preferences from the dominions were denounced in 1897, in spite of the fact that the trade of the United Kingdom with Germany was £47,952,000 in 1897 (£21,694,000 exports from the United Kingdom to Germany and £26,258,000 imports from Germany) and with Belgium £29,118,000 (imports from £20,886,000 ; exports to £8,232,000) *i.e.*, a trade of £77 millions in all, while the trade with Canada in 1897 was £24,390,000 (£5,172,000 exports to Canada and £19,218,000 imports from Canada). This meant that the United Kingdom was willing to risk the possible retaliation of her greatest European customer, Germany, to open the way for preferential relations which might lead to closer union.   This was done in response to a demand from the colonies at the Ottawa Conference of 1894. The United Kingdom continued to receive the most favoured nation treatment from Germany by a law passed by the Reichstag which could be repealed at any time.   Her commercial relations with Belgium were henceforth regulated by an " Exchange of Notes " terminable at three months' notice.   But the position was precarious.

From this time onwards the self-governing colonies began to give preferences to the goods of the mother country. Canada began in 1897, and was followed by South Africa and New Zealand in 1903 and by Australia in 1908.*   In Australia and South Africa the preference was granted by means of a reduction on the ordinary rates of duty, whilst in New Zealand a similar effect is aimed at by imposing a surtax on certain classes of goods when they are of foreign manufacture.   In Canada there are three tariffs—the General Tariff, the Intermediate Tariff for most favoured nations other than the British Empire, and the British Preferential Tariff.   This threefold scheme has now been adopted in the new Australian tariff of 1920.   British Guiana and some of the West Indian Islands made preferential arrangements with Canada in 1912.   In

*For variations and figures, Dominions Commission Interim Report, Cd. 7210, Cd. 7505, Cd. 8457.   For preferences for the Empire in Trust see p. 353.

1920 a much wider agreement was made and the islands concerned now include the Bahamas, Barbados, Bermuda, Jamaica, Trinidad, the Leeward and Windward Islands, as well as British Guiana and British Honduras.*

The amount of the preference in the four years preceding the war was computed to be as follows :

	Canada Million £	Australia Million £	New Zealand Million £	South Africa. Million £
1910 -	1.303	.972	.536	.539
1911 -	1.376	1.071	.725	.538
1912 -	1.667	1.266	.715	.349
†1913 -	1.573	1.244	.760	.555

In addition to this the colonies have arranged tariff reductions in favour of each other on certain specialities not covered by this British preference. Agreements exist between South Africa and Australia and between both of these and New Zealand. South Africa extended the British rebate to Canada in 1904. Canada and New Zealand extend the British tariff to other parts of the Empire and Canada made special arrangements with the West Indies in 1912 and 1920, and extended the preferences to Newfoundland. She also made a reciprocal tariff arrangement with Australia. Thus a system of preference with the mother country is supplemented by a system of preferences between the dominions themselves showing how the common connection with the mother country leads to mutual arrangements between the constituent parts of the Empire. In addition, subsidized mail services exist between New Zealand and Canada and between Canada and South Africa and between Canada and the West Indies (1920 agreement) under contracts which contain provisions intended to foster the exchange of produce between the dominions concerned.‡ Ships are in fact the shuttles which weave the weft of Empire.

*Cmd. 864. Bauxite for aluminium is brought from British Guiana to be smelted in the Saguenay District of Quebec. United Empire, December, 1925, p. 755.

†Dominions Commission, 1917, Cd. 8642, pp. 14-15.

‡In Canada the freight rates have to be approved by the Canadian Minister of Trade and Commerce and cannot be varied except with his consent. He also has power to fix maximum freight rates. Dominions Commission, Cd. 8462, p. 120.

It is interesting to notice that in the mail contract the Union of South Africa has stipulated that the Union Castle Line shall carry pedigree stock from the United Kingdom to South Africa free of charge. *Ib., p.* 68.

The Dominions Commission emphasized this fact very strongly, and said : " It has not, however, been adequately realized that the rates of freight which may be charged on goods to and from the dominions are in many cases a more important factor in the question of the development of inter-imperial trade than tariffs and tariff privileges on the present scale." " Freight and tariff rates being what they are, it is not too much to say that improvement in the cost of sea transport is amongst the most important problems which confront the statesmen of the Empire to-day."*

While the colonies gave preferences to the mother country the proposal to give tariff preferences here was defeated at the polls in 1906 as the United Kingdom would not alter her free trade policy. Nevertheless, preferences in other respects existed.

Under the Colonial Stock Act of 1900 (62 & 63 Vic. c. 62) colonial government stock was made trustee stock, thus enabling the colonies to borrow at a rate of interest otherwise unobtainable by a new country. The increase of the large amount of trustee stock depreciated the English Consols and other trustee securities, and so, while the colonies were enabled to borrow at lower rates, holders of existing trustee securities suffered. These investments, comprising about £650 million, had been supplied by Great Britain about one per cent. cheaper than Great Britain was prepared to lend it to countries outside the Empire. " This meant a saving to the colonies and India of at least £10,000,000 a year, which constituted a handsome preference."†

This policy has been still further developed in the arrangements with regard to taxation. Money invested by British subjects in a colony was subject to income tax within the colony and also subject to income tax within the United Kingdom. There was, therefore, a system of double Income tax within the Empire. The hardship of this was mitigated in the Finance Acts of 1916 and 1918‡ but only as a temporary expedient.

*Cd. 8462, p. 127-128.

†Paper by Sir Edgar Speyer on " The Export of Capital " quoted in *The Times*, May 28th, 1911.

‡An arrangement was also made with regard to the Excess Profits Duty whereby the income taxpayer paid the highest tax to which he is liable in either Great Britain or the Dominions, but not both taxes

By a conference between the representatives of the self-governing dominions and the mother country in 1919, it was agreed that there should only be one income tax and that the colonies should take their share from that and the United Kingdom take the remainder, provided it was not less than half the income tax she would have received in nominal circumstances.* There is thus a loss to the mother country in revenue of money invested within the Empire. As persons who invest their money in a colony only pay at the flat rate of 6s., which is divided between Great Britain and the colony concerned, Great Britain gets less than if they invested it in a foreign country when she would get the whole of the 6s., and the investor would have to pay tax in the other country as well. There is therefore a considerable preference given to capital invested within the Empire.

As a result of the war against Germany the mother country modified her position, in 1919, with regard to tariff preferences. On certain classes of commodities (cinematographs, films, clocks, watches, motor cars and musical instruments) there was a rebate of one-third on goods of colonial origin. The Chancellor of the Exchequer, Mr. Austen Chamberlain, spoke of the rebate of a third as "the General Empire rate."† On other goods such as tea, cocoa, coffee, sugar, dried fruits, tobacco and motors, the rebate was one-sixth. On wine there was a reduction, foreign wine being charged 1s. 3d. and 3s. duty, and colonial 9d. and 2s., according to strength. On foreign spirits, however, the plan adopted was that of levying a surtax of 2s. 6d. gallon extra.

The effect of the preferences was estimated to amount to a reduction of £2½ to £3 million on goods of imperial origin. It is interesting to notice that India would especially benefit on the question of tea.

Further efforts were made to increase the English hold on the colonial trade in accordance with the new constructive imperialism.

One of Mr. Chamberlain's first acts was to send a letter

on the same items, the division of the proceeds being agreed upon by the respective Governments.

*Report of the Royal Commission on Income Tax. Cmd. 615, 1920. Appendix I., p. 168.

†*Hansard*, 30th April, 1919, p. 194. Vol. 115. When these McKenna duties were abolished in 1924 the preference naturally disappeared also.

to all the colonies (November, 1895)* to find out the extent of the foreign competition in colonial markets and the reasons for its existence. The most important foreign imports of the colonies during the years 1884, 1889 and 1894 were set forth and the officials in thirty-one colonies and India were asked to send back patterns of competing goods. An exhibition was held to which manufacturers were invited in order to see where the foreign superiority lay. The result was to stimulate considerable interest in catering better for the colonial market. In 1897, Mr. Ritchie, when President of the Board of Trade, developed this idea for the whole of British foreign trade. A committee was appointed† to enquire into the best means of acquiring and disseminating trade information, and they recommended the establishment of what became the "Commercial Intelligence Branch of the Board of Trade." It was opened in 1899 and gave information as to the colonial as well as foreign trade. Four trade commissioners, specially attached to the colonies, were appointed in 1908; one to Canada and Newfoundland, one to Australia, one to New Zealand and one to South Africa, under whom were twenty-three local trade correspondents. Their business is to send back information of any contracts for railways, tramways, electric lighting, power installations, mines or harbour works, collect specimens of competing products, and advise on new openings for trade. They return at frequent intervals to keep in touch with the traders at home.‡ In 1917, sixteen trade commissioners were appointed, two of whom were allotted to India, one to the Straits Settlements and one to the British West Indies. The service was thus extended to the Empire in Trust.

In addition the colonies were allowed to use the British consular service in foreign countries for purposes of obtaining information. British consuls, therefore, have some care of the commercial interests of the Empire, not merely of those of the United Kingdom.§

*Cd. 8449 (1897).

†Cd. 8962, 8963 (1898).

‡Speech by Mr. L. Harcourt, *Hansard*, 8th, May, 1913, *p.* 2264, Vol. 23. They had just made a collection of competing hardware, tools and hollow ware.

§Dominions Commission, *op. cit.*, *p.* 147.

The new imperial trade commissioners have also received special instructions to represent the Empire in matters of trade, and not merely Great Britain, and are prepared to act for any of the dominions that wish to avail themselves of the services of the commissioners.*

An Advisory Committee was established in 1900 for the purpose of assisting the work of the Commercial Intelligence Branch of the Board of Trade. This Committee included representatives nominated by the Governments of Canada, Australia, New Zealand and South Africa, another instance like the Pacific Cable Board, of the Empire in Alliance working together as a whole for promotion of common interests in trade.

In 1904, the first Statistical Abstract of the British Empire was issued, in which the British Empire was treated as an entity exporting to and importing from the rest of the world certain quantities of merchandize.

Although Mr. Chamberlain went out of office in 1903, the spirit of his work still continued. The Liberal party which returned to power in 1906 was pledged not to interfere with the fiscal system of the country but was prepared to promote unity in other ways. In opening the Conference of 1907, the Prime Minister, Sir Henry Campbell Bannerman, said: " Gentlemen, freedom does not necessarily mean letting things drift." He went on to say: " You will not judge of the feeling entertained towards you by acclamations and festivities alone, although of these there will be abundance, but by the mutual spirit of friendship, the desire to stretch every point that can be stretched in order to meet the views of each constituent part of the Empire, the desire, equally strong, I hope, to avoid prejudicing in any way the interests of each other, and over and above all, you will be inspired and invigorated by our common pride in the great beneficent mission which the British people in all parts of the world are, as we believe, appointed and destined to fulfil."†

There has been no more striking phenomenon of recent years than " the growth in the English mind of the conception

---

*Instructions to the Imperial Trade Commissioners in Memorandum laid before the Imperial War Conference, 1918.   Cd. 9177, *pp.* 249-251.
†Colonial Conference, Cd. 3523, *p.* 6.

of the Empire as a *whole*, of which the United Kingdom and overseas dominions are parts."*

The policy of drift being abandoned by both parties, it naturally followed that some effort should be made to develop the resources of the Empire as a whole.

A Commission was appointed in 1912 to report on the natural resources of the self-governing dominions and the development of such resources " whether attained or attainable : upon the facilities which exist or may be created for the production, manufacture and distribution of all articles of commerce in those parts of Our Empire : upon the requirements of each such part . . . in the matter of food and raw materials and the available sources of such : upon the trade of each such part of Our Empire with the other parts, with Our United Kingdom and with the rest of the world . . . and generally to suggest any methods consistent always with the existing fiscal policy of each part of Our Empire, by which the trade of each part with the others and with Our United Kingdom may be improved and extended."†

It is interesting to notice that the Commission reported, in 1917, in favour of the creation of an Imperial Development Board " to deal with the scientific development of the resources of the Empire, with the deepening of its harbours on a co-ordinated plan, with the improvement of its shipping, mail and cable services, the preparation and publication of its statistics and other matters of joint interest to the whole." An Imperial Shipping Board was accordingly set up in 1920. It reports to all the governments of the Empire, but is subject to no one part of the Empire and to no one government. It deals with all questions relating to imperial shipping that are brought before it, and has been very successful in settling outstanding difficulties.‡   It is the first really Imperial body and is due to the fact that the Commission brought out the overwhelming importance of rapid and cheap sea communications to the prosperity of the Empire and advocated some control of freight rates to counteract any harmful discrimination by shipping rings in favour of foreign goods.

*Ashley, " British Dominions," *p.* VII.
†Reference, Dominions Commission, p. *iii*.
‡Report by Sir H. Mackinder to the Imperial Economic Conference, 1923.   Cmd. 2009, *p.* 293.

" The War has abundantly demonstrated that the life of the Empire depends upon its sea communications. Whatever the existing magnitude of the ocean-borne commerce between the United Kingdom and the Dominions and whatever the prospects of its development in the future, producer, manufacturer and merchant alike are concerned, and vitally concerned, with securing cheap, regular and efficient transport for their goods, and consequently with the progressive improvement of the Empire's shipping facilities. We emphasize this point for we feel that in discussions as to the best means of fostering trade within the Empire, its importance has been obscured by other factors affecting the exchange of merchandize and in particular by the prominence given to fiscal legislation. . . . If, therefore, it is possible to devise some means of permanent betterment of sea routes within the Empire, a powerful impulse will have been given to Imperial trade while the strength and cohesion of the Empire will be notably increased."[*]

The War accentuated the enormous importance for defence of the economic resources of the Empire.

Certain materials were almost wholly produced within the Empire, nickel, asbestos, jute, mica, palm oils, palm kernels and plantation rubber, and form for the Empire a valuable means of economic defence and commercial negotiation.

In certain other products the production within the Empire was, before 1914, enough to satisfy the demands of the Empire, such as butter, wheat, cheese and wool. In the case of wool, the Empire produced about forty to forty-five per cent. of the world's total supply and a larger proportion if the high grade merino wool alone is considered. The Empire also produces sixty per cent. of the world's output of gold.

Many of the specialities produced within the Empire were, however, sent outside for treatment before 1914. The nickel and asbestos were worked up in the United States, zinc or spelter was reduced in either Germany or Belgium. Tungsten, a necessary ingredient for hardening steel and for electric light filaments, although produced in Burma and Australia, was worked up in Germany. Monazite sand, found in South India, was also utilized in Germany for making incandescent

[*]Dominions Commission, *p.* 108.

mantles. In the case of tungsten the United Kingdom bought up, during the War, the whole production of the ores in Australia and New Zealand and the control of the monazite deposits in South India was transferred to British control and one of the seven directors has to be nominated by the Secretary of State for India.*

In July, 1918, an Imperial Mineral Resources Bureau was set up consisting of five representatives of the Dominions, one representative of India, one nominated by the Secretary for the Colonies and six members eminent in mining and metallurgy. To this the British Government contributes £10,000 a year and the Dominions another £10,000. Its function at present is to collect and publish intelligence regarding the mineral resources of the Empire. Here, again, is another instance of common co-operation in the great economic alliance and an insurance against the foreign control of imperial mineral wealth.

Again, in the Imperial Bureau of Entomology, founded in 1913, to deal with insect pests in the Empire, there is another common meeting ground, and the Dominions contribute towards its work as well as the Imperia Government. It was decided at the Imperial Conference of 1918 to establish a new Imperial Bureau of Mycology, working on the same lines as the Entomological Bureau, to deal with fungoid pests.

In 1924 a new departure was made. £1,000,000 was set aside by the British Government to promote inter-imperial buying and selling. An Imperial Economic Committee was created resembling the Imperial Shipping Committee which likewise contained representatives from all important regions of the Empire and reported to all the Governments of the Empire. They have reported as to the importance of publicity, labelling, better packing and grading, swifter transport and the need for scientific research into the preservation of food.†

Thus there has evolved a great mechanism for the economic development of the whole, centring in the United Kingdom as a common meeting place.

" I see quadrangles and corridors of empty grey-looking offices in which undistinguished-looking little men and little files of papers link us to islands in the tropics, to frozen wildernesses gashed for gold, to vast temple-studded plains, to forest worlds and mountain worlds, to ports and fortresses

*Dominions Commission, p. 69.
†Reports Cmd. 2493, 2499 (1925).

and light-houses and watch towers and grazing lands and corn land all about the globe."*

It is interesting to notice that the increased interest in the Empire after 1895 resulted in a diversion of British emigrants from foreign countries to the British Dominions. In the period 1891-1900, only 28 per cent. went to the Colonies and the rest principally to the United States ; in the period 1901-1912 63 per cent. migrated to places within the Empire and in 1913 78 per cent. of the whole remained under the British flag.†

Thus, from the time of Chamberlain, there has been a continuous development along many lines of a closer approximation of the self-governing dominions and the mother country which have tended to make the Empire a real economic alliance. Preferences in tariff, taxation and capital, direct postal and telegraphic communications, conferences and agreement on matters of common economic interest and the conscious development of imperial resources have all tended to create closer ties and a common working system, while the tendency towards a still closer approximation of the two Empires is developing. In addition, the great development of steamship communications which has been the outcome of private initiative, has been the vital link between the whole. Signs are, however, not wanting to show that the colonies are prepared to subsidize steamship communications as in the case of Canada and the West Indies. The Empire was built up as the result of maritime effort, yet it exhibits in Canada, in Australia, in South, East and West Africa, and in India, a vast continental expansion, and here the railway has played the decisive part in promoting emigration inland and making the interior resources available for the whole.

III.—THE EMPIRE IN TRUST.

While Chamberlain did so much to promote the closer union of the British Commonwealth of Nations or the Empire in Alliance, he was equally active in the development of the dependent or tropical Empire. It was relatively easy for

*Quoted by Sir A. Lyttleton in " British Dominions," ed. Ashley, p. 22.
†Dominions Commission, p. 85.

him to work up a certain amount of enthusiasm for the self-governing dominions akin to British in race and institutions, but he realized and emphasized the importance of the great tropical possessions of the Crown when few other people regarded the African possessions as anything but malarious regions inhabited by backward coloured races, and the West Indies were looked upon as derelict sugar islands always requiring assistance because of hurricanes or some other untoward incident.

In a famous speech on the Estimates (August 22nd, 1895)* Chamberlain defended West Africa as a very valuable possession and went on to say that these tropical possessions were England's "undeveloped estates," in which it would be necessary to sink national capital.† He considered that such a development was sheer self-preservation.

"It is only in such a policy of development that I see any

---

*" But if my right honorable Friend wishes from me a further declaration of policy I am not sorry that an opportunity has been given to me to make it. I regard many of our colonies as being in the condition of undeveloped estates and estates which never can be developed without Imperial assistance. It appears to me to be absurd to apply to savage countries the same rules which we apply to civilized portions of the United Kingdom. Cases have already come to my knowledge of colonies which have been British colonies perhaps for more than a hundred years in which up to the present time British rule has done absolutely nothing and if we left them to-day we should leave them in the same condition as that in which we found them. How can we expect therefore either with advantage to them or to ourselves that trade with such places can be developed. I shall be prepared to consider very carefully myself and then, if I am satisfied, to confidently submit to the House any case which may occur in which by the judicious investment of British money those estates which belong to the British Crown may be developed for the benefit of their population and for the benefit of the greater population which is outside." *Hansard*, August 22nd, 1895.

†This view was also maintained by Sir D. Morris in 1911 in a lecture on the West Indies, published in " The British Dominions," edited W. J. Ashley. " The productions of the tropics are in increasing demand as they are becoming more and more necessary to the inhabitants of temperate countries. . . It is not beyond the mark to state that our commercial supremacy may largely depend on our maintaining control of them. It is estimated that there are three million square miles of British territory within the Tropics. This area is producing commodities of the estimated value of £230 million. A large share of these are received in this country and they contribute materially to the prosperity and welfare of our people." *p*. 168.

solution of these social problems by which we are surrounded. Plenty of employment and a contented people go together and there is no way of securing plenty of employment for the United Kingdom except by developing old markets and creating new ones."* He also considered that Britain owed a duty to these regions simply because they belonged to her and that the development of their resources was plainly part of her duty. It was, however, contrary to all tradition that the Government of England should sink money in the development of colonial possessions. Nevertheless, from this time onwards England has assisted her tropical possessions in three ways :

(1) By directly financing railways and other permanent works, such as harbours.
(2) By encouraging institutions for the study of health in the Tropics.
(3) By encouraging scientific agriculture and the spread of agricultural knowledge.

(1.) In 1899 (62 & 63 Vict. c. 36) the Treasury wa authorised to advance to certain Crown colonies by way of loan £3,351,820 at 2¾ per cent., to be repaid within fifty years. A large part of this sum was spent on the West African railways (Gold Coast £578,000 ; Lagos £792,000 ; Sierra Leone £310,000). Further, £110,000 was given for the completion and equipment of a railway in Jamaica and £500,000 for railways in the Malay States. A further £98,000 was allotted to the Accra harbour works, and £43,500 to harbours on the Niger coast. In West Africa the Government itself made the line. "These railways have been constructed," said the official report in 1905† " through dense tropical forest, in a deadly climate which, in spite of every precaution in accordance with improved principles of malaria prevention, caused constant changes in the staff of every grade ; amid difficulties arising from heavy rainfall, from scarcity and inferiority of labour, from conditions under which cargo had to be landed as on the Gold Coast, by surf boats and lighters on an open roadstead, while native revolts and military operations have interrupted and delayed the work."

The results were quite astonishing. The development of large parts of Africa had been held back by the tsetse fly,

*Speech, *Hansard*, August 23rd, 1895.
†Cd. 2325.

which not merely communicated the sleeping sickness to human beings but destroyed the beasts of burden, leaving only man who had become immune, for porterage. The result was that in West Africa the natives carried on very little trade as they had no means of getting the products to the place of exchange. Their principal article of export was something that would walk to the market itself, viz., slaves. With the abolition of the slave trade exchange sank to insignificant proportions and the natives contented themselves with gathering some palm kernels and conveying them to the coast for barter.*

Their whole system of agriculture was primitive in the extreme and their trade was confined to wild forest grown products. With the coming of the railways new possibilities of exchange were opened out, the labour previously absorbed in porterage was set free for cultivation, the export of cocoa which had begun in 1891 extended rapidly and the Gold Coast became the premier cocoa-producing country in the world.† The natives began to make money out of a cultivated article, not mere self-grown forest produce. Instead of practising the primitive system of communal agriculture they began to regard their cocoa plantations as individual property and passed from communal to private ownership. With railways the country became safe and people left the walled towns and began to settle in the country districts.‡ The motor car added to the new mobility of goods and West Africa with its cocoa and palm oil, ground nuts and palm kernels, became an increasingly good customer for British goods. In this case the railway broke down the isolation of centuries imposed by the tsetse fly and created a new commerce and a new system of land ownership.

In a discussion at the Royal Colonial Institute, Mr. P. A.

*Sir H. Clifford, " The Gold Coast," *Blackwood's Magazine*, January, 1918, p. 51.

†	Cocoa exported.	Value.
1891	80 lbs.	£4
1901	960 tons	£42,827
1911	35,261 ,,	£1,613,468
1916	72,161 ,,	£3,847,720

Figures quoted by Sir Hugh Clifford, *op. cit., p.* 61-62.

‡J. Astley Cooper, " Recent Developments in West Africa," Royal Colonial Institute Proceedings in United Empire, August, 1910.

Renner, a native, said he had seen in the few years he had lived on the Gold Coast the improvement which had been brought about by the railways. It had so astonished the natives as to make them almost worship the white man. Previously tribal spite and feuds were so great that the people of one village would not visit those of another. Now they would find the men of the South in the North and those of the East in the West.*

The same spirit animated the Foreign Office after 1895 with regard to the construction of a railway from Mombasa to Victoria Nyanza, known as the Uganda Railway. "The obligation tardily accepted by the late government is assumed with readiness by the present government."† Goods cost £180 a ton to move by native porters, who were very few as the population was scanty; when the railway was built the price dropped to £17 a ton. By March, 1903, £5,384,370‡ had been furnished for building the railway, which was a State enterprise. Mr. Harcourt, in his speech on June 12th, 1912, said that the profit on its working had risen from £56,000 in 1906 to £134,000 six years later. A further loan of £250,000 was provided out of the surplus of 1910-1911 by Mr. Lloyd George, when Chancellor of the Exchequer, in order to help the growth and export of raw cotton in that region and a further loan of half a million was forthcoming in 1912. Railway construction was also pushed on by Mr. Harcourt in Nyasaland and West Africa. A loan of three million pounds Egyptian, promised in 1913 towards a railway in the Sudan, was largely spent in irrigation for cotton-growing.

The new constructive imperialism has its roots in railways and is bound up with the growing need of England for raw materials, especially cotton.

(2.) It is, however, no good building a railway for mass traffic if the white man cannot live in the region to superintend the production required or if the labour supply is continually decimated or destroyed by epidemics or malaria. The development of some means of making the tropics healthy for men to live and work in, was a necessary complement of railway building.

*Proceedings, *op. cit., p.* 550.
†Hansard, 4th ser. XXXVI. 1290; LXXXII. 309, 297.
‡Colonial Office List.

Chamberlain realized this and addressed a letter in 1898 to the leading medical schools of the United Kingdom urging the importance of encouraging the study of tropical medicine.

A School devoted to the study of tropical medicine was opened in London in 1899, and was subsidized by the Government and the Crown Colonies. Another school was started in Liverpool, in 1899, by the West African merchants there as a result of the awakened interest in tropical medicine The general result was to train medical men in the special diseases of the tropics, to start scientific enquiry into the causes of tropical diseases and to collect and disseminate the knowledge and research carried on by medical persons in tropical areas. Permanent research laboratories were started in the Federated Malay States, in Ceylon, British Guiana, the Leeward Islands and Nigeria. In the Colonial Office the Colonial Secretary was assisted by a permanent advisory committee on medical and sanitary matters connected with tropical Africa.

The general results of this crusade to make the tropics healthy have been remarkable.

West Africa used to have the reputation of a deadly climate and was known as the " White Man's Grave." A case occurred where the colonial governor who signed the despatch, the secretary who wrote it and the clerk who copied it had all died before it reached the Colonial Office from the Coast *

With steps taken to exterminate the mosquito the mortality dropped ; better men were willing to go to tropical regions ; the administration gained and the natives themselves felt the benefit of the preventive measures. The deaths among native troops in West Africa declined by 75 per cent.†

In the Federated Malay States in 1901 as many as 334 patients were admitted to the hospital as serious cases with malaria. Drainage specially adapted to destroy the mosquito breeding places was undertaken and the works were completed in 1902. The admissions to hospital then dropped to an average of 29. There was, however, an increase in the surrounding undrained districts which showed that the mosquito had not lost its virulence. In 1901, the sick leave granted

*Bruce, " Crown Colonies and Places," Vol. I., *p*. 403, 1910.
†Bruce, *op. cit.*, *p*. 438. See also " Health Problems of the Empire." *p*. 52.

to 176 government employees was 1,026 days ; in 1903 only
71 days' sick leave was required by 226 government
employees.*

Mr. Harcourt stated, in 1912, that in West Africa, in spite
of outbreaks of plague, yellow fever and other pestilence,
the death rate of European officials during the last nine
years had fallen from 56 to 25 per 1,000, and that the improvement was progressive. Research has also been undertaken
into sleeping sickness, yellow fever, beriberi, and other
tropical scourges. The result of the researches of British
tropical medicine was seen when the American Government
began its campaign of mosquito reduction in Havana and
extended its work to the Panama Canal.†

The new sanitation having made it possible for men to
live in the tropics, those regions have gained an added value,
trade has been stimulated and the wealth of the world and
its interdependence still further increased, the tropical areas
have come into their own and are regarded as valuable assets
to the Empire as a whole.

(3.) The third great line of development which owes its
impetus to Chamberlain's vision of what might be done with
the Empire in Trust was in the direction of the encouragement
of scientific agriculture and the destruction of insect and
fungoid pests that preyed on plants.

A Commission sent to the West Indies to investigate

*Bruce, *op. cit.*, *p.* 442.
The decline of malaria at Ismaila is very striking:

1902	-	1,551 cases.
(Anti-malarial operations)		
1903	-	214 cases.
1904	-	90 ,,
1905	-	37 ,,

Bruce, *op. cit.*, *p.* 439.

†Sir R. Ross, in a letter to *The Times* on July 24th, 1920, speaks of
his " very lucky observation of August 20th, 1897, which provided
the key to the whole mystery and even helped to unlock the yellow
fever problem in 1900. The sanitary measure of mosquito-reduction
was fully described in 1900, before Gorgas used it ; but he was able to
employ it and other measures against yellow fever as well as against
malaria. The whole work has been an international one in which
the British have taken a considerable part. But the British work has
been almost entirely due to the initiative of private medical men.
On the other hand, Gorgas worked with the whole support of the
American State behind him."

the depression reported in 1897 that their condition was
desperate. They depended on sugar ; their industry was
ruined by the bounties given by foreign governments to beet
sugar and any improvements the planters might introduce were
offset by further bounties. Chamberlain set to work to get
the sugar bounties abolished and the United Kingdom joined
the sugar convention in 1902. By agreeing to prohibit sugar
from bounty-giving countries the United Kingdom stopped
the bounty system as far as the important sugar producing
countries were concerned. If they could no longer get a free
market in the greatest sugar consuming country—Great
Britain, it was not worth their while to continue the bounty
system. The prosperity of the West Indies revived from
this time onwards and the financial situation of the Islands
was said to be better in 1910-1911 than in any time during
the previous fifty years. This was partly owing to the
abolition of the bounty system which gave a guarantee that
improvements in production would not be swamped by
further increase in the bounties. It is hopeless for individuals
to pit themselves against the bottomless purse of foreign
governments. Other factors, however, contributed. A tariff
war between Germany and Canada had the effect of shutting
out beet sugar from Canada, and the West Indian cane sugar
filled the gap.

A great deal of the growing prosperity of the West Indies
was due, however, to two other measures initiated by the
Colonial Secretary. He obtained a grant of £40,000 a year,
of which Jamaica paid half, to subsidize a line of steamers
to enable the fruit of the West Indies to find better markets.
As in Africa he tried to develop better communications. In
this case the object was to stimulate the growth of alternative
products and lessen the dependence on sugar as the staple
crop.

In 1898, an Imperial Department of Agriculture was
established in the West Indies, supported by Imperial funds,
and Sir Daniel Morris was transferred from Kew to act as
technical adviser. As a result of the application of science
to agriculture in this part of the tropics great improvements
were made in the sugar cane and the new canes yielded from
ten to twenty-five per cent. more than the old. Cotton,
once the great staple of the West Indies, was reintroduced

from the Carolinas in 1901 and an exceptionally fine variety of the high-class Sea Island cotton, indispensable for British fine cotton spinning was successfully evolved.*

Insect pests have been successfully combated and the yield of the sugar cane has been still further increased owing to the discovery of a parasite which preyed on a moth which bored into the sugar canes and caused great annual loss.†

Above all, the Department was successful in organizing agricultural education and in stimulating and educating a race of men willing to adopt new ideas and try experiments. The Department also furnished a bond of unity and a great thread of imperial communications between the Islands. In an article in *The Times* (May 24th, 1910) on the work of the Department, the writer said : " Some of the Islands do not hesitate to declare that the Imperial Department of Agriculture has been their salvation, having lifted them out of poverty into a condition of comparative affluence—as in Antigua and St. Kitts by the development of the sugar industry and in St. Vincent by the encouragement of cotton growing. . . . More than one-half of the total sugar production of the West Indies is being raised from varieties of cane developed by the Department." The cost to the Imperial Government was between £11,000 and £12,000 a year.

The brilliant success of science applied to tropical agriculture in the West Indies led to similar institutions being set up in other tropical areas belonging to the British Crown and the men trained in the West Indies went on to those

*Report of the Empire Cotton Growing Committee, Cmd. 523, 1920, *p.* 18.

†The story was told by Sir Daniel Morris in a lecture on the West Indies, printed in the " British Dominions," edited Ashley, *p.* 188, as follows :

" In connection with the experiments with sugar cane an interesting instance of the value of science for practical purposes was brought out in the case of a destructive pest known as the moth borer. For two hundred years this had caused immense damage on the sugar estates, but where or how it laid eggs and started its attack had never been ascertained." After an enquiry extending over less than a year conducted by a Cambridge entomologist, Mr. Lefroy, " not only discovered the eggs of the moth laid on the back of the leaf of the sugar cane, where they had escaped notice, but he also found a friendly parasite that could be utilized to keep the pest in check and enable the planters to save a considerable portion, if not all of the heavy annual loss hitherto sustained by them."

other regions, to spread the gospel, raise new crops and transform the conditions of tropical agriculture.* Similar Departments were set up in India, the Federated Malay States, British East Africa, the Gold Coast, North and South Nigeria and Egypt.

A great impulse was given to scientific agriculture in the tropical and semi-tropical areas by the growing world shortage of raw cotton. The causes of this shortage were many. It was partly due to insect pests. The cotton boll worm and the cotton boll weevil and the pink boll worm were so destructive in both Egypt and the United States that the loss was said to involve a quarter to a third of the crop,† and its ravages were increasing, especially in the United States. It was stated at the Imperial Entomological Conference in 1920 that the loss to the United States was £41 millions in one year and the loss in Egypt in 1917 was £17 millions.

The demand for cotton in the world was increasing owing to the greater beauty of the fabrics made of cotton, while new uses were continually being found for cotton, such as the casing of motor tyres. There was a growing attempt by other countries to manufacture cotton, with an increasing scramble for raw material. The supply varied considerably with the seasons and a frost in the United States, or the failure of the Nile to rise would seriously affect the staple industry of Lancashire. It became urgent for Great Britain to try and develop further cotton supplies or her greatest export industry would suffer irreparable losses.

The tropical and semi-tropical areas began to acquire a new value as possible cotton producing areas and the two favourite methods advocated to increase cotton production were improved transport facilities and better methods of cotton growing, so as to secure an increased yield. Thus Chamberlain's methods of constructive imperialism, viz., railways and science, were pursued as vigorously by his successors to get cotton. Mr. Lloyd George, when Chancellor of the Exchequer, devoted no less than half a million to the

*Two men trained in the West Indies went to the Imperial Department of Agriculture in India, two to the Federated Malay States, one to British East Africa, one to Fiji and three to Indian provincial departments. Sir D. Morris, *op. cit.*, *p.* 190. The Imperial College of Tropical Agriculture was set up at Trinidad in 1923 to train men for Agricultural service throughout the British Tropics. It also took over the work of the Department.

†Empire Cotton Growing Committee Report, *p.* 18.

provision of additional railway facilities in Uganda to secure the cotton crop. The Imperial Government in 1910 promised a grant of £10,000 a year for five years towards the experiments to be conducted by the British Cotton Growing Association, and it was empowered in 1923 to levy 6d. on each bale of cotton used by the spinners to be devoted to Empire cotton growing. A similar cess is levied in India, which is the largest cotton-producing country within the Empire. Her lint is short stapled and is at present unsuitable for English use on fine piece-goods. In Uganda, where cotton was almost unknown in 1903, a cotton of good quality is being produced; the export in 1924 is expected to reach over 100,000 bales. To develop the cotton possibilities of the Sudan the British Government guaranteed between 1919 and 1923 loans of £9,500,000 for railways and irrigation there. The 42,000 bales of good cotton produced in 1924 are expected to reach a million in fifteen years' time. Nigeria, especially the Northern provinces, is also a promising field, from which the 11 bales of 1914 had become 31,500 by 1921. Thus both the British and Indian governments have become interested in cotton production.

In 1911, the Department of Agriculture commenced its work in Egypt. As the chief product of Egypt was cotton, naturally cotton attracted the attention of the Department. Once the scientists had developed better varieties of seed it was the business of the Department to see that the fellahin got the seed. The task was first of all to evolve the strain and then distribute it. In addition to that a crusade against the cotton worm was necessary. Both these tasks were undertaken. The Government supplied the seed on credit and took payment after the cotton was sold. Instruction on the cotton worm and methods for keeping it under was given regularly at the Mosques after Friday prayers.‡ The local officials had to organize the crusade and persons who were slack in notifying and destroying the pest were punished by being made to go and help to destroy the insect in other people's fields. Parasites of the cotton worm were discovered in India and fostered to keep the worm under in Egypt as

*Empire Cotton Growing Committee, *p.* 38.
†*Op. cit., p.* 39.
‡Reports on Egypt in 1911, Cd. 6149, and for 1912, Cd. 6682.

they had done in India. Experiments were conducted with
a view to producing a cotton which matured earlier before
the boll worm had time to hatch.

Cotton growing has provided one of the chief motives for
the development of scientific agriculture in the colonies and
India. It has made for the creation of transport facilities
and irrigation schemes. It has caused peasant cultivators to
grow a product for exchange instead of depending on self-
sufficing agriculture and is one of the greatest stimuli to
colonial trade.

The movement for the establishment of scientific agricultural Departments had already spread to India. In 1905,
the Department of Agriculture was started. In addition
to the Imperial Department there are local agricultural
experimental stations in every province. Tobacco, wheat,
indigo, fruit, sugar, jute, flax and silk are all receiving scientific
attention. A large amount of energy is also being devoted
to the increased production of raw cotton for which India
offers special facilities in an abundant labour supply.*

The reports of all the tropical dependencies are full of
the new movement for re-fashioning the conditions of tropical
agriculture by advice, experiment, the provision of seed and
plants, marketing facilities, instruction and demonstration.
In no direction has English influence been more marked
than in the new health crusade in the tropics and the new
science of tropical production. The knowledge of her scientists
has increased the output of the staple articles of the world
and will still further increase it. Her scientific men have also
made the tropics a place in which the white man can live
and direct production ; her engineers have built the railways
which enable the products to be marketed and have irrigated
and drained the lands in Egypt and India and in other places
and have added large tracts to the fertile areas of the earth.
They have confined rivers, built bridges, harbours and roads
all of which made for increased production and increased
exchange of commodities. The great undeveloped estates
are in process of rapid development for the benefit of their
own inhabitants, the British Empire and the world. " It

*The India Cotton Committee place the normal crop in India at
4½ million bales of 400 lbs. each, cultivated on 22½ million acres.
Empire Cotton Growing Report, p. 42.

is the man of science who is to decide the fate of the tropics ; not the soldier or the statesman with his programmes and perorations, but the quiet entomologist. He is the man who of all others strikes the popular imagination the least and gets less of popular prestige but he has begun a fascinating campaign for the sanitary conquest of those enormous tracts of the earth. Before long he will have added their intensely fertile soil almost as a free gift to the productive resources of the human race."*

But the scientist must depend upon the administrator to devise the methods that shall make scientific research effective and it is, in the end, on the members of the colonial services that the constructive imperialism of modern times has rested and will rest. In speaking of their work " with knowledge and gratitude," Mr. Harcourt said : " They spent a great period of the best of their lives, on very moderate emoluments, in distant and often deadly lands—lost to their friends, removed from public appreciation in the obscurity of the jungle, but if they erred never spared from blame. They reaped few rewards except the advantage of the native, the credit of the Service and their own good name ; but they had at least the testimony of the civilized world to their probity and humanity. The Empire owed more than it would ever pay to her exiled and strenuous sons."†

The Empire in Trust has also developed its preference system. Following on the preferences given by Great Britain in 1919, Malta and Cyprus instituted, in 1920, preferences on the import of British manufactures. Jamaica also gave a preference on cotton-piece goods made from cotton produced within the Empire.‡

A new feature of the Imperial tariff system is the preference on raw material for imperial use.

As early as 1903 the Straits Settlements gave a rebate of the export duty on tin if it were to be smelted in the United Kingdom. In 1916 Nigeria followed suit and imposed a duty of 3¾ per cent. *ad valorem* on tin exported outside the Empire. In the same year an export duty of £2 per ton was imposed

*Lord Robson, quoted by Sir D. Morris, in " British Dominions," *p.* 169. †*Hansard,* June 12th, 1912.

‡For details *see* " Board of Trade Journal, 1920." As a result of the Canada-West Indies agreement of 1920 the preference given to Canada has been extended to the British Empire by Trinidad and Guiana. "Board of Trade Journal, December, 1920.

on palm oil from the Gold Coast, Sierra Leone, Gambia and Nigeria if it were utilized in foreign countries. Hides from India got a rebate of two-thirds of the duty on export if used within the Empire. The preference on palm oil, intended to last for five years, was repealed in 1922 owing to the after-war depression.

One of the most interesting of recent developments has been the changing position of India in the economic system of the British Empire. Before the War she formed part of the Empire in Trust ; during the War she began to join the Empire in Alliance. She was admitted, as we have seen, to the Imperial War Conference on equal terms with the other Dominions. When the English preference system was inaugurated in 1919 she obtained a preference on tea. One of her great grievances lay in the fact that in 1894 she had not been allowed to protect her own cotton manufacture. A duty had been imposed on cotton imports. The British Government insisted that it should be balanced by a corresponding excise duty on Indian cottons so that Lancashire might not be adversely affected. In 1917 this was changed. India offered £100 million towards the expenses of the War and proposed to raise part of the revenue to pay the debt by increasing the import duty on cotton piece goods. In spite of vigorous protests from the English cotton industry she was allowed to do this, the officials in India warmly approving.*

With the permission to impose duties even against Manchester, India virtually gained command of her own tariff and this was definitely confirmed in 1921. She thus attained in tariff matters the same freedom as the Dominions and like them has become increasingly protectionist, not merely in cotton, but in iron and steel, a new departure in 1924. A High Commissioner for India with economic as well as political duties was appointed in 1920, thus showing how India tended to approximate still closer to a self-governing dominion.

One continual source of trouble had been the opposition of the self-governing colonies to the migration of Indians as settlers. The people of India claimed that, as British subjects, they ought to be free to migrate within the Empire. The white population of the self-governing dominions were afraid of the introduction of cheap coolie labour with a lower

*Quoted by Mr. A. Chamberlain, 14th March, 1917, *Hansard*, p. 1150.

standard of living. The anomaly lay in the fact that although the self-governing Empire could exclude Indians, the government of India could not exclude colonials and they might even occupy the highest posts. Another grievance lay in the fact that Indians already se⁺tled in other parts of the Empire were not allowed to bring in their wives and children and were differently treated as regards the inhabitants of the other dominions in the matter of land holding and the power of acquiring land or setting up in trade. In the case of Canada, Japanese were more favourably treated than Indians.*

The Imperial War Conference of 1917 agreed that each country composing the British Commonwealth, including India, should enjoy complete control of the composition of its own population and might restrict immigration from any of the other communities.† India was thereby empowered to exclude colonials, if she wished. British citizens in any British country, including India, were, however, to be admitted into any other British country for visits of pleasure or commerce and Indians already permanently domiciled in other British countries were allowed to bring in one wife and minor children. " Indians in their outlook upon the Empire are at present powerfully swayed by two ideas. They are proud of the fact that they are British subjects and their country an integral part of the Empire. They wish to claim their Imperial privileges and they do not understand why, on the ground of race, they are unfairly excluded from large tracts of the Empire and worse treated in some matters than Asiatics who do not belong to the Empire. . . . Thus sentiment and imagination enter largely into the controversy."‡

Although the question of immigration was settled, a fresh controversy soon arose as to the status of Indians born or domiciled within a Dominion. The Indians claimed equality, while certain provinces of South Africa declined to grant equal rights in either trade, land-ownership, residence or franchise. While the friction in colonies of European descent was a rule the tropical possessions were permanently short of labour ever since the freeing of their slaves. The negro, with certain exceptions, satisfied his needs by short periods of work

*Cd. 8566, p. 160.
†Cd. 9177, p. 195.
‡Note on Emigration from India, Imperial War Conference, 1917, p. 161.

and saw no reason for doing more. Indian labourers were imported after 1837 to fill the gap and now such labourers are to be found in large numbers in Mauritius, British Guiana and Trinidad, in Fiji and Natal, in smaller proportions in Jamaica and some other West Indian islands and again in large numbers in Ceylon, the Malay Peninsula and North Borneo. While, therefore, they were unwelcome in one part of the Empire they were eagerly desired as a labour force in another.

The coolies were recruited on behalf of the colonial governments and were assigned to planters. They were indentured for a term of five years as a rule and then were expected to work for a further period as free labourers, after which, in some cases, their return fare was paid, or they became permanent residents and often became land-owners. In this way the population of Mauritius, Trinidad, British Guiana and Fiji contained a large proportion of Indians. Their departure from India was supervised by a Protector of Immigrants and they were under a Protector in the colony to which they went. Should their treatment be unfavourable in the colony to which they went, the Indian Government used to refuse leave to recruit for that colony and thus paralysed the flow of labour until conditions were improved.

Commissions of inspection were sent out from time to time by the government of India itself to supervise the treatment of the emigrant coolies. Most of them had prospered and improved their position and had added to the wealth of the countries that received them. The development of tea-planting in Ceylon and Assam would have been impossible without them ; they were all important in rubber-planting in Malaya and had helped the sugar production in the West Indies. Great difficulties arose over the admission and status of Indians in British East Africa, now Kenya. In a Crown colony the decision as to exclusion and general treatment of Indians rested with the British government itself. Indians had penetrated inland to build the Uganda railway, Indian traders had followed and in 1923 the 9,651 whites were outnumbered by 22,822 Indians, who claimed the franchise. Yet even here restrictions were imposed in 1923 on the ground that England was a trustee for the Africans who numbered 2½-3 millions. The continued immigration of Indians would, it was held, deprive the African natives of their chance of rising into the minor clerical and skilled posts. " His Majesty's government regard themselves as exercising a trust on behalf of the African population and they are unable to share or delegate this trust, the object of which may be defined as the protection and advancement of the native races." (Cmd. 1922, p. 10). Thus the emigration of Indians gave rise to a definite statement as to England's attitude towards Africans. The Indian government was so incensed that in 1921 it refused to allow its subjects to emigrate to any colony except Ceylon and Malaya, and this permission was withdrawn in 1923. Only the Viceroy in Council may allow emigration in exceptional cases. The new ease of migration thus created great friction between the colonizing races of the Empire.

The characteristic change in British commercial policy after 1895 is a reaction from world economics to imperial economics. Whereas in 1830 the *Westminster Review*** could speak of " the colonial dominion which has ever been the bane and curse of the people of this country," at the beginning of the twentieth century the evolution of the United Kingdom into a great federated Empire was regarded by many as self-preservation both for the United Kingdom and for its dominions and dependencies. All nations have missionary aspirations ;† the great thing is to be large enough to preserve your own type from other missionary nations each of whom believes not merely that its type is the best, but that it would be the best for others.‡

The British Empire is an Empire of diversities and would never attempt to impose one type on its extraordinarily varied parts ; uniformity would mean the death of its spirit. It therefore must be strong enough to resist attempts at proselytising by others if it wishes to have free play for its varying individualities. Moreover, the growth of a nation into an Empire is but part of an evolutionary principle. To refuse to obey it was to become stagnant. As the manor enlarged into the district or province containing both town and country,

---

*Quoted Mills, The Colonization of Australia, *p.* 21.
†*Cf.* Count Mouravieff, quoted Drage, Russian Affairs.
" I believe that Russia has a civilizing mission such as no other people in the world, not only in Asia but also in Europe. . . . We Russians bear upon our shoulders the New Age ; we come to relieve the ' tired men.' "
‡*Cf.* Prothero, German Opinion and German Policy before the War (1916).

A A

so the provinces or districts merged into the nations in their turn. Those countries that still kept their provincial restrictions, like Germany and Italy, remained backward until they went over to the national stage in the nineteenth century. In the twentieth century with the railways the nations have become Empires and to refuse to take the necessary economic or political means to draw the Empire closer is to perpetuate such local differences that the small areas become a prey to ambitious neighbours anxious to extend their markets, food areas or raw material supplies.

A large Empire makes for prestige and therefore for peace as it is not lightly attacked. Small or weak States are a great temptation to their neighbours as were Poland, Denmark, Belgium and the Balkan States.*

The transition to the imperial stage is said to be self-preservation in that it increases the man power of the whole, introduces into the older parts new and virile elements which prevent stagnation by proving to be a fount of perpetual youth and energy while it is of inestimable benefit to the new countries to be closely linked up with the revelations of an ancient civilization and storehouse of knowledge. That the British Empire should hold together is not mere political self-preservation but is self-preservation from the economic standpoint. When an Empire contains, as does the British, the bulk of the wool, rubber, jute, edible oils, nickel and gold of the world within its borders it has an unrivalled economic weapon for bargaining with other governments or with trusts, it has an insurance in peace and war. With

---

*Chamberlain at Colonial Conference of 1902 :

" I want you to consider for a moment what is the present position of the smaller nations with whom in population you may compare yourselves. What is the position of such nations in Europe as Greece, the Balkan States, or Holland, or the South American Republics ? Why, gentlemen, they are absolutely independent nations and accordingly they have to bear burdens for their military or naval defences or for both as the case may be, to which yours bear no proportion whatever. I point out to you, therefore, that in the clash of nations you have hitherto derived great advantages even from a purely material standpoint from being a part of a great Empire. But the privileges we enjoy involve corresponding obligations. The responsibilities must be reciprocal and must be shared in common. I do not think that any Empire is on a sure foundation which is not based upon recognized community of sacrifice."

the growing shortage of raw materials in the world it has a valuable asset in its quasi-monopoly of these articles.   From the point of view of the United Kingdom it is a great advantage to be able, if necessary, to supply the greater part of her food requirements from within the Empire.   As a basis for shipping and for its profitable employment, the widespread distribution of colonies is of inestimable advantage. To the dominions and dependencies the connection with the richest country in the world has been of great assistance in getting cheap and abundant supplies of capital which has hastened their rapid and prosperous development.

A diversified Empire, such as the British Empire, means further development of the faculties.   No place is far off to the English, and nothing can happen in the world without affecting the Empire at some point.   The conscious cooperation of all parts which is bound up with developed communications is self-preservation from the point of view of safty from aggression, from the point of view of the adequate development of the economic life of all parts and from the point of view of having an ideal or enthusiasm without which all else is vain.   His Majesty, in replying to the address from the Imperial War Conference of 1917, said : " The value of the Empire lies not in its greatness and strength alone but in the several contributions that each of its diverse parts with their varying circumstances and conditions makes to the one general stock of knowledge and progress."*

*Cd. 8566, *p.* 163.

PART VII

THE EFFECT OF THE DEVELOPMENT OF MECHANICAL TRANSPORT ON BRITISH AND IRISH AGRICULTURE.

SYNOPSIS

I.—THE EFFECT OF THE DEVELOPMENT OF MECHANICAL TRANSPORT ON ENGLISH AGRICULTURE.

1—The differences between the agriculture of Great Britain and that of European countries.
- (a) Great Britain a food importing urban country after 1850.
- (b) Disappearance of the peasantry in Britain.
- (c) Absence of serfdom and serf problems.

2—Periods.
- (a) 1793-1850. The victory of the large farm.
    - 1—The French Wars hastened the creation of the large farm. Result, an agricultural revolution consisting of improved cattle breeding, more corn fields, abolition of the open fields, enclosure of wastes.
    - 2—Causes of the disappearance of the small tenant and owner as the typical feature of English agriculture. Inferior methods, fluctuations in prices, loss by enclosure of the wastes, the expense of resorting the strips, incapacity to adopt improvements, lack of capital, loss of bye-employment, high poor rates. The large farmer was more efficient, import of corn difficult owing to lack of transport, import of meat impossible except salted. Encouragement by the Government of large farmers for the sake of food supply for growing town population.
    - 3—Agricultural labourers recruited from the small farmers : abundant and superfluous
- (b) 1851-1873. The National Market created by the Railway and the "Good Years."
    - Farming was very prosperous owing to the development of the railways in Great Britain which developed a national market for food products. Foreign competition undeveloped. Scientific farming and the adoption of machinery. Women ousted from field work. The agricultural labourer went to the towns.

(c) 1874-1894.   The World Market and American Competition.
Fall in the prices of meat and wheat. Loss of capital
by landlords and farmers, reduction of rents,
general uncertainty. The reversion from
arable to pasture. Growth of food imports.

(d) 1895-1914.   Agricultural   reconstruction   and   social
experimentation.

1—Great Britain concentrated on dairy farming,
market gardening and cattle breeding.

2—Decline in the numbers of agricultural labourers.
" Rural depopulation."

3—The attempt to establish a peasantry by Government aid and compulsion where necessary.
Development of agricultural education and
co-operation.

4—Increasing State intervention in agriculture. The
Board of Agriculture, 1889.   The Development Grant, 1909.

II.—THE EFFECT OF MECHANICAL TRANSPORT ON IRISH AGRICULTURE
AND THE RELATIONS BETWEEN GREAT BRITAIN AND IRELAND.

1—The attempted Anglicisation of Ireland.   Elizabeth to Charles II.
2—The suppression of competition in Ireland.   1660-1783.
3—The equal treatment of Great Britain and Ireland. Laissez-faire.
1801-1870.

The overpopulation of Ireland, the decline in numbers,
emigration, free contract in land.

4—Constructive policy for Ireland, 1870-1913.

(a) Improvement of tenures.

1870. Fixity of tenure and compensation for eviction.
1881. Land Commission to fix " fair rents."
1885-1891. Facilities for land purchase.
1903. Wyndham Act. Greatly extended facilities
for purchase with the intent to transfer the
land of Ireland from landlord to tenant.
1909. Financial breakdown. Further facilities and
increased advances of capital.

(b) Improvement in methods.

The Irish Co-operative Movement.
The Department of Agriculture, 1899.
The Congested Districts Board, 1891.

I.—THE EFFECT OF THE DEVELOPMENT OF MECHANICAL

TRANSPORT ON ENGLISH AGRICULTURE.

WHILE the development of mechanical transport provided new outlets for British industry abroad and
created new branches of industry at home, it also stimulated
the growth of food in other countries and facilitated its

import into England. The agriculture of both Great Britain and Ireland had to be radically reconstructed in consequence of the world competition and there was here as elsewhere an abandonment of the laissez-faire attitude as far as the United Kingdom was concerned. The result was that the United Kingdom became the land of social experiment in agriculture. One of the most remarkable land transferences of modern times was being carried out in Ireland and a great scheme to restart a peasantry with State aid was being attempted in England prior to the outbreak of the War of 1914.

The fundamental difference between the history of English agriculture and that of the other Great Powers in the nineteenth century lies in the fact that Great Britain had become the urban country, *par excellence*. While even in Germany, the most urbanized of the Great Powers forty-eight per cent. of her people still lived in the country in 1900, in England only twenty-three per cent. did so.* Instead of exporting food stuffs like the United States and Russia, she imported them in increasing quantities before 1914 and paid for them with her manufactures. She did not attempt to feed herself by increased yields as did Germany and France. Seventy-eight per cent. of her wheat and forty per cent. of her meat were imported during the years 1901-1913.† She was the chief market for the raw material and food producing countries, illustrating very forcibly the interdependence of the new world economy brought about by the industrial and transport revolutions.

While the continental countries were occupied with freeing

---

*In the United Kingdom 71.3 are urban. Agricultural Ireland pulls the figure down from 77 per cent. in England.

†" Looking at the populous areas of Lancashire and Yorkshire (with a population of nearly ten millions between them), one is probably justified in suggesting that to feed them with American grain, imported in bulk and milled in large mills, is, in view of the relative expenses of production and distribution, as scientific a method of providing them with bread as any other. In this connection it is interesting to quote the experience of the North Eastern Railway, which in one year carried 265,222 tons of foreign grain from two ports, Hull and Newcastle, while the 265,893 tons of English grain carried by them the same year had to be collected from 467 separate points." Report to the Board of Agriculture for Scotland on Agricultural Credit in France, *p. 9.*

their serfs, England had no serfs to free and, therefore, no compensation to arrange. Having got rid of her small farmers as the typical feature of her agriculture by the first quarter of the nineteenth century she did not have to face the task of training an ignorant people to farm properly. Nor had she to teach her big farmers better methods, English agriculture was the model for the world up to the period of the great depression.

While other countries protected their farmers from the great drop in prices that ensued after 1873, England remained free trade and therefore all the grain and meat which were prevented from entering other countries came to the United Kingdom, and prices fell still further. She had deliberately subordinated her agriculture to her industry in 1846 when the repeal of the Corn Laws took place and for the sake of cheap food for her preponderant town population this country was prepared to rely on imported food stuffs and let English agriculture take its chance of surviving.

In yet another feature the agriculture of Great Britain differs from that of the continent. The small farmer was deliberately sacrificed to the large one, as being the more efficient type. Towards the end of the nineteenth century a reaction took place with the object of re-establishing a peasantry in this country. This could only be done by Government action and a certain measure of compulsion. Both were applied with the result that, prior to 1914, England was carrying out a great social experiment—the creation of small holdings in large numbers. The reaction from laissez-faire in this respect is very marked.

The outstanding events of English agriculture during the nineteenth century are the disappearance of the peasant farmer as the typical cultivator of English land, the change from a home-grown to an imported food supply, the transition from arable to pasture farming after 1880, and the attempt to revive the small farmer by Government pressure, County Council ownership and compulsory purchase of land.

(a) The Victory of the Large Farm.

The history of English agriculture during the nineteenth century divides itself into four periods. Between 1793-1850 we get the disappearance of the small farm and the rise of the large one and the establishment of free trade in food stuffs. From 1851 to 1873 we get the " good years " in English agriculture. From 1874-1894 the agricultural depression set in, due to the grain from America and meat imports from both America and Australasia and the consequent drop in prices. After 1894 we witness an agricultural reconstruction lasting to the outbreak of the European War.

During twenty-two years of the first period (1793-1850) Great Britain was at war with France (1793-1815) and the effect of the French wars was to hasten a movement which was also taking place both in Germany and France during the eighteenth century, namely, the creation of the large farm. In England it had begun as early as the sixteenth century but it proceeded at an accelerated pace in the eighteenth century. Great improvements were being carried through in English agriculture in the last half of the century. Breeds of cattle were developed which put on far more weight and did it in a quicker time than the old stocks. It was no longer necessary to kill off the cattle at Martinmas and salt the meat in for the winter as roots were grown and fodder crops developed which would keep the cattle alive in the lean months whereas previously they would have starved. The yields of corn were increased by growing clover, which stored up the nitrates in their roots. When ploughed into the ground in the following year they added this valuable chemical to the soil and the next grain crop flourished accordingly. It was possible to grow clover or roots in the fallow year and thus keep all land under cultivation. Where land was enclosed the farmers made money. The new canals and roads opened up new markets and England had a considerable corn export up to 1776. The typical picture of John Bull represents him as a prosperous farmer, not as a captain of industry.

A large part of England lay in the open fields in strips and here no improvement was possible unless the strips were re-sorted into compact farms, the cattle when herded together caught disease ; rotation of crops, clover and turnips were out of the question when the beasts were turned over the fields after harvest and the scattered acre strips were uneconomical units to cultivate.

There were thus two strongly contrasted systems—that of

the large man on a compact farm producing wheat and meat of good quality, and the little common field farmer with poor yields and lean beasts eking out a living on the margin of subsistence, depending on his own weaving or his wife's spinning to provide part of their needs. His ability to hold on was, as elsewhere in Europe, largely dependent on the existence of commons or waste where he could cut fuel or timber and turn out his animals during the hay or corn harvest when the cultivated fields were in crop.

The peasant farmers were either copy-holders, paying a nominal sum per year and the best beast or chattel on inheritance, small free-holders or small lease-holders. The last could be ejected, the two former could not. The interesting thing is that all three disappeared alike.

During the war period British towns were growing and the urban population needed more food. Import of corn from the two places where import was usually possible, viz., Dantzig and Odessa, became very uncertain. This was all the worse as the years 1797-1801 were characterized by bad harvests in this country as were also 1810-1813. The very greatest difficulty was experienced in providing the people with food. The large farmer grows grain and meat more economically than the small one ; he can afford better seed, better implements and better cattle. He can fatten his animals on patent foods and can afford to wait for his sales till prices rise. The small farmer cannot use his horses as economically as a large one unless he can fill in some days carrying coal. He can, however, produce profitably fruit and vegetables, poultry and eggs, and can economize in labour by employing his family, but to make a small farm pay really well intensive culture of a high degree is needed and that was something unknown to the common field farmer of the late eighteenth century. Even if he had known it he could not have practised it if he held his land in strips with the right of common pasture for all animals after harvest.

On the one hand then, there was the big farmer able to carry out improvements and produce the food needed by the nation ; on the other hand, there was the small man, socially valuable but economically inferior. Was England going to starve the urban population to let him live ? Moreover, with the pressure on the food supply it was impossible to tolerate the amount of waste lands or commons that existed. " A great portion of the unstinted common lands remain nearly as Nature left them ; appearing in the present state of civilisation and science as filthy blotches on the face of the country, especially when seen under the threatening clouds of famine which have now repeatedly overspread it."* They must be broken up and cultivated. They were, however, an integral part of the economy of the small man. He was, as a matter of fact, threatened on three sides. If he sold his surplus, his poor grain and miserable scraggy beasts fetched a low price in competition with the good wheat and fat bullocks of the large farmer. If he were to enclose strips he had to pay for the cost of the enclosure surveys and hedge his land ; his commons were at the same time being taken in for arable and he lost not merely his pasture rights but he had to buy coal to replace the fuel gathered freely from the wastes, and this happened at a time when he was losing his weaving and his wife her spinning, as both were being taken over by the factories, more especially the spinning.

The years 1793 and 1815 were prosperous times for large farmers. The price of wheat rose to unprecedented heights† and the new methods paid well and were nationally valuable. It was said of Coke, of Norfolk, that he saved his country with a ploughshare where the sword would have failed. Every encouragement was therefore given to enclosing by the Government. Nor was the common field farmer altogether unwilling to enclose. He could not make it pay as he was going on, the only hope seemed to be better methods and those could only be tried on the compact farm. Enclosing had been continuing all through the eighteenth century.

*Marshall (W.), " The appropriation and Inclosure of Commonable and Intermixed Lands," 1801.

† 339 of 1912 Return to House of Commons.

Average Annual Price of British Wheat, per quarter.

	s.	d.			s.				d.
1800	113	10	Highest price,		141.5	Bread (4-lbs.)			15.3
1801	119	6	,,	,,	161.2	,,	,,		15.5
1805	89	9	,,	,,	103.3	,,	,,		13.1
1809	97	4	,,	,,	113	,,	,,		13.7
1810	106	5	,,	,,	121	,,	,,		14.7
1811	95	3	,,	,,	111.8	,,	,,		14
1812	126	6	,,	,,	160	,,	,,		17
1813	109	9	,,	,,	126.11	,,	,,		15.7

The persons interested in the area to be enclosed would get
a private Act of Parliament* empowering them to re-sort the
land of the parish or manor as the case might be, and the
parish would be re-divided and the objectors would have to
submit to the Act of Parliament. Unfortunately for the
small farmer the change came at a time when there was a
diminution of family earnings, uncertain seasons, very high
poor rates and wild fluctuations in prices. The Government
engaged in fighting Napoleon, with its cloth barred out of
Europe by the continental system, with its ships being sunk
in large numbers,† was in no position to initiate co-operative
societies, provide credit or train teachers to instruct the small
man in the new methods, even had they known these modern
expedients, which they did not. It had primarily to get on
with the war and spent the fifteen years after the war nervously
watching France lest she should break out again. It was
quite obvious that the big farmer produced more for other
people ; he had a larger surplus and all the experts were
unanimous as to the superiority of the large farm.‡ Landlords, too, preferred the big tenant. He needed fewer repairs,
was able to pay high rents and pay them regularly ; he was
not prostrated by the death of a cow as was the small tenant,
with consequent remission of rent. There was no doubt
that, where possible, small tenancies would be thrown into
large farms. The small owner, however, equally disappeared.
Times were against him and he could not hold on. The more
enterprising sold their lands and either started cotton factories
as did the Peels, or rented large farms, utilizing the money
received for the small farm to stock the large one. Others
sank to the rank of agricultural labourers. The French
peasant was saved by the French Revolution ; the German
was partly saved by the paternalism of the Government,
though East of the Elbe " times " were against him too, and
he disappeared there in large numbers. But these countries

---

*The notice that a petition for enclosure had been presented to
Parliament had, in 1774, to be fixed on the church door for three
Sundays so that objectors should not remain ignorant and could be
heard against the proposed change. Prothero, " English Farming,
Past and Present," *p.* 250,

†Forty per cent. of English shipping was destroyed between 1793
and 1815. W. R. Scott.

‡*See* List of authorities quoted, Prothero, *p.* 303.

were predominantly agricultural. In England an industrial town people needed food and concentrated on the most economical method of producing it, viz., large farmers. No constructive effort on behalf of the small man was put forward because this country firmly believed at that time in laissez-faire. Her population as in France in 1789 and Russia in 1905 was increasing beyond her capacity to feed it; large imports were not then feasible. Hence the peasant paid the price.

It is possible that the peasant owners might have survived had not the years between 1815 and 1830 witnessed a great agricultural depression;* prices fell, all farming became sheer speculation and under such circumstances the small man was at a special disadvantage, as he can rarely survive two bad years. He cannot hold on for prices to rise, he cannot afford to wait. The result was that, whereas in 1913 farms of one to fifty acres numbered 292,720 as against 143,166 larger than fifty acres, they did not occupy sixteen per cent. of the total acreage, i.e., there are actually even now more small farms than large ones, but the large ones occupy eighty-four per cent. of the area.†

The large farm being thus definitely established as the typical feature of English agriculture, it proceeded to new

	Average price of Wheat. Per quarter.		Highest.		Lowest.		Bread.
	s.	d.	s.	d.	s.	d.	d.
1815	65	7	74	1	55	10	10.3
1816	78	6	107	6	54	0	11.7
1817	96	11	118	7	76	6	14.3
1818	86	3	94	0	81	9	11.8
1819	74	6	83	0	67	0	10.3
1820	67	10					10.2
1821	56	1					9.3
1822	44	7	52	5	39	4	8.3
1823	53	4	64	5	41	2	9
1824	62	11	69	9	55	8	10.4
1825	68	6	71	11	63	5	10.8
1826	58	8	63	4	55	10	9.2
1827	58	6	68	8	50	9	8.9
1828	60	5	76	7	51	3	

Note the drop in price and the fluctuations between the highest and lowest prices in 1815, 1816 and 1817. Compare these prices with those on p. 114 for 1800-1813.

†Agricultural Statistics, 1913, cd, 6597.

triumphs. Improvements in drainage were introduced in the thirties ; chemical manures, such as nitrates and phosphates, were put into the soil thereby raising the yields to a record, nor were prices of wheat especially high although wheat was shut out by a high tariff ; prices both of wheat and bread actually dropped considerably in the period miscalled " the hungry forties."*

It was possible to carry through these improvements because there was an abundance of agricultural labour. It seems to have been inefficient, but there were plenty of men. They were very loth to emigrate, so appalling were the conditions in the emigrant sailing ships of those days. Their position was, however, hopeless. They had lost the chance of getting a little farm and rising in the world. They had not enough capital for a large farm. They had lost the free fuel and grazing for a cow or goat on the waste and they subsisted on doles from the Poor Law varying according to the size of the family. Their wages between 1824 and 1851 averaged 9s. 6d. to 9s. 7d. a week.†

In 1846 the manufacturers obtained the repeal of the Corn Laws, the idea being that they could extend their sales abroad if the continent could pay for manufactures in corn. As the English population was increasing it was obvious that the price of food would rise and higher wages be demanded. Food imports, it was hoped, would keep prices and wages down. The manufacturers were joined by other interests and succeeded in repealing the laws against the import of

*Return, *op. cit.*

	Average price.		Bread.	
	s.	*d.*	*d.*	
1839	70	8	10	
1840	66	4	10	
1841	64	4	9	
1842	57	3	9.5	
1843	50	1	7.5*	This is the first time bread
1844	51	3	8.5	touched 7d. for the
1845	50	10	7.5	century.
1846	54	8	8.5	
1847	69	9	11.5	
1848	50	6	7.5	
1849	44	3	7.0	
1850	40	3	6.8*	This is the first time it touched 6d.

†Bowley, Wages in the Nineteenth Century.

both meat and wheat. Nor did those who voted for the repeal believe that a large import would take place. Railways and steamships were quite undeveloped and Cobden showed that the English farmer had 10s. a quarter protection in sheer distance, that being the cost of freight and insurance of wheat from Dantzig to these shores. Dead unsalted meat, Macculloch pointed out in 1842, could not be imported. The main object of the repeal in the minds of many of its supporters was to remove any suspicion that the industrial classes were in any way being sacrificed to the agricultural interests. Import might, it was thought, steady prices in bad years. As a matter of fact, a corn import did develop and was mainly in the hands of Germans who brought it from Dantzig and Greeks who did the same from Odessa. Of American competition there was no sign for a quarter of a century after the repeal. When it came American grain centred in Liverpool and the wheat import fell into English hands. It seems unlikely that the free traders would have been anything like so drastic had they realized the possibilities of mechanical transport thirty years later.

(b) The National Market created by the Railway and the "Good Years."

Between 1851 and 1873 the free traders were brilliantly justified. Never did English farming prosper better. Foreign competition was scarcely felt and wheat prices fell very little.* In the sixties Russia was just reconstructing after the Crimean War and freeing her serfs and was in no position to compete while her railways had not been built ; Germany, the other great exporting country, was engaged in fighting Denmark in 1864 and Austria in 1866, and was preparing for the war with

*Quinquennial Average Price of Wheat, 1841-1875. Report of the Agricultural Sub-Committee (Ministry of Re-construction), 1918, Cd. 9079.

		s.	d.
1841-1845	-	54	9
1846-1850	-	51	10
1851-1855	-	55	11
1856-1860	-	53	4
1861-1865	-	47	6
1866-1870	-	54	7
1870-1875	-	54	8

Bread averaged in cost 8.13d.

France. The Civil War was absorbing the United States and her mass export had not yet developed. Hence England had an effective protection in distance and wars. There was a greater consumption of meat by artisans which reacted favourably on English agriculture, the prices of all animal products being higher than the general rise in prices; the railways widened the agricultural market and cheapened manures; machinery came in and enabled the harvests to be saved quicker and in better condition.

Agricultural machinery, however, economised labour, especially women's labour, and they began to disappear as agricultural workers. They went into service in the towns. The steam thresher also did away with a good deal of winter work, such as threshing out the corn in the barns with the flail.* As far as corn growing was concerned agriculture became a seasonal employment.

The agricultural labourers' position would have been worse had not the railways opened up a large prospect to them as navvies, porters and railway officials while the pick of them went into the new police. There was also an endless demand for iron workers and coal miners, if they cared to enter these trades. Emigration was a very different thing in the new iron steamer after the fifties, and new prospects were opened up by the discovery of gold in California in 1849, and in Australia in 1848. The agricultural labourer began to leave the land, a process hastened in the next period.

(c) *The World Market and American Competition.*

Between 1874 and 1894† there fell on England the full

*This was very unhealthy owing to the dust.

†Return, *op. cit.*

	Wheat prices.		Bread.	
	s.	d.	d.	
1875	45	2	6.8	
1877	56	9	8.1	
1884	35	8	6.2	
1885	32	10	6.3	
1886	31	0	5.6*	This is the first time bread
1887	32	6	5.7	cost 5d.
1888	31	10	6.	
1889	29	9	6.	
1890	31	11	6.2	
1893	26	4	5.8	
1894	22	10	5.5*	The lowest year.
1895	23	1	5.1	
1896	26	2	5.5	

force of the American wheat exports. Other countries dyked up with tariffs, which they increased as occasion required ; only England remained free trade. The chilled beef followed the wheat and the frozen mutton from Australasia coming in in the eighties, completed the rout. The loss of capital and profits in English agriculture between 1875 and 1905 was put by Sir Inglis Palgrave at £1,600 million.* There was a destruction of confidence and enterprise, the wheat area began to contract, land was allowed to revert to rough pasture,† improvements in agriculture such as drainage, etc., were arrested. The English farmer became the prey of international forces and never knew where the drop would stop. Argentina seemed to be ready to take

The fall in the case of beef between 1876-1895 was between thirty and forty per cent. Report of the Royal Commission on the Agricultural Depression, p. 46, Cd. 8540, 1897.

AVERAGE PRICE OF BEEF, PER STONE, 8-lbs.

	1. Inferior Quality.	2. Second Quality.	3. First Quality.	Index No. 1.	2.	3.
	s. d.	s. d.	s. d.			
1876-1878	4 5	5 6	6 0	100	100	100
1884-1886	3 9	4 9	5 3	85	86	87
1893-1895	2 8	4 0	4 7	60	73	75

SHEEP, PER STONE, 8-lbs.

	Inferior.	Second.	First.	Index No.		
	s. d.	s. d.	s. d.			
1876-1878	5 5	6 5	6 11	100	100	100
1884-1886	4 9	5 6	6 0	88	86	87
1893-1895	3 9	5 1	5 9	69	79	83

Depreciation of about twenty per cent. in first and second qualities and thirty per cent. in the inferior.

Ib., p. 49.

There was a decrease of 9.8 per cent. only in pork, bacon and hams. p. 50. For causes, see p. 142.

*Journal of the Royal Statistical Society, 1905. Estimate of Agricultural losses in the United Kingdom during the last thirty years, 1872-1904.

†Total area of land in England and Wales, 1916, Cd. 8240 :—

37,137,564 acres.
10,965,707 arable.
16,087,393 grass.

Acreage under wheat, 2,170,170 acres, 1916.
   ,,   ,,   ,,   4,213,651   ,,   (1856-1857).

the place of the United States should the latter's food exports be consumed at home, and Canada became increasingly important as a corn grower. The growth in the value of the food imports can be seen from the following table :

GROWTH OF IMPORTS.

From   Agricultural Statistics, published 1912.   Cd. 6385.

VOL. XLVI.   PART IV.

Annual Average during seven years.

	1856-1862 £	Per head s.	d.	1905-1911 £	Per head s.	d.
Wheat and Flour ...	17,876,000	12	6	48,104,000	20	3
Potatoes - -	174,000	0	1½	1,570,000	0	8½
Meat - -	3,584,000	2	6	48,042,000	21	7
Butter and Margarine -	3,217,000	2	3	25,783,000	11	7
Cheese - -	1,249,000	0	10	6,902,000	3	1
Eggs - -	408,000	0	3	7,247,000	3	3
Raw Fruit - -	839,000	0	7	9,073,000	4	1
Nuts - -	367,000	0	3	1,444,000	0	7¾
Vegetables, other than Potatoes - -	121,000	0	1	2,477,000	1	1½
	£27,835,000	19	4½	£147,642,000	66	3½

AGRICULTURAL STATISTICS, 1913.

VOL. IV.   Cd. 7551.

SOURCES OF SUPPLY.

Wheat Supplies, 1913.

Per cent.

Australia supplied	8.7	of the total import.
Canada ,,	22.5	
India ,,	15.3	
	46.5	
Argentina ,,	12.3	
Chile ,,	.6	
Germany ,,	.9	
Russia ,,	4.1	(In 1910, 24.3)
U.S.A. ,,	34.8	
Total Foreign	53.5	

Thus nearly half came from within the Empire.

B B

Meat Supplies, 1913.
Per cent.

Australia supplied	,	15.1	of the total import.
Canada	,,	1.6	
New Zealand	,,	10.9	
Argentina	,,	38.1	(In 1907, 19.8%)
Denmark	,,	11.5	
Netherlands	,,	4.3	
U.S.A.	,,	12.8	(In 1907, 41.8%)
Uruguay	,,	3.4	
Others	,,	2.3	

Argentina had taken the place of the United States as the chief source of the imported meat supply.

" It would be difficult to paint in too black colours the depression of the last quarter of the nineteenth century affecting, as it did, first the corn lands of the South and West and later the pasture districts of the other parts of the country. Bankruptcies among farmers increased to an alarming extent ; many lost their whole capital. . . . In spite of the reduced incomes and the depreciated value of their estates remissions of rent by landlords up to fifty per cent. were common. . . . In some cases land became derelict. . . The effect of the depression was felt with peculiar severity by yeomen and small occupying farmers. Labourers, having less to lose, suffered perhaps to an even greater extent than their employers. While the remuneration for every other class of labour was steadily increasing, the wages of the agricultural labourer actually declined."*

The large imports of wheat often took the form of flour and the English flour mills were adversely affected. Corn mills were set up at the ports, such as Liverpool and Hull, to grind the imported wheat and to get cheap water carriage for the offals which were increasingly in demand in Ireland for pig feeding and in Denmark and Holland. There was in consequence a wholesale destruction of the local corn mills which had been dotted all over the country at ten or twelve mile intervals.

(d) Agricultural Reconstruction and Social Experiment.

Between 1895 and 1914 great changes took place in English agriculture. The English farmer began to concentrate on growing those things where he still held a sort of national monopoly, where he would not, in fact, be a prey to harvests

*Report of Agricultural Reconstruction Committee, *p.* 11, Cd. 9079, 1918.

in Saskatchewan or Entre Rios, California or Ukraina, all to him incalculable factors. The result was that England began to assume the aspect of a dairy farm and market garden. The very drop in price of wheat and meat had caused people to have surplus money for other things, such as fruit and vegetables. These began to be eaten in increasing quantities ; the consumption of milk per head doubled, and strawberries became the temporary joy of the boy in the street, while the growth of jam factories provided a ready market for the surplus fruit crops. The result was an increase of cows and dairying, an extension of the area devoted to fruit growing and an extension of flower, vegetable and fruit culture. The improvement in the breeds of English live stock was also maintained as the chilled meat was inferior in flavour to home bred meat ; the price of the latter was higher and it still paid to produce first quality meat. The quantity of wheat grown declined and the imports rose, as will be seen by the following table, which is in striking contrast with the rise in the wheat and rye production in Germany during the same period.*

In 1863-1864 the production of wheat in the United Kingdom was 44,805,120 cwts.

FISCAL BLUE BOOK, 1909.

p. 176.   Thousand Cwts., Wheat.

Annual Average.	Home Production.	Foreign Imports.
1880-1884	41,225	73,418
1885-1889	39,598	39,295
1890-1894	34,621	90,815
1895-1899	31,604	96,836
1900-1904	27,136	108,036
1905-1908	30,993	112,278

Germany.   Thousand cwts.

	Production.	Imports.*
1880-1884	46,549	11,910
1885-1889	51,198	9,317
1890-1894	56,417	19,238
1895-1899	62,458	28,650
1900-1904	69,270	38,072
1905-1908	73,236	44,039

Rye.   Thousand cwts.

1880-1884	109,997
1905-1908	195,444

*" The main value of the tariff policy to German agriculture was the sense of security which it created in the farmer. It was the conviction

It is interesting to notice that the production of meat in England increased, although the imports increased but the amount eaten per head declined.

MEAT IMPORTS.

AGRICULTURAL STATISTICS, IV., CD. 7551.

Year ending June 4th.	THOUSAND CWTS.		PERCENTAGE.		LBS.
	Home.	Imports.	Home.	Imports.	Per head.
1900-1901	29.330	20.936	58.3	41.7	136.3
1912-1913	31.087	21.104	59.6	40.4	127.6

During this period the decline in the number of agricultural labourers continued and there was in some counties a labour famine. The young men left the land and went to the towns or emigrated to America or Canada.

The decline may be seen from the following figures :*

1881	•	983,919
1891	•	866,543 − 117,376
1901	•	689,292 − 177,251

This decline gave point to an agitation to re-establish small holdings. It was said that the labourer was leaving the land because there was no incentive for him to remain. He could not get a small farm, therefore he went to America. It was held to be socially desirable to keep men in the country. It was thought that the country physique was better and re-enforced the town population with fresh and vigorous stocks. There was a superstition that a third generation of Londoners never existed. It was an admitted fact that the country population got some of the best posts in the towns. The country born are to be found in preponderant

---

that he was essential to the community and that the community would not permit his land to go out of cultivation rather than the prospect of receiving an extra two marks per 100 kilos for his wheat after the year 1906 that stirred the German agriculturist of the new century to make an effort ; and it may be added it was the knowledge that his grain was not wanted and that his fellow country-men did not depend upon his exertions that led the British farmer at the same period to cut down expense and reduce or at least fail to increase the productivity of his land." Middleton, Cd. 8305, p. 34.

---

*Report on the Decline of the Agricultural Population, 1906, Cd. 3273, p. 7. The figures vary somewhat in different authorities. It all depends what one calls " an agricultural labourer." There is no doubt about the marked diminution in the labour available.

numbers in the Civil Service, the scholastic profession, the police, railway service, gas workers, carters, shop assistants. It was not the country migrant who was found to be unemployed in the enquiry undertaken in 1906,* but the townsman. Hence the country population was held to be worthy of special encouragement. "Times" were now as much in favour of the small farmer as they had previously been against him. He does not produce wheat and meat as satisfactorily as the large man but he does grow fruit and vegetables even better; spade culture is more profitable for these than the plough. The labour difficulty of milking, which involves Sunday work, was obviated in the case of the small farmer who could utilize his family, and the generation before 1914 did not worry much about the cleanliness of its milk.

While a great agitation was carried on to revive small holdings it became obvious that it would have to be carried out by some public authority. Landlords, unless the circumstances were very exceptional, would not be willing to establish them. Small tenants are far more trouble than large ones, and moreover, land in England was not vacant: no good landlord would disturb a sitting tenant for the uncertain returns and very certain bother of small men. Houses, too, were a great difficulty. It was almost impossible to put up a house for a small farmer at any price that he could afford to pay if he were to give an economic rent. The cost of labour and building materials were the obstacles.

In 1908 the Small Holdings Act came into operation. It provided that County Councils should acquire land in

---

*Memorandum by Mr. Wilson Fox on Country Born in Large Towns in Report on Agricultural Settlements in British Colonies, Cd. 2978, 1906, *p.* 29. His conclusions were:

1. " That the major part of London poverty and distress is home made and not imported from outside."
2. " That the country-men who migrate to London are mainly the cream of the youths of the villages.
4. " That they usually get the pick of the posts . . . and in general all employments requiring special steadiness and imposing special reliability.
6. " That the country immigrants do not, to any considerable extent, directly recruit the town unemployed who are, in the main, the sediment deposited at the bottom of the scale as the physique and power of application of a town population tend to deteriorate."

every county and sell or let it to suitable small holders. If they could not acquire the land they wanted by negotiation they were given the power to take it compulsorily and submit the price to arbitration.   Should the County Councils refuse to act, the Board of Agriculture could act in default.   Government grants were given to the County Councils for legal and other preliminary expenses and loans were advanced at low rates to the County Councils for the general purposes of the Act.   These local authorities were to let or sell the land at such a rate that there should be no loss.   They need not make a profit.   The Government appointed certain Small Holdings Commissioners to help the local authorities and see that the Act was put into execution.

As will be seen from the following figures, much has been accomplished but it has not been a revolution in tenures.

ANNUAL REPORT.   SMALL HOLDINGS COMMISSIONERS, 1915.   Cd. 7851.
1908-1914.

Number of applicants, 46,660 and 96 Associations.
Land applied for, 782,286 acres.
Applicants approved, 27,667, of whom 18,486 obtained holdings.
Land acquired, 195,499 acres.
       (a) purchased, 138,405 for £4,549,068.
       (b) leased, 57,094, rents, £71,221.
Let to 12,584 individuals.
       506 acres sold to 50 persons.
       8,436 acres let to 63 Co-operative Societies who sub-let to
         1,451 members.
       3,580 persons provided with 47,500 acres by private land-
         owners.
Board of Agriculture leased 182 acres at Bournemouth for £370 a year
    let to 16 tenants and one Society.
In 1914, 32 per cent. of applicants were agricultural labourers.   Of
    2,100 applicants in 1914, only 587 asked for houses to be provided
Number of new houses erected, 774.
Compulsory orders, 491 for 35,588 acres.
Loans sanctioned, £5,255,553, and Government advanced £207,179
    for preliminary expenses.

People already on the land had their holdings enlarged ; those who had a house often got a bit of land in addition, but there was very little done towards increasing the number of people actually living in the country.   The cost of building is almost prohibitive even for a County Council, which has not to make a profit—only avoid a loss.   A Departmental Committee, sitting in 1913, reported that the cheapest single

house that could be put up would be one costing £183 and
that contained no " parlour." Fancy a farmer's wife without
a parlour !* Useless as that appendage is, it is a social
necessity. The result was that only 774 new houses have
been built by the local authorities for small holders between
1908 and 1914. The young man who wanted to marry
and who was driven to the towns or the colonies to get house
or land would not be much encouraged to stay in the country
by the possibility of getting one of the 110 new houses provided
annually on the average all over England.

It was obvious, too, that giving a man a farm does not
make him a good farmer. The economy of the small man
should be totally different from that of the large man. To
prosper he must farm highly intensively. A man with four
hundred acres making £2 an acre profit may live comfortably.
A small man on thirty acres will not keep a wife and family
and pay rent on £60. He must, therefore, farm better than
the big man. The only object lesson, however, that he
knows is the large farmer. How can he get instruction ?
The State has to provide it. Efforts were being made in
this direction prior to the outbreak of war in 1914.†

Instruction is, however, not enough. The small man must
be able to sell his produce, hence it is necessary for him to
combine for purposes of sale. He will also do better if he
co-operates for the purposes of purchasing his agricultural
manures and using machinery in common. He needs credit
to tide him over the year. Farming is mainly outlay in the
Spring and returns in the Autumn. He has to live till the
Autumn.

Without co-operation in purchase and sale, instruction
and credit, it is unlikely that the small holder will be a success.

*Cd. 6708, Plan II.
†" As regards the small holders themselves their two great needs
are education and co-operation. . . . As a rule the small holder
is not in a position to avail himself of the courses of instruction given
at Agricultural Colleges and it is necessary, therefore, that instruction
and advice should be brought to his very door. It should be the aim
of the County Councils to supply this need, both through the agency
of travelling advisers and instructors and by the practical illustrations
of the results that may be obtained by a small holder through the
application to his business of the findings of agricultural research."
Annual Report of Small Holdings Commissioners, Cd. 6157, 1912,
p. 22.

Meanwhile the big man does not need all this assistance. The small man is socially desirable but nationally expensive. It is an interesting fact that there is scarcely any demand by English small holders to purchase land. Their tenure is secure and they prefer to rent land so as to leave if it does not pay* or if profitable move on to something bigger. They prefer to sink their capital in stock and not in land.

The movement to re-establish a peasantry was checked by the war of 1914. With the submarine menace the question became one of increased corn growing in this country. Celery and rhubarb are not much use when the country lacks wheat. The future fate of the small holder will depend on the extent to which Great Britain will revert to her former position as a wheat importer. If the small holder is to be further developed he will have to be housed, taught and organized into co-operative societies with State assistance.

One effect of the Great Depression in England was to interest the State actively in agriculture. Up to that time it had pursued a policy of laissez-faire. A Member of Parliament described the position, in 1881, as follows:

"If one wanted to get any information about the diseases of cattle he was referred to the Minister who looked after art and science, education and religion; if one wished to hear something about agricultural statistics or corn returns he was referred to the Minister whose main duty it was to look after railways and ships; while the President of the Local Government Board, to whom they looked for information with regard to public highways, roads and bridges, had for his main duty to look after paupers."†

In 1889 the Board of Agriculture was created. Its functions have been mainly to give free advice and spread knowledge. It has undertaken a good deal of research into insect pests and has stamped out certain cattle diseases. It administers the Acts intended to protect the farmer with regard to the contents of fertilizers and food stuffs. In 1909, a "development grant" was specially allotted to agriculture. The

*"During the past year 99 tenants in England and Wales have given up their holdings at their own request and 20 tenants have received notice to quit from the Councils," p. 12, Small Holdings Report, Cd. 6157, 1912.

†Hansard, 3 Ser., Vol. CCLXI., p. 442.

Treasury were empowered to advance money for improvements in forestry, for research in agriculture, for rural industries and for improved transport. Agriculture was thus directly subsidized.

England was, in 1914, organized into twelve agricultural provinces, each with its expert advisers on agricultural matters.* There has also been a considerable development of State aided agricultural education.† The tendency to direction from above was considerably developed by the war which witnessed compulsory corn growing, the fixing of minimum wages in agriculture and a guaranteed price for wheat production.

II.—THE EFFECT OF MECHANICAL TRANSPORT ON IRISH AGRICULTURE AND THE RELATIONS BETWEEN GREAT BRITAIN AND IRELAND.

The most marked reaction from laissez-faire in the direction of State leadership and Government alteration of existing tenures is to be found in Ireland which had become in the thirty years prior to the war the classic country of economic experiments by the State.

(1.) From the reign of Elizabeth to that of Charles II. England was occupied in trying to anglicise Ireland. Geographically she commanded the sea approaches from the Atlantic to both the English and the St. George's Channels, and lay strategically right across England's coming and going on the highways of the world. Ireland was not, as James I. said, England's " back door," but her front door. A Roman Catholic Ireland was, therefore, an ever present menace to England's security, for if the Irish were not strong enough to interrupt English commerce the country, nevertheless, constituted an excellent base for Spanish and French attempts against England.

The object of the sixteenth and seventeenth century rulers was to colonize the country and make it English in religion, land tenure and agricultural methods, language and administration. Although England and Scotch settlers were

*Leaflet issued by the Board, No. 279, " Technical Advice for Farmers," 1915.

†As to the extent and organization of English Agricultural Education in 1913-1914, see Annual Report of Education Branch, Board of Agriculture, Cd. 7450, 1914.

" planted " in Ireland with a view to leavening the Irish and creating an English type of civilization, England's object was not wholly attained. By the end of the seventeenth century the communal land tenure was altered to freehold, the language had become English, the country was divided into counties, the English judicial system was set up in place of the jurisdiction of the clan chiefs, a money economy was introduced, but the great mass of the Irish, except in Ulster, still acknowledged the Roman obedience and were alien to English habits of thought. Nor, owing to the chronic insecurity caused by rebellions, repressions and consequent confiscation of land, did the methods of agriculture improve as they were improving in England during the seventeenth century.

(2.) Between 1660 and 1783 the English policy altered. If Ireland was not to be anglicised she must at least be made harmless. Those years may be termed the period of the suppression of competition of Ireland with England. The result was a series of penal laws which created a great gulf between the two religions in economic status. They were designed to make the Protestant interest economically the stronger. The possession and disposal of land was regulated in such a way that the Protestants became the chief land owners. Being in the nature of settlers in a hostile country, they did not develop their land along intensive lines, as did the corresponding English proprietors in the eighteenth century. The Irish landlord did not even erect buildings. He let the bare land, the Irish tenant made it into a farm with his " improvements." The other side of the shield is to be found in the fact that there was no ousting of the small farmer by improving landlords in the eighteenth century as there was in England.

Certain commercial restrictions were placed on Ireland as on the other English colonies after the Restoration. They were intended to prevent Irish industries competing with English in such a way that the English cloth trade should be injured or the cattle industry of England affected. On England rested the main burden of defence and nothing, so seventeenth and eighteenth century statesmen considered, must be allowed to impair her financial capacity. Hence the Irish were not allowed (1699, 10 & 11 W. III. c 10) to export cloth other than frieze except to England and between England and Ireland there was a heavy duty which restricted the export of cloth. Nor were they allowed to send their cattle to England, between 1663 and 1759. If, however, the woollen industry was penalized, linen was encouraged by bounties.*

(3.) After a brief period in which the Irish Parliament became independent (1783-1801) the two kingdoms were united again and a third period, in which Ireland was treated on exactly the same footing as England ensued, which lasted to 1870.

The equal treatment of Ireland and Great Britain was perhaps not the best possible arrangement for Ireland at the beginning of the nineteenth century. A backward, over-populated agricultural country, such as Ireland was in the first half of the nineteenth century needs very different treatment from a rich industrial country like Great Britain. In all countries the peasantry in the nineteenth century have needed State guidance, but England believed firmly in laissez-faire. She allowed her own peasantry to disappear without making any effort to save them and it was not likely that she would have taken special measures to preserve Irish peasants even had she known how to do it. The agricultural experts believed in the large farm ; they considered it better that the Irish peasant should go to the United States and leave room for what they deemed the only efficient type of agriculture—the large holding. The famine of 1845 merely confirmed the experts in their opinion that the country was wretchedly under-cultivated, that small farmers were bad farmers and that the best thing for Ireland was to place no obstacles in the way of emigration and clear away all legislative hindrances to " high farming."† It was impossible for the soil of Ireland to feed nearly eight and a quarter million people, and there were no industries sufficiently developed to absorb the population and permit them to buy their food with manufactures as England was doing. Sir Horace Plunkett has calculated

*Bonn : " Die englische Kolonisation in Irland " considers that the effect of the commercial restrictions on Ireland has been greatly exaggerated and that they scarcely injured Catholic Ireland at all, especially as the native manufacture of frieze was not affected. Vol. II., *p.* 230.

†The population of Ireland was 8,175,124 in the Census of 1841. It was 4,337,000 in 1915.

that only three million people can now live on the land of Ireland with a reasonable standard of comfort. Although it comprises twenty million acres, a quarter of them are bog, barren mountain and waste.* How much the less, then, could nearly treble that number exist with their backward methods of farming in 1845 ? In a like case, the French and Russian peasants had agitated for more land and had made revolutions to obtain it. The Fenian agitation arose out of similar agricultural conditions but Ireland is too small a country for more land to be any real alleviation and intensive farming, so the Government thought, could not be carried through by small farmers.

Emigration was the only remedy for an over-population of that kind. It is interesting to notice that the intense hostility felt towards the English caused the Irish to go mainly to England's revolted colonies, the United States, and not appreciably to the colonies that had remained within the Empire. While the result was to create in the United States a party intensely hostile to Great Britain it has produced a homogeneity in the other colonies which has probably done something to cement the new British Empire of the nineteenth century.

Meanwhile England wished to introduce large farms into Ireland to promote the sinking of capital in the land and better yields and the opportunity was given her in the fact that about one-seventh of Irish land passed into the bankruptcy courts. By two Acts, the Encumbered Estates Act of 1848 and Deasy's Act of 1860, land sales were facilitated and the way cleared for free contract in land. In thirty years no less than £50 millions' worth of land changed hands. Large farms cannot develop in a country covered with small farms without ousting small farmers, and the growth in the larger farms and the decline in the smaller may be seen from the following figures :

	1 to 5 acres.	Per cent.	5 to 15 acres.	Per cent.
1841	310,436	44.9	252,799	36.6
1851	88,083	15.5	191,854	33.6
1871	74,809	13.7	171,383	31.5
1891	63,464	12.3	156,661	30.3
1901	62,855	12.2	154,418	29.9

*Ireland in the New Century (1905), p. 50.

	15 *to* 30 *acres.*	*Per cent.*	30 *acres and upward.*	*Per cent.*
1841	79,342	11.5	48,625	7.
1851	141,311	24.8	149,090	26.1
1871	138,647	25.5	159,303	29.3
1891	133,947	25.9	162,940	31.?
1901	134,091	26.0	164,483	31.9

(1901)

Since 1841, reduction in holdings :

1 to 5 acres	247,581
5 to 15 acres	98,381
	345,962

Increase of holdings :

30 acres and upwards	115,858
15 to 30 acres	54,749
	170,607*

That the disappearance of the small farmer was not even more rapid was due to the fact that the remittances from America enabled many of them to give exorbitant rents and it did not pay the speculators and middlemen who had bought a large proportion of Irish land to turn out people who would often pay as much as £10 an acre for agricultural land without buildings. On the other hand, if a tenant had to leave, the very fact that he had to leave behind the "improvements" and buildings he had made himself, constituted a peculiar hardship when eviction took place. In England, the bulk of the small owners had been able to sell their land at good prices, while the disappearing tenants had an alternative in factory work or canal and road making. In Ireland it was land or the United States. Prior to the abolition of the duties on food stuffs grown at home between 1842 and 1846, Ireland had supplied Great Britain with young cattle for fattening and dairy produce. The free trade measures threw open Ireland's best market to the competition of all the world. The situation as far as Ireland was concerned was not much affected till 1878, transport was not sufficiently developed.

The Irish peasant could not always make his farm pay even in the "golden age" (1850-1873); his methods were primitive and his rent high. Hence evictions followed which

*Bonn : Modern Ireland and her Agrarian Problem. *p.* 46.

gave rise to the Fenian agitation and forced England to abandon laissez-faire in Ireland before it was even contemplated in England.

(4.) In 1870 began the fourth period in Irish land history which may be styled that of the constructive treatment of Ireland. In that year Gladstone fathered a Land Act which guaranteed fixity of tenure where it had previously existed under a custom known as the Ulster Tenant Right, thus abolishing the idea of free contract in these cases. Where evictions took place compensation had to be paid, the object being to make them so expensive that the landlord would not resort to that procedure.

In 1878, however, the depression set in, prices of dairy produce and cattle dropped and all Irish rents based on the good years seemed to be on too high a level. In Great Britain the landlords reduced rents by as much as fifty per cent., but in Ireland evictions increased and the Land League was formed. Rents were not paid, the boycott was resorted to and land agents were shot.

In 1881 another Land Act was passed to satisfy the demand for the three F's : Fixity of tenure, Free sale and Fair rents. On demand of either landlord or tenant a newly created body, the Land Commission, was to fix the rent for the ensuing fifteen years. When that term had expired they could be fixed again for another fifteen and so on.* This was a drastic interference with the liberty of the landlord to make the best bargain he could and marks a striking departure from the idea of letting the Irish land system freely work itself out along English lines to large farms. In this way reductions of rent amounting on the average to 20.7 per cent. were carried through for the " first term " rents and when they came up for the " second term " a further cut on the first term rents of 19.3 per cent was made, while the " third term " showed a reduction of 9.2 per cent. on the rents as fixed for the second term.

1881-1916.   Cd. 8481, 1917.

			Originally	Reduced to
First term,	381,687 rents fixed		£7,523,816	£5,968,174
Second term,	143,394	,,	,, £2,571,983	£2,074,512
Third term,	5,007	,,	,, £84,558	£76,799

*In 1886 a similar Act was passed for Scotch crofters, the term being seven years. (49 and 50 Vict., c. 29.)

The land question was, however, by no means settled. The fixing of fair rents did not conduce to peace and progress. It resolved itself into a battle between landlord and tenant every fifteen years and was destructive of good relations between them. Men farmed badly to get a reduction of rent at the next term and Ireland was overburdened with a mass of solicitors, valuers, land commissioners and surveyors to administer the Acts and conduct the disputes. As no uniform standard of what constituted a " fair rent " was adopted there were considerable dissimilarities of procedure which caused much resentment.

A tentative attempt at buying the landlord out was made in 1885 by the Ashbourne Act, when £5 million was provided by the Government of the United Kingdom to enable the tenant to purchase his land. The tenant was to pay the advance off in annual instalments over forty-nine years. Another £5 million was advanced in 1888.

Balfour provided £23 million more in 1891 for the purpose of land purchase. By the end of the century, however, the landlords were refusing to sell. They had been paid in stock reckoned at par, the actual value of which had fallen below par, and there was a loss they would not face. The tenants, anxious to buy, clamoured for compulsion to sell and it seemed as if there would be further agrarian difficulties. Landlords and tenants met in conference in 1903 and drew up a scheme acceptable to both which was embodied in the Wyndham Act of 1903. The result was the provision of a much larger scale of money for land purchase. The amount was estimated to reach £100 millions. The landlord was to be induced to sell by being paid in cash and by receiving a bonus over and above the selling price. This was estimated to cost £12 million.

The £100 millions was found to be inadequate and further grants were authorized in 1909 of sums estimated to reach £83 million more. These sums were in addition to the £33 million already advanced under the Balfour and Ashbourne Acts.

The amount of the instalments to be paid under the 1903 Act was $3\frac{1}{4}$ per cent. on the purchase price for $68\frac{1}{2}$ years, at the end of which period the Irish tenant would own the land. In 1909 the interest on future advances was raised to $3\frac{1}{2}$ per cent. A special body, the Estates Commissioners, was created in 1903 to deal with the land sold in whole estates, while the Land Commission dealt with small portions of land.

Under this system land was rapidly being transferred to the tenant when the European War brought the advances to an end. The Government of the United Kingdom was borrowing at five and six per cent., and could not advance money at three per cent.(the half per cent. in the $3\frac{1}{2}$ is sinking fund). The question immediately before the War was that of bringing compulsion to bear on landlords who would not sell.

GENERAL RESULTS OF LAND PURCHASE ACTS, 1885-1913.

*Advances made by Land Commission :

Under Acts, 1881-1896 - -	£23,380,229
Under 1903 Act - - -	£2,073,786
Advanced to Congested Districts Board for purchase of estates - -	£2,295,079*
†Sales arranged for by Estates Commissioners -	£90,932,824
Already advanced - - - -	£56,887,014
To be advanced - - - -	£32,553,711
Cash provided by tenants - - -	£ 527,529

The result was that half the land of Ireland had changed hands, but it was a change of tenures, not methods.

The great difficulty in Ireland remained, viz., the backward nature of Irish farming. It is no use encouraging a man to own land if he does not know how to use it. Special measures were taken against his subdividing his land and he might only mortgage it for a sum not exceeding ten years of his instalments and then only with the consent of the Land Commission. But the Irish small farmer needed instruction and guidance. Small farmers also need co-operation to enable them to surmount the disadvantages of their smallness. Sir Horace Plunkett started an Irish Co-operative Movement in 1889 and the Irish Agricultural Organisation Society was so successful that a Government Department was created alongside of the Society to continue the work and carry out research. The result was the creation in 1899, of the Department for Agriculture and Technical Education, of which Sir Horace Plunkett was made the chief. Its functions were to aid agriculture in every possible way. Its peculiarity was that it worked with representative advisory bodies and through local authorities and committees.

*Report of Irish Land Commissioners, Cd. 6979, 1913.
†Report of Estate Commissioners, Cd. 7145, 1913.

It was bureaucracy tempered by constant contact with democracy.

Another State experiment was tried for the very poor regions of the West of Ireland, known as the Congested Districts. In 1891, Mr. Balfour set up a special body—the Congested Districts Board—the task of which was to raise the general condition of the people in these " rural slums." They dealt with about 3½ million acres and half a million of people. The Board was given £55,000 a year and its task was to raise holdings to an economic size, to start bye-employments, such as carpet, lace and tweed making, to provide cheap seed, improve breeds of cattle, make roads, build harbours, start fishing and transplant population. Their functions as to the encouragement of better agricultural methods were handed over to the Department in 1904. In Congested areas the Board had the right of compulsorily acquiring land should they deem it necessary. In 1909 its annual income was raised to £166,000 to continue the work of raising the standard of life in these regions and their efforts seem to have been attended with marked success.*

In 1906 the Government of the United Kingdom provided £4 million for labourers' cottages in Ireland.

After the separation of the Irish Free State from the United Kingdom the former Government undertook all administration connected with Land Purchase within its area as from April 1st, 1923. The British Government, however, will guarantee future issues of capital for the completion of land purchase and continue to pay the interest on the advances for past purchases, being recouped the amount thereof by the Free State. In Northern Ireland, which remains part of the United Kingdom, Land Purchase finance is still a British Government service, though certain administrative functions in connection therewith have been transferred to the Northern Government. The net result appears to be that Irish land has been rapidly passing into the hands of the Government of the United Kingdom which has transferred the land of Southern Ireland to the Free State. After the instalments have been paid, which will take over a century to complete, the land will be re-transferred to peasant owners who meanwhile must be taught

*A Congested Districts Board was started for Scotland in 1897 to deal with the specially poor regions there. (60 and 61 Vict., c. 53.) Its functions were transferred to the new Scotch Board of Agriculture in 1911.

to farm better. It is an experiment in temporary land nationalization, land transfer and agricultural education on a huge scale. Although the instalments were paid with great regularity in a time of rising prices the Irish peasant will have to learn to get more out of the land he holds if he is to stand the strain of the lean years when they come. He cannot meet them any longer by insisting on rent reductions. His instalments are purchase price, not rent due to a landlord.

The framework for improved agriculture exists in the co-operative societies which have brought out local leaders and have aroused a sense of common interests. They are anxious to help the farmer to buy, sell or grade his produce or organize credit. A Department was created to instruct the landed classes and help them to farm better, while a special body was formed with the interesting constructive task of raising the condition of those who have managed to exist where a European goat would starve.* "Peasant rescue," so notable a feature of continental governments has also been operative in Ireland.

It is interesting to notice that the effect of the over-population of Ireland was to lead to a great emigration to the United States which was also re-enforced by German peasants freed from serfdom. The result was that the United States obtained a population which helped on the rapid settlement of the prairie lands. But the United States would not have developed as rapidly as she did had Great Britain not been a manufacturing country and free trade, and thereby provided the great market for the grain and meat exports of the Middle West and the cotton of the South, while the carriage of the American grain and cotton contributed in no small degree to the building up of the British Mercantile Marine prior to 1914. No country can now live to itself ; it is action and reaction all the time.

A great deal of discussion took place in Germany before the War as to the relative advantages and disadvantages of the agrarian and industrial State. Some German writers considered Great Britain to be in a parlous condition because she had sacrificed her agriculture to her industrial development.† As a matter of fact the British Empire is still one of the great agricultural entities of the world and Great Britain is the pivotal exchange point of a great wheat producing, meat producing, wool producing commonwealth. This island is in reality the commercial and industrial centre of a vast agricultural community widely scattered over large portions of four continents.

*Bonn, *op. cit., p.* 23. †Wagner, *Agrar und Industriestaat.*

CONCLUSION

IF one were to summarize the economic effect of English technique on the world during the nineteenth century it would be true to say that Great Britain's inventions altered the whole of the methods of industrial production and that no country remained uninfluenced thereby. But her influence did not stop at industry ; she revolutionized the agriculture, the distribution of the population, the industrial code, the sanitation, the labour movement and the commerce of the globe.

The British reaction upon the world during the nineteenth century has been stupendous. By producing and exporting large quantities of coal the necessity for reserving certain areas in every locality as fuel areas was removed, commons were broken up and food areas increased. The application of machinery to agriculture enabled the crops to be saved in better condition because it shortened the duration of the harvest and minimized the risks of weather and this again added to the food supply.*

The opening up of the interiors of North and South America by the railway and the linking up of the world by the steamship, both of which were worked out to a successful issue in Great Britain, temporarily relieved the world from the fear of famine. By lessening the famines, Great Britain lessened the plague and pestilence which were their invariable consequences. New outlets at home and abroad were made possible for the European peoples just struggling out of serfdom and rapidly increasing in number. In the new industrial system copied from England they found new occupations in which they were trained to a large extent by British foremen ; Britain's demand for food and raw materials made it worth while for the growing populations to emigrate and open up new continents. Agricultural

*See Prothero, " English Farming, Past and Present," pp. 369-370 for English inventions of agricultural machinery. Machinery in agriculture was only a logical outcome of machinery in industry.

machinery assisted their development by enabling these continents to be cultivated in spite of the shortage of labour in a new country. This was true of both America and Australia. The invention of the railway set free an immense amount of labour in Africa which had been used in porterage. Mechanical transport made it possible to specialize production to a high degree and enabled the coal and manufactures of England to be exchanged for the food products of the world. The savings of the British people, invested, as they were, all over the world, caused a far more rapid development of its resources than would otherwise have been possible.

The factory system did not create the evils of overwork and child labour, it brought them into prominence and possibly somewhat intensified them ; nor did it create the insanitary state of the towns which had existed since the Middle Ages. It was the good fortune of this country to be the pioneer of much needed industrial and sanitary reforms which have been adopted all over the world.

The British developments of trade unionism and co-operation helped to mould the labour movement all over the world.

The British inventions of cables, railways and steamships made the whole world one great trading area and ushered in a revolution in commerce and international trade. While Great Britain influenced all production with her machinery, she revolutionized all distribution by her developments in mechanical transport and this again ushered in the second stage of the industrial revolution which witnessed the spread of machinery to practically every important trade, the organization of international businesses, international combines, and international labour movements. With this there was an increased national rivalry to obtain control of the raw material and food producing areas since the resources of the world could be easily developed or exploited from any one centre. The result was the emergence of a new national economic imperialism due to the desire to get a control or monopoly of commodities limited in amount which cut across the new internationalism created by the increased possibilities of exchange. There was a reaction from world economics to imperial economics, the aim being self-sufficiency within large areas and yet, on the other hand, there was the increasing interdependence of the world.

APPENDIX

Some Books for further reference on

PART II.

THE INDUSTRIAL REVOLUTION

GENERAL :—

    Cunningham : " Growth of English Industry and Commerce," Vols. II. and III.

    Ashley : " Economic Organization of England."

    Mantoux : " Revolution Industrielle," 1906.

    Fay : " Life and Labour in the Nineteenth Century," 1920.

    Hammond : " The Town Labourer, 1760-1832," 1917.

       ,,   " The Skilled Labourer, 1760-1832," 1919.

COTTON :—

    Baines, E. : " The History of the Cotton Manufacture in Great Britain," 1835.

    Chapman, S. J. : " The Lancashire Cotton Industry," 1904.

    Daniels : " The Early History of the Cotton Industry."

    Schulze Gaevernitz : " The Cotton Trade in England and on the Continent " (Translation), 1895.

    French, G. J. : " Life and Times of Samuel Crompton."

    Unwin : " Samuel Oldknow and the Arkwrights."

WOOL AND WORSTED :—

    Clapham, J. H. : " The Woollen and Worsted Industries," 1907.

    Heaton : " Yorkshire Woollen and Worsted Industries," 1921.

    James (John) : " The History of Bradford," 1841.

    James : " History of the Worsted Manufacture in England," 1857.

    Lipson : " History of the English Woollen and Worsted Industries," 1921.

HOSIERY :—

    Felkin, W. : " An Account of the Machine Wrought Hosiery Trade," 1845.

IRON AND ENGINEERING :—

    Ashton : " Iron and Steel in the Industrial Revolution."

    Smiles : " Industrial Biography, Iron Workers and Tool Makers," 1863. (Many later Editions.)

       ,,   " Lives of the Engineers," 1861-1862. (Many later Editions.)

       ,,   " Lives of Boulton and Watt," 1865.

IRON AND ENGINEERING :— *(contd.)*

Fairbairn : " Iron, its History, Properties and Processes of
Manufacture," 1861.

      "   " Life, partly written by himself," 1877.

      "   " Treatise on Mills and Millwork," 1861-1863.

Bell, Sir Lowthian : The Iron Trade and its Allied Industries in
" The Reign of Queen Victoria," Vol. II., edited
Ward, 1887.

Bell, Sir Lowthian : " The Iron Trade of the United Kingdom,"
1886.

Lord : " Capital and Steam Power," 1750-1800.

Report on Artisans and Machinery, 1824, Vol. V. ; 1825, Vol. V. ;
1841, Vol. VII.

COAL :—

Boyd, R. N. : " Coal Pits and Pitmen," 1895.

Galloway, R. L. ; " Annals of Coal Mining, 1898," 1904.

      "   " Papers relating to the History of the Coal
Trade."

Report on the Condition of Women and Children in Mines," 1842.
Vol. XV.

ROADS AND CANALS :—*See p.* 399.

CONDITIONS IN FACTORIES :—

Hutchins and Harrison : " History of Factory Legislation,"
1911. (Bibliography).

Dunlop : " English Apprenticeship and Child Labour, a History."
1912.

Report on Women and Children in Factories, 1833. Vols. XX.
and XXI.

HOME WORK :—

Report on the Woollen Manufacture, 1806.

Reports on Hand Loom Weavers, 1835, Vol. XIII. ; 1839, Vol.
XLII. ; 1840, Vols. XXIII. and XXIV. ; 1841,
Vol. X.

Report on Frame Work Knitters, 1845, Vol. XV.

Gaskell : " Artisans and Machinery," 1836.

Lloyd, G. H. I. : " Cutlery Trades."

TOWN LIFE :—

Buer, M. C. : " Health, Wealth and Population in the Early
Days of the Industrial Revolution," 1926.

George, D. : " London Life in the Eighteenth Century."

Hutchins : " The Public Health Agitation," 1909.

Jephson : " The Sanitary Evolution of London," 1907.

Reports on the Sanitary Condition of Towns and of the Labouring
Classes, 1840, Vols. XXIII. and XXIV. ; 1844,
Vol. XVII. ; 1845, Vol. XVIII.

Trade Unions :—

    Webb, S. & B. : " History of Trade Unionism." (Bibliography).

Poor :—

    Report on the Administration and Practical Operation of the Poor Laws," 1834. Vols. XXVII. and XXVIII. (Since reprinted).

Wages :—

    Bowley, A. L. : " Wages in the United Kingdom in the Nineteenth Century," 1900.

Encyclopædias (containing valuable references to early processes and inventions) :

    Rees : " The Cyclopædia or Universal Dictionary of Arts, Sciences and Literature," 1819.

    " The Penny Cyclopedia," 1833, etc.

    Ure : " A Dictionary of Arts, Manufactures and Mines," 1839. (Many later editions).

    " The Dictionary of National Biography," 1908.

Original Documents :—

    Bland, Brown and Tawney : " English Economic History," Select Documents.

    Smart : " Economic Annals of the Nineteenth Century."

Part III.

NINETEENTH CENTURY POLICY AND PROGRESS

General :—

    Dicey : " Law and Public Opinion in England during the Nineteenth Century," 1905.

    Porter : " Progress of the Nation from the Beginning of the Nineteenth Century " (new edition, ed. Hirst).

    Tooke and Newmarch : " History of Prices and of the State of the Circulation from 1793." (1838-1857).

    Layton : " Introduction to the Study of Prices."

    Page (Editor) : " Commerce and Industry, 1815-1914, Vol. I. " Historical Review, Vol. II., Statistical Tables.

    Giffen : " Economic Enquiries and Studies," 1904.

COMMERCE AND COMMERCIAL POLICY :—

Leone Levi : " History of British Commerce."

Bowley : " England's Foreign Trade in the Nineteenth Century,"
1905.

Fuchs, C. J. : " Trade Policy of Great Britain and her Colonies."

Macculloch : " Dictionary of Commerce."

Ashley, W. J. : " The Tariff Problem."

Cunningham : " The Rise and Decline of the Free Trade Movement," 1905.

Robertson, D. : " A Study of Industrial Fluctuation," 1915.

Nicholson, Prof. J. S. : " The History of the English Corn Laws."

Rathgen : " Englische Handelspolitik am Ende des 19ten
Jahrhunderts."

Schulze Gaevernitz : " Britischer Imperialismus."

Fuchs : " Englische Getreidehandel und seine Organisation."

Rees : " Short Fiscal and Financial History of England," 1815-
1918.

OFFICIAL :—

Report on the Depression of Trade, 1886.   Vols. XXI., XXII. and
XXIII.

Customs Tariffs of the United Kingdom, 1800-1897. C.8706
(1897).

Report on Food Supply and Raw Material in Time of War, 1905.
Cd. 2643.

Report on Commercial and Industrial Policy after the War, 1918.
Cd. 9035.

Report on Co-operation in the American Export Trade, Vols. I
and II., 1916, issued by the Federal Trade Commission.   (This official publication examines into
the causes of the successful prosecution of foreign
trade by other countries, notably Great Britain,
as the greatest of the commercial nations).

STATISTICS OF TRADE AND SOCIAL CONDITIONS :—

Fiscal Blue Books : Cd. 1761 (1903).
                    Cd. 2337 (1904).
                    Cd. 4954 (1909).

Statistical Abstract of the United Kingdom (annual).

Food and Raw Material Requirements of the United Kingdom,
1915. Cd. 8123.

Bowley : " Statistical Studies relating to National Progress in
Wealth and Trade since 1882."

    „   " Elements of Statistics," 1907.

STATISTICS OF TRADE AND SOCIAL CONDITIONS :— (contd).

Bowley: " Manual of Statistics," 1915.

„ " The Effect of the War on the External Trade of the United Kingdom : An Analysis of the Monthly Statistics, 1906-1914."

Stamp, J. C. : " British Incomes and Property."

Public Health and Social Conditions, Statistical Memoranda and Charts. Cd. 4671, 1909.

LABOUR AND SOCIAL CONDITIONS :—

Webb : " History of Trade Unionism."

„ " Industrial Democracy."

Hobhouse, L. T. : " The Labour Movement."

Beer : " History of British Socialism." (Translated).

Hovell and Tout : " History of the Chartist Movement."

Potter, B. : " The Co-operative Movement in Great Britain," 1899.

Simon, Sir J. : " English Sanitary Institutions."

Brassey : " Work and Wages," 1872.

Tawney : ed. " Studies in the Minimum Wage."

Schloss : " Methods of Industrial Remuneration."

Beveridge : " Unemployment."

OFFICIAL :—

Report of Commissioners on Employment of Children and Young Persons in Trades and Manufactures not already regulated by law. 1863, Vol. XVIII ; 1864, Vol. XXII. ; 1865, Vol. XX. ; 1866, Vol. XXIV.

Report of the Royal Commission on Labour, 1892, Vols. XXXIII.-XXXVI. ; 1893, Vols. XXXII.-XXXIX.

Report on Truck, 1871, C. 326. 1908, Cd. 4442.

„ Sweating, 1888, Vols. XX. and XXI. ; 1889, Vols. XIII. and XIV.

„ Home Work, 1908, 246.

„ Trade Disputes and Trade Combinations, 1906, Cd. 2825.

Report of Royal Commission on Mines, 1909, Cd. 4820.

„ an Enquiry by the Board of Trade into Working Class Rents, Housing and Retail Prices, together with the Standard Rates of Wages, 1908, Cd. 3864.

Report on Physical Deterioration, 1904, Cd. 2175.

„ the Poor Laws and Relief of Distress, 1909.

„ Compensation for Injuries to Workmen, 1904, Cd. 2208. 1920, Cmd. 816.

INDUSTRY AND INDUSTRIAL ORGANIZATION :—

Macrosty : " The Trust Movement in British Industry."

Levy, H. : " Monopoly and Competition."

Marshall, A. : " Industry and Trade, a Study of Industrial Technique and Business Organisation," 1919.

" British Industries," ed. Ashley, W. J.

" British Industries under Free Trade," ed. Cox, H.

OFFICIAL :—

Report on Trusts, Cd. 9236, 1919.

    „   Co-operation in the American Export Trade (an Examination of Foreign Trusts, etc.).

    „   the Textile Trades after the War, Cd. 9070, 1918.

    „   the Iron and Steel Trades after the War, Cd. 9071, 1918.

    „   the Coal Trade after the War, 1918, Cd. 9093.

Coal Conservation Committee, Cd. 9084, 1918.

Report of the Engineering Trades after the War, 1918, Cd. 9073.

    „   the Controller of the Department for the Development of the Mineral Resources in the United Kingdom 1918. Cd. 9184.

    „   on Bank Amalgamations, Cd. 9052, 1918.

FINANCE :—

Buxton : " Mr. Gladstone as Chancellor of the Exchequer."

    „   " Finance and Politics."

Mallett, Sir L. : " British Budgets, 1887-1888 to 1912-1913."

BIOGRAPHIES OF REFORMERS AND OTHERS :—

Hodder : " Life of the Seventh Earl of Shaftesbury."

Podmore : " Life of Robert Owen."

Wallas : " Life of Francis Place."

Morley : " Life of Cobden."

Helps : " Life and Labours of Mr. Brassey, 1805-1870."

Milner and others : " Life of Joseph Chamberlain."

PERIODICALS AND DICTIONARIES :—

Journals of the Royal Statistical Society.

Economic Journal.

Dictionary of Political Economy.

Conrad's " Handwörterbuch der Staatswissenschaften."

Bibliographies :—Published by U.S.A. Library of Congress, edited A. P. Griffin.

List of Books with reference to Periodicals on Trusts (1907).
On Labour, particularly relating to Strikes, 1903.
Industrial Arbitration, 1903.
Child Labour, 1906.
Iron and Steel in Commerce, 1907.
Employers' Liability and Workmen's Compensation, 1911.
Working-men's Insurance, 1908.

Parts IV. and V.

THE COMMERCIAL REVOLUTION, RAILWAYS AND SHIPPING.

General :—

Sax : " Die Verkehrsmittel in Volks und Staatswirtschaft," 1918
Cohn : " Geschichte und Politik des Verkehrswesen."
Acworth : " Elements of Railway Economics."
" State in relation to Railways," Papers by various writers. Royal Economic Society.
Hadley : " Railroad Transportation."

Roads :—

Webb, S. & B. : " Story of the King's Highway."

Canals :—

Jackman : " Transportation in Modern England."
Smiles : " Brindley " in " Lives of the Engineers."
Priestley : " Historical Account of the Navigable Rivers, Canals and Railways throughout Great Britain," 1831.
Royal Commission on Canals, Cd. 4979 (1909).
Pratt : " History of Inland Transport."

Railways :—

Francis : " History of the English Railway ; its Social Relations and Revelations, 1820-1845 " (1851).
Lewin, H. G. : " British Railway System. Outline of its Early Development to the year 1844."
Cleveland Stevens : " English Railways, their Development and their Relation to the State," 1915.
Ross : " British Railways."
Cohn, G. : " Untersuchungen über die englische Eisenbahnpolitik, 1873-1883."

OFFICIAL :—

> Report of Royal Commission on Railways, 1867, Vol. XXXVIII. (Historical Introduction).
>
> Report on Amalgamations, 1872.
>
> „ Railway Agreements and Amalgamations, 1911, Cd. 5631.
>
> Report of The Board of Trade Railway Conference, Cd. 4677, 1909.
>
> Report on the Railway Conciliation and Arbitration Scheme, Cd. 5922, 1911.
>
> „ Accounts and Statistical Returns rendered by Railway Companies, Cd. 4697.

PERIODICAL :—

> Archiv für Eisenbahnwesen.

SHIPPING :—

> Fairbairn : " Treatise on Iron Shipbuilding, its History and Progress," 1865.
>
> Lindsay, W. S. : " The History of Merchant Shipping," 1876.
>
> Kirkaldy, N. W. : " British Shipping, its History, Organisation and Importance."
>
> Sargent : '' Seaways of the Empire," 1918.
>
> Jones, G. M. : " State Aid to Merchant Shipping." Studies of subsidies, subventions and other forms of State aid in the principal countries of the world. U.S.A. Bureau of Foreign and Domestic Commerce, Special Agents Series, No. 119 (1916).
>
> „ " Navigation Laws." Comparative study of the principal features of the Laws of the United States, Great Britain, Germany, Norway, France and Japan, 1916. U.S.A. Bureau of Domestic and Foreign Commerce. Special Agents' Series, No. 114 (1916).

OFFICIAL DOCUMENTS :—

> Committee on Steamship Subsidies, 1901, Vol. VIII. ; 1902, Vol. X.
>
> Royal Commission on Shipping Rings, Cd. 4668 (1909).
>
> Dominions Commission, Cd. 8462 (1917).
>
> Memoranda and Tables of the Chief Harbours of the British Empire and certain foreign countries and as to the Suez and Panama Canals, Cd. 8461.
>
> Committee on Shipping and Shipping Policy after the War, 1918, Cd. 9092.

Appendix

IBLIOGRAPHIES :—Published by U.S.A. Library of Congress, edited A. P. Griffin.

List of Books with references to Periodicals on Mercantile Marine Subsidies, 1906.

Ditto, on Government Ownership of Railroads, 1904.

Ditto, on Railroads in relation to the Government and Public, 1904.

Ditto, Railroads in Foreign Countries, 1905.

PART VI.

THE COMMERCIAL REVOLUTION AND THE NEW IMPERIALISM.

Egerton : " Short History of British Colonial Policy."
Lucas, Sir C. P. : " The British Empire."
    " Greater Rome and Greater Britain."
Bruce, Sir C. : " The Broadstone of Empire."
Ireland : " Tropical Colonisation."
  " The Far Eastern Tropics."
  " Colonial Administration in the Far East."
Curzon, Lord : " The Place of India in the Empire."
Drage : " The Imperial Organisation of Trade."
Ashley, W. J. (Editor) : " The British Dominions."
Johnson, S. : " Emigration from the United Kingdom to North America."
Mills, R. C. : " The Early Colonisation of Australia."
Chamberlain, J. : " Speeches, edited C. W. Boyd, 1914.
Milner, Lord : " The Nation and the Empire," 1913.

REPORTS :—

Return of Differential Duties in favour of Colonies, 1823-1860, Cd. 2394 (1905).

The Dominions Commission, especially Cd. 8462, 1917. Final Report.

Statistical Abstract of the British Empire. Annual.

Report on the Moral and Material Progress of India. Annual.

Reports of all Crown Colonies and Protectorates, Published annually.

Reports of the Colonial Conferences, 1887, 1894, 1897, 1902, 1907, 1911.

REPORTS :—(contd.)

> Proceedings of Imperial War Conference. 1917, Cd. 8566 ; 1918, Cd. 9177.
>
> Report on Emigration from India, 1910, Cd. 5192.
>
> Colonial Office List.
>
> Report of the Empire Cotton Growing Committee. Cmd. 523.
>
> Report of East Africa Commission, 1925. Cmd. 2387.
>
> Proceedings of the Royal Colonial Institute and the Empire Review.

BIBLIOGRAPHY :—Published by U.S.A. Library of Congress, edited A. P. Griffin.

> List of Books with References to, Periodicals on " The Theory of Colonisation, Government of Dependencies, and Related Topics," 1900.
>
> Ditto, on " British Tariff Movement under Chamberlain," 1906.

PART VII.
BRITISH AGRICULTURE IN THE NINETEENTH CENTURY

GENERAL :—

> Prothero : " English Farming, Past and Present."
>
> Levy, H. : " Large and Small Farms."
>
> Bedford, Duke of : " Story of a Great Agricultural Estate."
>
> Nicholson, Prof. : " Rent, Wages and Profits in Agriculture."
>   ,,   " The English Corn Laws."
>
> Rew, Sir H. : " An Agricultural Faggot."
>
> Hall, A. D. : " Agriculture after the War."
>
> Hasbach : " A History of the English Agricultural Labourer " (Translation from German), 1908.
>
> Dunlop : " The Farm Labourer. The History of a Modern Problem." 1913.
>
> Curtler : " The Enclosure and Redistribution of our Land."

OFFICIAL :—

> The Reports on the Agricultural Depression : (i.), 1880-1882 ; (ii.), 1895-1897.
>
> Report on Small Holdings, 1906.
>   ,,   our Food Supply in Time of War, 1906.
>   ,,   Fruit Culture, 1905.
>
> Reports of the Small Holdings Commissioners. Annual.
>
> Report on Equipment of Small Holdings. Cd. 6708, 1912-1913.

OFFICIAL :— *(contd.)*

Agricultural Statistics dealing with : Prices, Acreage, Crops and Live Stock ; Imports and Exports, Foreign and Colonial Statistics. Annual.

Agricultural Output of Great Britain, 1912, Cd. 6277.

Agricultural Policy (Reconstruction Committee), 1917. Cd. 8506 (3*d.*).

Report on the Decline of the Agricultural Population of Great Britain. Cd. 3273, 1906.

Report on Migration from Rural Districts, 1912. Cd. 6277.

Middleton : " Recent Development of German Agriculture," Cd. 8305.

Rural Education Conference. Cd. 6871.

ARTICLES :—

Palgrave : " Estimate of Agricultural Losses." Journal of Royal Statistical Society, 1905 (May).

Fox, W. : " Agricultural Wages in England and Wales during the last Fifty Years." Journal of Royal Statistical Society, 1903.

Eversley : " Decline in the Agricultural Population." Journal of Royal Statistical Society, 1907.

IRELAND

Bonn : " Die englische Kolonisation in Irland."

,,   " Modern Ireland and her Agrarian Problem." (Translation).

Barker : " The Irish Question."

Murray : " The Commercial and Financial Relations between England and Ireland."

Plunket : " Ireland in the new Century."

Staples-Smith, Gordon O'Brien : " Rural Reconstruction in Ireland."

Report of the Bessborough Commission on the Landlord and Tenant Act, 1881-1882.

Report on the Irish Land Acts, Fry's Commission, 1898.

Reports of the Irish Land Commissioners, Estates Commissioners, Congested Districts Board, the Department of Agricultural and Technical Education. Annual.

Report on the Financial Relations between Great Britain and Ireland, 1896.

Report on Land Purchase Finance, 1908. Cd. 4005.

APPENDIX

THE RAILWAY ACT OF 1921*

A NEW era in railway history was begun with the reorganisation of the British railway system after the War. The Act of 1921 hastened the process of amalgamation, provided a new experiment in the control of railways by the State and set up a new method of settling wages and other disputes in which railway workers were involved.

There were still in 1921 no less than 214 separate railway undertakings in Great Britain.† Of these 121 were to be combined into four groups, so as to create larger units. These larger units were expected to show great economies in working, in the building of railway equipment and in the handling of traffic. The highest group of railways outside the four combines was formed by the urban and suburban railways of the London area, worked mainly by electricity and confined chiefly to passenger traffic.

The idea was that each of the four groups was to be a unit which would give economic working.

The railways were amalgamated into (1) a Southern, (2) a Western, (3) a Midland, North-Western and West Scottish and (4) an Eastern, North-Eastern and East Scottish group. The old historical names such as the London and North-Western dating from 1846, the Midland from 1844, the Great-Northern from 1845, and the London and South-Western from 1839, all disappeared and only the Great Western retained the name it inherited from 1835.

*Grouping under the Railways Act, 1921 ; Acworth *Economic Journal*, March, 1923. Communications (Resources of the Empire Series). W. T. Stephenson, *Railways of the United Kingdom, p.* 173.

†The present number is eighty-eight of which thirty-three are not working Companies. Further of the fifty-five, four form the Underground Group and ten are joint lines, the property of the Big Four.

According to the Act the companies were to settle their own terms of amalgamation and present them to an Amalgamation Tribunal. Should they fail to agree the Tribunal would itself undertake the task of combining the companies. As a matter of fact the amalgamation of over a thousand millions sterling of capital and of powers and lines which had grown up as we have seen in the most haphazard fashion, was carried out by the companies themselves, under the stimulus of the Act and approved by the Tribunal.* Competition was not, however, abolished. In the amalgamation each of the big four absorbed other companies which had extensions into the territory of its rival, and as long as two companies run to Exeter and Plymouth, or two lines serve London and Manchester, so long will competition exist. The competition really centres round the towns and the industrial centres, so that while the bulk of the territory is non-competitive, the bulk of the traffic is still competitive, as Sir William Acworth pointed out. A limitation of this competition by agreement seems inevitable in the future.

Instead of having to apply to Parliament whenever such an agreement is arrived at, as was necessary in the olden days, with the result that such an agreement was usually refused in order to maintain competition, a much simpler and cheaper procedure has been substituted in an appeal to the Ministry of Transport, which is empowered to sanction such a proceeding by the issue of an Order. Thus the Ministry of Transport has control over all future agreements regulating competition.

It is interesting to see how Parliament, after doing its best for over three quarters of a century to keep alive and stimulate competition, has now completely abandoned that attitude and has itself forced on amalgamations and made the path easy for further cessation of competition.

*The chief exception was the Caledonian Railway.

DD

The benefit of the savings which it is hoped will ensue under the new grouping is to go as to 20 per cent. to the Company making the saving, and 80 per cent. to the customers in reduction of rates.

The old statutory maximum rates, above which the companies could not go, but below which they might vary their rates, subject to control if they wished to move them upwards, are now a thing of the past. New schedules of rates have been drawn up after examination and will be fixed by a new body, the Railway Rates Tribunal. This is a business body, the function of which is to fix rates that shall actually be paid. It is to fix the charges at such an amount as shall yield to the companies the standard net revenue of 1913, with sundry named additions, provided they work with efficiency and economy. The companies may only lower the standard rates by exceptional rates varying between 5 and 40 per cent. of the standard rate. The right of appeal by the trader, either for or against an exceptional rate, is maintained, and the Tribunal has to review all the exceptional rates and standard charges at intervals. The whole system of returns, accounts and statistics has been overhauled and more information is now available as to the cost of working the railways and the sources of their revenues.

The vast task of classifying the goods and of fixing the new standard rates was thrashed out between the Railway Rates Committee and the Traders Co-ordinating Committee. It then went to the Railway Rates Advisory Committee and on to the Railway Rates Tribunal, which in 1926 is concerned with the final fixing of the rates.

In addition to these changes a new body was set up to deal with labour questions—the National Wages Board. On it are represented the Railway Companies, the railway workers and the railway customers, i.e., the public, under an independent Chairman. Below this body each company has a series of Councils and there is a Central Wages Board on which the Companies and the employees alone are represented. The final appeal lies, however, to the National Wages Board. Thus the wages and conditions of employment are no longer left to the companies to decide, but in the last resort the public, too, is brought in and its wishes considered, so that the exploitation of the public by an industry in which there can be no foreign competition is avoided. There is, however, no compulsion on the companies or the employees to obey the award, and strikes have taken place by bodies of workers who refused to accept the decision of the National Board. How important labour remuneration is in the costs of working may be seen from the fact that 120 millions sterling, or 52 per cent. of the total receipts of the railways in 1924 was absorbed by salaries and wages, and 47 millions, or 20.5 per cent. went to the remuneration of capital.*

Thus the Act of 1921 is yet another attempt to provide an alternative to state railways.†

*Ry. Returns, 1924.
†I am indebted for assistance on the technical points to my colleague, Mr. Stephenson, who is in charge of the Railway Department at the London School of Economics.

INDEX

A

AFRICA, 14; Exports to, 30; Railways in 186-187, 343-345; Scramble for, 322. (*See also* South Africa, West Africa, Egypt).

AGRICULTURAL LABOURERS, decline of, 376; and Industrial Revolution, 64, 367, 369, 371; in Ireland, 389; Wages of, 369. (*See also* Peasant Farmers).

AGRICULTURE, Acreage of Arable, 372; Board of, 154, 380; as Bye-Occupation, 101; and Canals, 246; Change in Methods of, 67, 118, 364-366, 369, 375; and Coal, 18; in Crown Colonies, 149, 349-352; Depression in, 142, 372, 374; Development Grant for, 380; Education for, 381; Effect of Abolition of Serfdom on, 9; Effect of Imports on, 13, 142, 363, 372; Features of British, 362; German, 375*n*; History of British, 360*ff*; Losses in, 372; Machinery in, 137, 371, 391-392; Methods of in Ireland, 388-389; Numbers employed in, 177, 376; Prosperity of, 138, 370-371; and Railways, 224; State Assistance to, 13, 153, 154, 380-381. (*See also* Small Holdings, Farming, Peasants, Free Trade, Corn Laws, Wheat, Meat, Famines and under separate Countries).

AIRE AND CALDER CANAL, 244.

ALSACE, 146, 164*n*.

AMALGAMATIONS (*See* Business Organization, Combines, Railways, Shipping).

AMERICA, 3, 4; (*See also* Colonies, United States, Argentina, Canada).

ANTWERP, 306*n*.

APPRENTICESHIP, 67, 93, 114; Abolition of, 126; Pauper, 92.

ARGENTINA, 229; Emigrants to, 230*n*; Food Exports of, 372; Wheat from, 202, 373.

ARKWRIGHT, 47, 48, 57, 72, 76, 99.

ARTISANS, Skill of British, 153.

ASBESTOS, 339.

ASIA and Railways, 187.

ASIATIC EMIGRATION, 231, 232, 355*l*.

AUSTRALIA, 179, 322; and Imperial Preferences, 332-333; Meat from, 374; Population of, 329; Wheat from, 202, 373; Wool from, 56, 57.

B

BANK OF ENGLAND, 22, 129.

BANKING, British, 34, 163, 166, 167; Joint Stock, 25, 129, 215; German, 145; French, 34.

BAUWENS, 76*n*.

BEEF, Price of, 372; Beef Trust, 209. (*See also* Meat).

BELL, SIR I., 143.

BENTHAM, J., 128.

BERTHOLLET, 59.

BESSEMER, 143, 186.

BILBAO, 41.

BIRMINGHAM, 31, 69, 73.

BLEACHING, 21, 59.

BLENKINSOP, 71.

BLINCOE, 92*n*.

BLOUNT, SIR E., 139*n*.

BOARD OF AGRICULTURE, 154, 380; of Health, 105, 126; of Trade, 265-266, 284.

BOULTON, 72.

BRADFORD, 52, 72, 80.

BRASSEY, 99, 138, 163*n*.

BRAZIL, Emigration to, 230*n*.

BREAD, Price of, 366*n*, 368, 369.

BRIDGWATER, DUKE OF, 28, 242, 243.

BRINDLEY, 99, 243.

BRITISH EAST AFRICA, 350.

BRITISH EAST AFRICA COMPANY, 324, 325.

BRITISH EMPIRE, Eighteenth Century, 316-320; Nineteenth Century, 118, 321*ff*; Advantages of Belonging to, 357-359; and Mechanical Transport, 193, 195, 196; Metal Resources of 339, 340; Products of, 196, 339; Raw Materials of, 339; Statistics of, 328; Trade of, 329. (*See* Colonies, Crown Colonies, Chartered Companies, Empire in Trust, Empire in Alliance, Shipping)

BRITISH GUIANA, 356.

BRITISH NORTH BORNEO COMPANY, 324, 325*n*.

BRITISH SOUTH AFRICA COMPANY, 324,

BRUNEL, 74*n*.

BUSINESS COMBINATIONS. (*See* Combinations.)

BUSINESS ORGANIZATION, 23-25; Types of, 25, 207, 208, 209, 210, 211; Control of, 212.

C

CABLE COMMUNICATIONS, 205, 211; Imperial, 331, 332.

CALEDONIAN CANAL, 244.

CALICO. (*See* Cotton.)

CANADA, 42, 178, 179, 318; Canada Company, 324; Emigration to, 229; and Imperial Preferences, 332, 333; Meat from, 374; Population of, 329; and Railways, 195, 196; and Shipping Subsidies, 333; Wheat from, 202, 373; and West Indies, 296, 332, 333.

CANALS, 28, 37, 79, 80, 240-253; Advantages of, 242, 245, 246, 247; Characteristics of, 245; Decline of, 249-251; Disadvantages of, 251, 252;

CANALS—*contd.*
Dividends of, 244 ; French, 248 ;
Mileage of, 243 ; and Railways,
249, 263, 272 ; Revival of, 252, 287,
288 ; Tonnage on, 249, 250 ; Royal
Commission on, 203 ; Scotch, 243.

CAPITAL, 82, 113 ; Accumulation of in
Great Britain, 34 ; British Investment of Abroad, 138, 139, 167 ;
in Railways, 213, 214 ; in Colonies,
334 ; Correctives of, 113 ; Freedom
for, 130 ; Organization of, 25 ;
(*See also* Business Organization and
Combines).

CAPITALISTS, 99.
CARDING, 49, 51.
CARDWELL'S ACT, 274.
CARRON, 298.
CARTELS, 208.
CARTWRIGHT, 55, 56.
CENSUS, 3. (*See* Population).
CHAMBERLAIN, AUSTEN, 385.
CHAMBERLAIN, JOSEPH, 148, 158, 325,
327, 329, 330, 337, 341, 342, 343,
346, 347, 348, 358.
CHARLOTTE DUNDAS, 298.
CHARTERED COMPANIES, 174, 323-325.
CHARTIST MOVEMENT, 133.
CHILDREN, 39, 53-63, 67-68, 103 ; Death
Rate of, 81 ; Education of, 96,
151 ; in Factories, 50, 91, 92 ; and
Factory Acts, 95 ; Feeding of,
151 ; French, 123 ; Home Work of,
91, 94 ; Labour Market for, 68 ;
Legislation for, 116, 151 ; in Mines,
63*n*, 71, 94, 96, 124 ; and Steam
Power, 93 ; Work of, 151. (*See
also* Education).
CHINA, Emigration from, 232.
CIVIL SERVICE, 155.
CLEMENT, 75.
CLOTH, 27, 38, 102*n* Manufacture of,
32, 39 ; Trade in, 38, 39.
COAL, 8, 10, 19. 137, 391 ; Bunker, 302 ;
and Canals, 28, 29 ; Export of,
80, 118, 141, 164 ; Freights of,
308*n* ; Importance of, 18, 21, 164-
165 ; International Comparisons,
161, 165 ; and Iron Smelting, 69 ;
Output of, 18, 71, 79, 141, 165 ;
and Railways, 19, 261 ; and
Shipping, 299, 300, 302, 312 ;
Tonnage Exported, 200 ; Trade in
London, 204 ; Transport of, 18th
Century, 70, 71, 261 ; Winding of, 71.
COALBROOKDALE, 69.
COAL MINING, 27, 70, 71, 116.
COAL MINERS, 71, 94.
COAT'S SEWING COTTON, 208.
COBDEN, 140, 186.
COCOA, 187, 344.
COKE, OF NORFOLK, 366.
COLBERT, 146, 176.
COLD STORAGE, 200.
COLONIZATION, Attitude towards, 179,
320, 321, 326, 329 ; by Chartered
Companies, 174, 323-325 ; Coloniza-
tion, Nineteenth Century, 14, 119,
149, 322*f* ; and Railways, 13, 14,
179 ; Sixteenth Century. 316*f*.

COLONIAL CONFERENCES, 147, 314, 330,
331.
COLONIAL STOCK ACT, 334.
COLONIES, and Bounties, 42 ; and Cable
Communications, 331, 332 ; and
Commercial Treaties, 332 ; and
Cotton, 42, 345, 350, 351 ; Customs Union proposed, 323 ; Emigration to, 229, 317, 341 ; Exports
from Britain to, in Eighteenth Century, 30 ; Freedom of Trade granted
to, 297 ; French trade with, 30 ;
and Immigration of Asiatics, 231,
255 ; and Industrial Revolution,
30, 42 ; Investment in. 334 ; and
Mechanical Transport, 315*ff* ; and
National Policy, 147-149 ; Nature
of Trade of, 329 ; Postal Facilities
to, 331 ; Preferences to and from,
13, 42, 146, 148, 297, 317, 320, 331-
333 ; and Shipping Rings, 310 ;
(*see also* Tariffs) ; Reaction in
Favour of, 147, 148, 315, 322, 323,
326, 329, 337 ; Rebellion of, 319,
320 ; in Seventeenth Century, 316
317, 318 ; Raw Materials from, 41,
42, 353 ; (*see also* Raw Cotton) ;
and Railways, 185 ; and Shipping,
13, 294, 295, 297, 333 ; (*see also*
Navigation Acts) ; and Slavery,
318, 320 ; Struggle for, 326 ; Trade
Commissioners for, 336 ; Unpopularity of, 320, 321, 357 ; (*see also*
Crown Colonies).
COMBINATION ACTS, 84 ; Repeal of, 128.
COMBINATIONS IN BUSINESS, 11, 152,
153 ; Advantages of, 211 ; Control
of, 212 ; Disadvantages of, 211 ;
Horizontal, 208 ; International,
208, 210 ; in Trade, 206*f* ; in United
Kingdom, 210 ; Vertical, 207 ;
(*See also* Trusts, Cartel, Shipping
and Railway Amalgamations).
COMMERCE, Change in Methods of, 204*ff* ;
Effect of Industrial Revolution on,
79 ; Revolution in, 10, 185-212 ;
Staple Commodities of, 199 ; Train-
ing for, 155. (*See also* Trade.)
COMMERCIAL INTELLIGENCE DEPART-
MENT, 155, 336, 337.
COMMERCIAL TRAVELLERS, 247.
COMMERCIAL TREATIES, 140, 332.
COMMONS, Enclosure of, 366.
COMPOUND ENGINE, 299, 300.
CONSTRUCTIVE IMPERIALISM, 148, 149,
326*f*.
CONGESTED DISTRICTS BOARD, 389.
CONSULS, 155.
CONTINENTAL SYSTEM, 102.
CONTROL STATIONS, 229 ; (*see* Shipping)
CO-OPERATIVE SOCIETIES, 113, 138,
154 ; in Agriculture, 379, 380 ;
in Ireland, 388, 390 ; Capital of,
114 ; Numbers in, 114 ; Whole-
sale, 114.
CORN, Import of, 190, 365 ; Laws,
Repeal of, 130, 131, 132, 369, 370 ;
Transport of, 186, 189, 190.
CORNWALL, 72.

CORT, 28, 69
COTTON Duties in India, 354; Gin, 47; Factories, Numbers in, 59; Children in, 125; Export of, 118; Famine, 140; Growing in West Indies, 348; International Comparisons, 160, 161.
COTTON MANUFACTURE, 27 and *n*, 31, 40, 41, 47, 79, 116; Inventions in, 21; Importance of, 59; Jennies in, 46, 48; in Germany, 146; Location of, 60; Machinery in, 29, 32, 43, 47*ff*; in Moscow, 164; in Russia, 188; Mule in, 49, 50; Power Looms, Numbers of, 55; Water Frame, 48; Water Power, 49, 50.
COTTON PIECE GOODS, 43, 44, 45, 48; Bleaching of, 59; Dyeing of, 60; Export of, 47, 141; Printing of, 44; Wearing of Printed, 44.
COTTON, Raw, 41, 42, 45, 46; and Colonies, 42; Competition for, 41; Egyptian, 149, 351; and French Wars, 102; Imports of, 40, 45, 46, 241; Indian, 352*n*; Production of, 187; Shortage of, 59*n*, 140, 350; Sources of Supply, 188, 196, 345, 350; from United States, 46; from West Indies, 149; Sea Island, 349.
COTTON SPINDLES, 161.
COTTON SPINNING BY MACHINERY, 47*ff*.
COTTON WEAVING BY MACHINERY, 53*ff*.
COTTON WORM, 350, 351.
COTTON YARN, 32, 46, 48, 53; Export of, 49, 141; Price of, 46, 101.
COUNTRY BORN IN LARGE TOWNS, 377.
CRAWSHAY, 80, 99.
CRIMEAN WAR, 105.
CROMPTON, 49, 50; Boyhood of, 91.
CROWN COLONIES, Applied Entomology in, 349, 350; Financial Assistance to 343; Indian Labour in, 354, 355; New Policy for, 343; Officials in, 353; Position of Colonial Secretary with regard to, 328*n*; Preferential System in, 353; Railways in, 343-345; Scientific Agriculture in, 348-349; and Tropical Medicine, 345*ff*; Value of, 342.
CREUSOT, 36, 37, 75.
CROMFORD, 48.
CROMPTON, 49, 50, 91.
CYPRUS, 353.

D

DALE, 50.
DANGEROUS TRADES, 149.
DANTZIG, 202, 365.
DARBY, 28, 69.
DAVEY, SIR H., 106.
DEATH DUTIES, 116, 158.
DEFOE, 27, 32, 38, 39.
DENMARK, Meat from, 374.
DEPARTMENT FOR AGRICULTURE AND TECHNICAL EDUCATION IN IRELAND, 388, 389.
DEPRESSION OF TRADE COMMISSION, 146, 156.
DOMESTIC INDUSTRY (*see* Home Work).
DOMINIONS COMMISSION, 195, 332*f*, 338.

DRINKWATER, 72.
DUTCH, 42; (*See* Holland).
DYEING, 21, 60.

E

EARNINGS, Family, 96; in France, 170.
EAST INDIA COMPANY, 40, 41.
EDUCATION, 96, 114, 126; Technical, 151.
ENGLAND, Position of in Seventeenth Century, 4; in Eighteenth, 5; in Nineteenth, 108-109; (*See* Great Britain).
EGYPT, 179; Cotton Growing in, 149, 351.
ELECTRIC POWER, 19.
EMIGRANTS, Asiatic, 231*f*, 355*f*; Conveyance of, 227, 229; Destination of, 228, 229, 341; Remittances by, 229; Statistics of, 229, 230*n*.
EMIGRATION, 371; Causes of, 226-229; to America, 11, 14; by Chartered Companies, 324; to Colonies, 317, 341; Control of, 230, 231; Control Stations for, 229; of Indian Subjects, 354; from Germany, 229; from Ireland, 384, 390; Outside British Empire permitted, 128; Problems of, 230, 231.
EMPIRE IN ALLIANCE, 327, 330-341.
EMPIRE IN TRUST, 328, 341*f*.
EMPIRE OF RULE, 328.
EMPIRE OF SETTLEMENT, 327.
ENGINEERING, 19, 20, 68; British Skill in, 163; Development of, 27; (*See also* Machinery, Railways, Shipping)
ENGINEERS, British, in France, 75, 76, 77; Chinese, 167; rise of, 20; Shortage of, 56, 72, 74; Skill of British, 75; Training of, 167.
EXPORTS, Comparative Statistics of, 159; to British Possessions, 149; Values of, 138, 141, 147.

F

FACTORIES, Children in, 50, 91, 92, 93; Irish in, 63; Lighting of, 94; Location of, 60; Numbers employed in, 59; Unpopularity of, 64, 91.
FACTORY ACTS, 93, 94, 95, 96, 103, 113, 125, 126, 133, 137, 152; Adults under, 132.
FACTORY Buildings, 60; Hands, 89, 90, 91; Inspectors, 89, 108, 125, 126, 151.
FACTORY SYSTEM, 79; Advantages of, 65, 84, 100; Growth of, 49*f*; and Employment, 90; (*See also* Machinery, Industrial Revolution).
FAIRBAIRN, 34, 75, 99.
FAIRS, 205.
FAMINES, 391, 201, 203; Irish, 383.
FARMERS, Large, 364, 366; in Ireland, 383; Peasant, 367.
FARMS, Numbers of, 368; Size of, 154; (*See* Small Holdings).
FAY, SIR S., 205.
FEDERAL TRADE COMMISSION, 166, 212.

FEDERATED MALAY STATES, Railways in, 343 ; Tropical Medicine in, 346.

FISHING INDUSTRY, 224.

in, 343 ; Tropical Medicine in, 346.

FISHING INDUSTRY, 224.

FLAX, 42 ; (See Linen).

FLOUR, Import of, 206.

FLOUR MILLING, 207, 374.

FLYING SHUTTLE, 54, 61.

FOOD STUFFS, Exports of, 10 ; Import of 177, 200, 362, 372, 373 ; within British Empire, 339.

FOURNESS, 72.

FRAME WORK KNITTERS, 58, 87n.

FRANCE, 2, 4, 5, 108, 109, 138 ; Area of, 258 ; Birth Rate of, 159 ; British Machinery in, 76, 77 ; British Workmen in, 75 ; Canals of, 248 ; Capital in, 34 ; Child Labour in, 128 ; Coal Output, 161, 165 ; Colonies of, 3, 5 ; Colonial Trade of, 35 and n ; Cost of Living in, 170 ; Cotton Statistics of, 160, 161 ; Export Trade of, 26, 30, 34, 35 ; Home Work in, 61, 64 ; Free Trade in, 132 ; Growth of Towns in, 219 ; Industrial Position of, 3, 19, 26, 33, 35, 37, 145 ; Iron in, 20 ; Iron Output of, 161 ; Labour in, 22, 33, 122 ; Machinery in, 36, 66 ; Model for England, 146 ; National Policy of, 171 ; Population of, 3, 26, 153 ; Railways in, 130, 161, 192, 253 ; and Raw Cotton, 40, 41, 43 ; Rivalry with England, 41, 42, 43, 145, 146, 147 ; Roads of, 35, 239 ; Shipping of, 160 ; Subsidies to Shipping, 305 ; Trade Statistics of, 159, 162 ; Wages in, 33 ; Wool Consumption, 161 ; Raw Wool, 39, 43.

FREE TRADE, and Agriculture, 363 ; Continental, 131 ; Deviations from, 335 ; Establishment of, 130, 131 ; Importance of, 132 ; Motives for, 120, 127, 131 ; Policy of, 116 ; and Trusts, 153 ; (See also Corn Laws, Navigation Acts, Tariff).

FRENCH REVOLUTION, 83, 102, 175, 176 ; Influence of, 6, 7, 8 ; Destructive Effect of, 34, 35.

FRENCH WARS (see War and Napoleon)

FRIENDLY SOCIETIES, Funds of, 169.

G

GERMANY, 2, 3, 7, 8, 11, 19, 24, 31, 102n, 138 ; Area of, 258 ; Banking in, 145 ; Birth and Death Rate, 159 ; Cartels in, 208 ; Coal Output, 10, 161, 165 ; Colonial Possessions of, 326 ; Commercial Treaties with, 147 ; Cotton Statistics, 160-161 ; Divisions of, 139 ; in Eighteenth Century, 4 ; Emigration from, 228, 229 ; Flax from, 41 ; Growth of Towns, 218, 219 ; Industrial Development of, 36, 37, 145. 188 ; Influence on Great Britain, 147, 156 ; Iron Manufacture of, 146, 188 ; Iron and Steel Statistics, 160-162 ; Locomotives in, 192 ; Mercantile Marine of, 160, 313, 305, 306, 307 ; Paternalism of, 173 ; Population of, 159 ;

GERMANY—contd.
Effect of Railways on, 7, 187, 191 ; Numbers travelling by Railway, Effect of Railways on, 7, 187, 191 Numbers travelling by Railway, 215 ; Rivalry with Great Britain, 117, 145, 146, 178, 310 ; Trade Unions in, 24 ; Wheat Imports, 375 ; Wheat Export to England, 373 ; Wheat Production, 372 ; Woollen Statistics, 160, 161.

GILCHRIST-THOMAS PROCESS, 144, 146.

GIRARD, 57.

GLADSTONE, 130, 132, 386.

GLASGOW, 59, 80, 103.

GOLD DISCOVERIES, 136.

GORGAS, COLONEL, 347n.

GOTT, 52, 99.

GREAT BRITAIN, Area of, 258 ; 1815-1914, 177f ; Capital of, 34 ; Causes of Success of, 162ff ; Change in Character of, 104 ; Colonial Possessions, 1815, 1819 ; Effect of Colonies on, ; Industrial Development of, 34, 35, 42 ; Colonial Trade of in Eighteenth Century, 35 ; Effect of Inventions on, 183, 391 ; Effect on the World, 108 ; Emigration from, 228 (See Emigration) ; Export Trade in Eighteenth Century, 26, 27, 30, 35 ; Import Trade, Eighteenth Century, 30, 35 ; Individualism of, 171ff ; Investments Abroad, 167 ; Loans, 213, 214 ; Prosperity of, 136, 137, 138 ; and Railways, 192, 235-290 ; and Shipping, 291-313 ; (See also United Kingdom) ; France, Rivalry with ; Germany, Rivalry with ; British Empire and Colonies).

GREAT POWERS, 2.

GUADELOUPE, 42.

H

HAMBURG, 46, 48.

HANDLOOM WEAVERS, Earnings of, 87, 119n ; Efficiency of, 100 ; Food of 87 ; Homes of, 84, 86.

HARGREAVES, 48, 76.

HARCOURT, L., 328n, 345, 353.

HARKORT, 188.

HIGHS, 48.

HOLLAND, 3, 4, 5, 36, 109, 146 ; Difficulty in Obtaining Raw Material, 46.

HOME WORK, 39, 56, 58, 61, 62, 63, 64, 82, 85, 87 ; Cheapness of, 119 ; and Children, 91 ; Disadvantages of, 65.

HOME WORKERS, 90, 96, 103 ; Earnings of, 64 ; (See also Hand Loom Weavers, Women, Children).

HORROCKS, 55.

HOSIERY, 57, 58.

HOUSING, 64, 67, 156 ; Rural, 377-379.

HUDSON, GEORGE, 270.

HULL, 207.

HUNTSMAN, 28, 69.

I

IMPERIAL BUREAU OF ENTOMOLOGY, 340.

IMPERIAL BUREAU OF MYCOLOGY, 340.

IMPERIAL DEPARTMENT OF AGRICULTURE 348, 349, 350n.

IMPERIAL FEDERATION, 323.
IMPERIAL MINERAL RESOURCES BUREAU, 340.
IMPERIAL PREFERENCE, Financial, 334 ; in Communications, 331, 332 ; (see also Tariff Preference, Shipping and Colonies).
IMPERIAL SHIPPING BOARD, 338.
IMPERIAL TRADE COMMISSIONERS, 148, 336.
IMPERIAL WAR CONFERENCES, 314, 331, 355, 359.
INCOME TAX, 116, 120, 131, 132, 137n, 158 ; Double, 334.
INDIA, 3, 5, 22, 146, 179 ; Changing Position of in the Empire, 354 ; Cotton Goods from, 43, 44 ; Discovery of, 2 ; Emigration from, 231 ; 354-357 ; (see also Colonies and Asiatic Labour) ; Exports to, 30 ; High Commissioner for, 354 ; Inclusion in Imperial Conference, 331 ; Population of, 328 ; and Preferential System, 335, 353 ; and Railways, 187 ; Reaction of on England, 149 ; and Scientific Agriculture, 350, 352 ; Trade of, 196 ; Tariff Autonomy, 354 ; Wheat from, 202, 373.
INDIVIDUALISM, British, 171.
INDUSTRIAL CHEMISTRY, 21, 59.
INDUSTRIAL REVOLUTION, Causes of in Great Britain, 26ff ; and Colonial Trade, 30, 42 ; Date of, 78 ; Effects of, 79ff, 101ff ; Features of, 107, 108 ; and Middle Classes, 98-100 ; Periods of, 22, 25 ; Slow Progress of, 61 ; Special Complications in Great Britain, 108 ; Spread of, 22, 25 ; and Transport, 22 ; World Effect of, 108, 391-392.
INLAND NAVIGATIONS (see Canals).
INSPECTORS (see Factory Inspectors) ; in France, 108n.
INSURANCE, for Accidents, 114 ; Sickness, 114 ; Health, 157 ; Unemployment, 150.
ITALY, 41 ; Emigration from, 229, 230.
IRISH LABOURERS as Factory Hands, 63, 86 ; as Railway Navvies.
IRELAND, 381-390 ; Commercial Restrictions on, 382 ; Congested Districts, 154, 389 ; Co-operative Movement in, 154, 388 ; Cultivated Area of, 384 ; Estates Commissioners in, 388 ; Department of Agriculture, 153, 154, 388 ; Emigration from, 228, 384, 390 ; English Influence, 382 ; Evictions in, 385, 386 ; Fair Rents in, 153, 386, 387 ; Famine in, 383 ; Financial Advances to, 174, 175 ; Improvements in Agricultural Methods, 388 ; Land Commission in, 386 ; Land Legislation for, 384, 386, 387, 388 ; Large Farms in, 384, 385 ; Land Purchase in, 153, 387 ; Linen Yarn from, 45, 46 ; Peasantry in, 383 ; Peculiarities of Land System, 382 ; Population of, 67, 383 ; Size of Farms, 384 ; Spinning in, 57 ; Wool from, 384.
IRON, 7 10, 19, 21 ; Comparative Statistics of 160 : Exports of, 141, 160 ; for Machinery, 29 ; Import of, 69 ; Industry, Migration of, 69.
IRON MANUFACTURE, 20, 26 ; and Coal, 128 ; in Germany, 146 ; and Hot Blast, 78 ; Inventions in, 28 ; Location of, 20 ; Effect of Railways on, 19, 188, 190 ; and Rails, 71 ; Smelting of, 69, 70 ; Ships, 137, 298 ; (See also Steel, Machinery, Engineering.
IRON MASTERS, 69.
IRON ORE, Import of, 143, 165 ; Smelting of, 69, 70.
IRON, PIG, Production of, 144n, 188.
IRON TRADE DEPRESSION in 143 ; and Railways, 135, 136 ; and Renewals, 19, 136, 137 ; (See also Germany, France, United States and Railways)
IRONWORKS, British, in France, 75.
ISMAILA, 347n.

J

JAMAICA, 312, 319, 321, 353.
JAPAN, 311, 326.
JENNY (see Spinning).
JOINT STOCK COMPANIES, 23, 25, 113 ; Capital of, 113, 130, 135n ; Formation of, 129, 130 ; Legislation for, 114 ; Numbers of, 113n ; and Shipping, 291, 304.

K

KAY, 54.
KENNET AND AVON CANAL, 247.

L

LABOUR, Change in Character of, 10, 89, 90 ; Combinations, 11, 25 (See Trade Unions) ; Condition of, 157 ; Comparison with French, 122, 170 ; International Comparisons, 24, 170 ; Improvement in Condition of, 168, 170 ; and French Wars, 122 ; Legislation for, 13, 114, 149, 150, 151 ; New Aim of Labour Legislation, 150 ; Movements, 25, 133 ; Effect of Labour Movement on Legislation, 157 ; Labour Party, 153, 157 ; and Railways, 25, 197, 281, 285, 286 ; Scarcity of, 32.
LAISSEZ-FAIRE, Policy of, 12, 106, 115, 178 ; Abandonment of, 117, 147 ; Exceptions to, 116 ; and Germany, 156 ; Limitations of, 117 ; and Railways, 156 ; Reaction from, 147f, 156, 184 ; Success of, 172 ; Theory of, 127 ; (See also Free Trade).
LAND (see Agriculture, Farms).
LAND PURCHASE (see Ireland).
LASCARS, 221.
LEEDS, 52, 71, 73, 80.
LEEDS AND LIVERPOOL CANAL, 247.
LEICESTER, 31.
LEPROY, 349n.
LEVANT, Cotton from, 41, 46.
LINEN, 32, 40 ; Factories, Numbers in, 59 ; Machinery in Spinning, 59 ; Power Looms for, 55, 58 ; Yarn, 41, 45, 46, 48, 57.
LINER, Shipping, 301.

LIVERPOOL, **28**, 103, 207, 346; Flour Milling at, 374.
LIVERPOOL AND MANCHESTER RAILWAY, 255, 261, 263.
LLOYD GEORGE, 345, 350.
LOMBE, 48.
LONDON, 3, 81, 103, 177; Trade, 144, 145.

M

MACADAM, 70, 238.
McCULLOCH, 199.
MACHINERY, in Agriculture, 137, 371, 391; Advantages of, 102; British in France, 75, 76, 77; in Germany, 171; Causes of, 28*f*; in Calico Printing, 60; Carding, 47; in Chemical Industries, 21; Cotton, 43, 45, 46-57, 118; Dearer than Hand Work, 119, 121; Difficulty in Manufacture of, 72, 74; Disadvantages of, 103; and Engineers, 10; Exports of, 76, 78, 129, 137, 141, 160; Flying Shuttle, 54; in France, 9; in Hosiery, 58; Hostility to, 46, 50, 51; and Labour Troubles, 65, 66; in Linen, 57, 58; in Net and Lace, 58; and Machine Tools, 74; Manufacturers of, 60, note, 74, 99; Material of, 19, 29; and Population, 66; and Power Loom, 54, 55; and Railways, 206; Renewal of, 137; on Ships, 300; Shortage of, 56, 69; Slow Progress of, 48, 50*f*; Social Effects of, 7, 10, 62, 82*f*; Spinning, 21, 31, 47, 48, 49; Spread of, 25; Transport of, 10; from United States, 75, 163; in Weaving, 53*f*; Water Power for, 49, 50, 72, 73; Wool Combing, 56; in Woollen Manufacture, 50*f*, 118, 119; (*See also* Steam Engine, Engineering and Industrial Revolution).
MACHINE TOOLS, 19, 56, 74, 75, 77, 78.
MALTA, 353.
MANBY, 298*n*.
MANCHESTER, 28, 31, 45, 54, 59, 60, 65, 73, 86*n*, 220; and Canals, 242, 246.
MALTHUS, 121.
MARINE ENGINE, 143, 194, 298, 299, 300, 301.
MARSHALL, A., 57.
MAUDSLAY, 99.
MEAT, Imports of, 142, 199, 200, 376; Packers' Trust, 209; Sources of Supply, 374.
MECHANICAL TRANSPORT, 180*f*; (*See* Railways and Steamships).
MERCANTILISM, 12.
MERCHANDISE MARKS ACTS, 155.
METALLGESELLSCHAFT, 208.
METALLURGICAL TRADES, Combination in, 207, 208.
MIDDLEMEN, Growth of, 99, 221, 222.
MILLING INDUSTRY, 207.
MINERS, Wages of, 151.
MINES, Children in, 63, 71, 94, 124; Difficulties of Reform, 107*n*; Inspection, 126; Work, 126; Ventilation, 72-94; Women, 62, 71*n*.

MISSISSIPPI, 189.
MONAZITE SAND, 339.
MORRIS, SIR D., 342*n*, 348.
MOSCOW, 164.
MOSQUITO (*See* Tropical Medicine).
MULTIPLE SHOPS, 208.
MUNICIPALITIES, Activities of, 140, 156; Debt of, 156; Reform of, 129; Relief Works, 150; Trading by, 114; (*See also* Towns).
MULE, for Spinning, 49, 50, 52.
MUSLIN, 44, 45, 49, 53.
MUTTON, Price of, 372.

N

NAPOLEON I., 8, 43, 101, 102, 239; and French Industry, 173, 176.
NATIONAL DEBT, 119, 178.
NATIONAL POLICY, Change in, 147*f*; Comparison of, 171*f*.
NASMYTH, 34, 66, 75, 99.
NAVIGATION ACTS, 3, 293-298, 303, 316; Repeal of, 131, 296-298; and Colonies, 316.
NAVVIES, 267, 268.
NEILSON, 70, 99.
NET MANUFACTURE, 58.
NEWCOMEN, 6, 70, 72.
NEW LANARK, 50.
NEW ZEALAND and Imperial Preference, 332, 333; Meat from, 374; Population of, 329.
NEW ZEALAND COMPANY, 324.
NICHOLSON, PROF. J. S., quoted, 121.
NICKEL, 339.
NIGERIA, 196; and Cotton, 351; Preferences in the Export of Raw Material, 353.
NINETEENTH CENTURY, Characteristics of, 1, 2, 9, 14; in Great Britain, 113; Periods of, 118.
NORWICH, 41, 104*n*.
NOTTINGHAM, 31.
NYASALAND, 345.

O

ODESSA, 202, 365.
OLD AGE PENSIONS, 144, 150.
OLD COLONIAL SYSTEM, 316.
OTTAWA CONFERENCE, 332.
OWEN, ROBERT, 125*n*.

P

PACIFIC CABLE BOARD, 332.
PALM OIL, 197, 353.
PATENT ACT, 197.
PATERNALISM in Germany and France, 171.
PEASANT FARMERS, 377; in England, 363 365, 366, 369; in France, 367; in Germany, 367; in Ireland, 384, 385; Rescue of, 390.
PEEL, SIR R., 60, 72, 99.
PLUNKET, SIR H., 384, 388.
POLAND, 202.
POOR LAW, 96, 121; Reform of, 129.
POPULATION, Belt, 20; English (in 1760), 26; (in 1811), 120; (in 1911), 178; of British Empire, 328, 329;

POPULATION—*contd.*
Comparative Statistics of, **158**;
Density of, 80; in Eighteenth
Century, 3; French, 26, **158**;
Grouping of, **217**; Growth of, 66,
67; Irish, 120, 383; Migration of,
225f; Scotch, 26, 120; Theory of,
120, 121; Town, 3, **217, 219, 220**.
PORTUGAL, 2.
PORTS, 165, 188, 216, **247**.
POWER LOOMS, 55, 56, **57**.
PREFERENCES (*see* Tariffs).
PRICES, Fall of, 200; International, 199;
and Railways, 224.
PROTECTION, 12; and Railways, 184,
197; and Trusts, 210; (*See* Tariffs).
PRUSSIA, Policy of, 171f; Railways of,
289; (*See* Germany).
PURITANISM, Economic Effect of, 174.

R

RAILS, Iron, 71; Steel, 186.
RAILWAY AND CANAL COMMISSION, 252,
276, 279.
RAILWAY ACTS (1844), 266; (1854), 274;
(1888), **279**; (1894),   ; Number
of Private, 373.
RAILWAYS, and Africa, 186; and Agriculture, 224, 360f; Amalgamations
of, 155, 157, 207, 269, 275, 284, 290,
404-405; and American Civil War,
189; in Asia, 187; and British
Commerce, 187-191, 192, 193, 196;
British Peculiarities of, 253; Building of, 129; Building of Abroad, **138**;
and Business Combinations, 24, 207,
210, 211; and Canals, 249, 250 and
note, 259, 263, 272; Capital of, 135,
214; Clearing House for, 270, 271;
and Coal Traffic, 261; and Colonial
Development, 13, 14, 149, 179, 185;
and Commercial Methods, 204, 205,
206f; Commercial Revolution
caused by, 10, 180f; Conciliation
Boards on, 285, 286; Control of,
13, 265f, 278; and Corn, 190-196;
Cost of British, 257; in Crown
Colonies, 343ff; Earnings of, 261;
Effect of, 2, 10, 180f; in Great
Britain, 134, 135, 136, 191, 192;
and Employment, 134; Equipment
Renewals, 23, 200; and Emigration,
226; and Famines, 201, 203; and
Finance, 212f; and Flour, 207; and
Food Supply, 142; French, 139,
253, 254; Gauges, 187n, 192, 266,
274; German, 187, 191; History
of British, 235-290; in India, 187;
and Industrial Revolution, 22, 25;
and Iron Manufacture, 19, 23, 70,
188, 190; and Local Government,
223, 226; and Machinery, 206;
Material of, 186; and Migration,
225; Mileage of, 270, 272; Members employed by, 134; National
Wages Board, 406; Objections
to, 255; and Parliament, 265;
Political Effect of, 184, 191, 226;
Preferences given by, 274, 276;
and Prices, 224; and Protection,
184, 197; Prussian, 289; Railway
Rates Tribunal, 406; Rates of
Charge, 186, 190, 260, 270, 277, **287**;

RAILWAYS—*contd.*
406; Revision of, 278, 280; State
Control of Rates, 154, 279, 281, 290,
406; Rates in United States, 190n;
Reaction from Laissez-faire, 156;
Rebates and Shipping, 305, 306;
Revolution caused by, 180-231;
Rivalry created by, 193, 197; and
Roads, 238; Royal Commissions
on, 266, 274, 275; and Russia, 188
(*see* Russia); Social Effects of, 11,
215; State Railways, 184; Arguments for and against, **288**, 289;
State Purchase of, **266**; Statistics
of, 161; and Steamer Competition,
272; Strikes on, 284, 285, 286;
Tonnage carried by, 249; in United
States, 201; and Town Growth,
217; and Trade Unionism, 24, 284,
286; and Trading Classes, 221; in
Tropical Colonies, 149; Trucks,
259; in United States, 183, 188,
189, 190, 191, 254; in West Africa,
343, 344; Wooden, 70; and Women
225; Workers on, 267, 284 (*see*
Transport Workers), 406.
RAINHILL, 264.
REEDEREI-VEREINIGUNG, 307.
REFORM, Social, 124f, **149**; Difficulties
of, 104-107; Era, 104.
RHODESIA, 196.
RICHLIEU, 176.
ROADS, 25, 37, 70; British Policy with
regard to, 236f; **240**; French, 239;
and Railways, **238**; Repair of, 236,
237, 238; (*See* Macadam, Turn-
pikes).
ROCHDALE CO-OPERATORS, **113**.
ROEBUCK, 59.
ROSS, SIR R., 347n.
ROYAL NIGER COMPANY, 324.
RUSSIA, 2, 4, 5, 7, 8, 10, 14, 20, 24, 173;
Area of, **258**; Coal and Iron in, 10;
Effect of Railways on, 7, 188, 213,
326; Growth of Towns in, 219;
and Japan, 326; Emigration from,
226; Wheat from, **373**.

S

SAILING SHIPS, Number of, 299 and note.
SANITATION, 80, 81, 105; Reform of,
126; Effects of, 178.
SCHOOLS, Dame, 95n.
SCOTLAND, 34, 37, 70, 80; Congested
Districts Board for, 389n; Popula-
tion of, 120; Growth of Towns in,
219.
SCOTT, SIR WALTER, 106.
SEAMEN, Numbers of, 221.
SERFDOM, Abolition of, 7, 8, 9, 10, 37 :
Consequences of, 173, 363.
SHAFTESBURY, LORD, 125, 126, 128, **129**.
SHEARMEN, 66.
SHEFFIELD, 31, 69.
SHIPBUILDING, British Capacity for, 193,
194, 311; Capital in, 301; Changes
in Technique of, 194.
SHIPPING, Abolition of Protection of,
131, 132, 296f; and American Civil
War, 138; British Supremacy in,
193, 194 and note, 301f; and Coal

SHIPPING—*contd.*
302, 312 ; and Colonies, 293, 294-297, 333, 334, 339 ; Combinations, 307-310 ; Comparative Statistics, 160 ; Control of, 298 ; Control Stations, 306, 307 ; Depression in, 142 ; and Emigration, 227-230 ; Freedom of, Reasons for, 291, 292 ; Freights, 308 ; German, 302, 305-307 ; and Industrial Revolution, 36 ; Influence on British Empire, 195 ; Liners, 194, 301 ; Losses in Napoleonic Wars, 296 ; in War of 1914, 311 ; and Machinery, 300*n* ; Political Importance of, 194 ; Protection of (*see* Navigation Acts) ; Rebates, 309 ; Reservation of the Inter-Imperial Trade for, 312 ; Rings, 143, 207, 293 ; Rivalry of American and British, 304, 305 ; of German and British, 305-307 ; Size of, 302 ; State Encouragement of, 305 ; Statistics of, 193, 194 ; Subsidies, 293, 305, 306, 312, 333 ; Subventions to (British), 303, 304, 312 ; of United States, 193, 194, 291-297 ; Tonnage of British, 302, 313 ; of World, 313 ; Trade Nature of British, 194*n*, 303 ; Tramps, 194, 301 ; and Suez Canal, 142, 194, 299 ; (*See* Steamships, Seamen, Navigation).

SILK, Exports of, 141 ; Factories, numbers in, 59 ; Manufacture, 44, 48*n*, 140 ; Power Looms in, 55 ; Raw, 40, 41, 42.

SHOPKEEPING, 208, 223.

SLAVERY, 320, 344.

SMALL HOLDINGS, 154 ; Act, 154, 377, 378 ; Figures relating to, 378 ; Movement, Cause of, 376, 377 ; State Assistance to, 379.

SMITH, A., 71.

SMYRNA, 41.

SOMERSET, 51.

SOUDAN, 345.

SOUTH AFRICA, 178, 326 ; Imperial Preference in, 332, 333 ; Population of, 329 ; and Shipping, 310, 333*n*.

SOUTH AUSTRALIAN COMPANY, 324.

SOUTH WALES, 80.

SPAIN, 2, 3, 4, 109, 146, 165.

SPELTER, 339.

SPINNERS, 31, 39 ; Scarcity of, 47 ; Wages of, 85*n*.

SPINNING, 39, 96 ; in Cottages, 51, 52 ; of Cotton, 46, 47*f* ; by Hand, 31 ; in Factories, 51, 52 ; Flax, 57 ; Jennies, 31, 46, 49, 50, 51, 61, 62, 76 ; Machinery for, 31, 47*f*, 57 ; Monotony of, 82 ; Mule, 66 ; Power, 48*f* ; Rates of, 46, 52, 57 ; of Wool, 51 ; Worsted, 52.

STATISTICAL ABSTRACT of Br. Emp., 337.

STEAM ENGINE, 29, 69, 72 ; and iron manufacture, 21, 69, 70 ; for Pumping Mines, 21, 27, 70 ; Number of, 73.

STEAM LOCOMOTIVE, 264.

STEAM POWER, 60 ; Advantages of, 61 ; Characteristics of, 17 ; and Children, 92 ; Effects of, 5, 6, 17, 73.

STEAMSHIPS, 10, 25, 71; on American Rivers, 189 ; Cold Storage, 301 ; and Commercial Methods, 205, 206 ; Effect of, 142 ; History of, 291*f*, 298*f* ; Material of, 143, 194, 298, 300 ; Numbers of, 299, 301, 302 ; and Pirates, 203 ; Tanks, 301.

STEEL, 25 ; Bessemer, 143 ; Exports of, 141 ; Manufacture of, 190 ; by Gilchrist-Thomas Process, 190 ; in Germany, 144 ; Output of, 188 ; Rail, 143, 144, 186 ; Combine, 210 ; Ships, 143, 194, 300.

STEPHENSON, G., 99, 262.

STOCKTON AND DARLINGTON RAILWAY, 262.

STRIKES, 66, 98, 284-286.

STRUTT, 48, 99.

SUEZ CANAL, 142, 194, 299.

SUGAR, 317 ; Bounties, 148, 156, 348 ; Production of, 149 ; in West Indies, 348.

SURREY IRON RAILWAY, 262.

SWEDEN, 165.

T

TAFF VALE CASE, 285.

TARIFFS, 12, 37.

TARIFF PREFERENCES, 147, 148, 297, 317, 320, 321, 331-334 ; Amount of Preferences, 333 ; on Raw Material, 353 ; Given by United Kingdom, 335 ; (*See* Protection).

TARIFFS, Reform of, 130 : (*See* Free Trade).

TAXATION, after 1815, 119, 120 ; Weight of, 115, 116 ; (*See* Income Tax ; Death Duties).

TELFORD, 258.

TENNANT, 59.

TEXTILES (*See* Cotton, Wool, Worsted, Silk, Linen).

THOMAS PROCESS, 144.

TIN, 353.

TOBACCO, 317, 318.

TOWNS, 73, 80 ; Growth of, 10, 11, 19, 216, 217 and note ; Overcrowding in, 64 ; Mortality in, 181 ; Population in, 3, 177 ; Sanitation of, 81, 105 ; Water Supply of, 80, 140, 220.

TRADE BOARDS, 151, 157.

TRADE, Commissioners, 336, 337 ; Depression of, 115, 117, 141*f* ; Disputes Act, 152 ; Government Assistance to, 155 ; Inter-Imperial, 329, 336, 337, 338 ; International Comparisons, 162 ; Prosperity of, 139 ; Revival of, 136 ; Statistics of, United Kingdom, 136, 141, 147, 158*f* ; Success of British, 163*ff* ; Routes, Change of, 144 ; (*See* Exports, Imports, Commerce, Board of Trade).

TRADE UNIONS, 11, 22, 24, 25, 83, 84, 89, 133, 137, 152 ; Effects of, 157 ; Growth of, 157 ; Importance of, 113 ; Legislation for, 117 ; Numbers in, 113 ; Railway Servants, 284, 286 ; and Women, 152 ; (*See also* Labour, Combination Acts).

TRAMP SHIPPING, 301.
TRANSPORT AND INDUSTRIAL REVOLUTION, 22, 25; Stages of, 182.
TRANSPORT WORKERS, 220, 221, 222, 223.
TROPICAL MEDICINE, 345, 346, 347.
TRUCK, 87-89, 103, 126; Acts, 151; and Railway Labourers, 268.
TRUSTS, 24, 25; (See Business Organization and Combinations in Business).
TUNGSTEN, 339, 340.
TURBINES, 300.
TURNPIKES, 28, 237, 238.

U

UGANDA, 187, 196; Railway, 345, 351.
ULSTER, 175.
UNEMPLOYMENT INSURANCE, 150.
UNEMPLOYED WORKMEN'S ACT, 150.
UNITED KINGDOM, as Imperial Clearing House 340; National Debt of, 119; Shipping Tonnage, 313; Trade Statistics of, 136, 141, 147, 158f, 196; (See Great Britain, British Empire, Ireland).
UNITED STATES, 8, 9, 19, 24, 138, 178; Area of, 258; Coal in, 10; Coal Output, 163 165; Combines in, 208-210; Civil War in, 189; Corn Export of, 190; Cotton Statistics, 160, 161; Divisions of, 139; Emigration to, 11, 14, 228-231, 341; Exports of, 196; Flour Exports, 206; Food Exports of, 10, 13, 142; Growth of Towns in, 218, 219; Individualism of, 174; Industrial Development of, 145; and Ireland, 390; Iron in, 10; Iron Manufacture of, 190; Iron and Steel Statistics, 160, 161; Population of, 158; Railways in, 7-10, 188-190, 197; Railway Statistics, 161; Shipping of, 193, 291, 295, 296, 304; Shipping Statistics, 160; Transport Workers in, 223; Steel Corporation, 207; Tonnage carried on Railways 201; Tonnage of Mercantile Marine, 193, 313; Trade Statistics of, 159; and Steamers, 189; Wheat from, 202, 373; Woollen Statistics, 160, 161.
USURY LAW, 130.

V

VIRGINIA, 42, 318.

W

WAGES, 33, 96, 163n; of Agriculture Labourers, 369; Course of, 168 169; Family, 52, 62, 63; Fund, 115, 120; in France, 123, 170; in Germany, 170; Home Workers, 64; on Railways, 406; Weavers, 86, 119; Minimum, 151; Rise of, 137; (See Truck; Trade Boards).
WAGGON WAYS, 262.
WAKEFIELD, 321.
WALES, Iron in, 69.
WALPOLE, 40.
WAR, between Great Britain and France, 78, 84, 101, 102, 115, 121; Civil

WAR—contd.
War in United States, 138, 140, 142; Continental (1864-1870), 138; Recovery from, 122; Thirty Years' War, 4.
WATER FRAMES, 48, 57, 76.
WATER MILLS, 93.
WATER POWER, 50, 51, 60, 61, 73.
WATER SUPPLY, 80, 140, 220.
WATT, 6, 8, 34, 59, 69, 70, 72; and Marine Engine, 298.
WEAVERS, 31, 39, 47, 64, 65, 68n, 94n; Earnings of, 53, 119; Factory, 86; Hand Loom, 53-57; Scarcity of, 53, 54; Superfluity of, 119, 121; Wages of, 86.
WEAVING, 56; of Cotton, 53; Machinery in, 53f; (See Power Loom).
WEST AFRICA, 179; and Chamberlain 342; and Railways, 343-345; Scientific Agriculture in, 350; and Tropical Medicine, 346, 347; and Tariff Preferences, 353.
WEST INDIES, 3. 18. 41. 42. 46. 179; 321; and Canada, 296, 332; and Cotton, 40, 41, 42, 46, 149, 348; and Indian Immigrants, 356; and Scientific Agriculture, 348, 349; and Shipping 296; and Shipping Subsidies, 333; and Sugar, 148, 348; (See Jamaica).
WEDGWOOD, JOSIAH, 80.
WHEAT, Acreage, 372; Arrivals of, 202; Freight of, 308n; Imports, 142, 201, 202, 370, 373, 375; Imported by Germany, 375; Price of, 366n, 368, 369, 370; Production of, 375.
WHITNEY, E., 47.
WILKINSON, J., 69, 72, 75, 99, 298.
WILKINSON, KITTY, 92n.
WOMEN, 39, 63n, 78; Earnings of, 51, 96; and Factory Acts, 126, 132; Effect of Railways on, 225; and Industrial Revolution, 96, 97, 100; Legislation for, 116; in Mines, 62n, 71, 94; as Spinners, 31; and Trade Unions, 113, 152; and Truck, 88; as Weavers, 53, 55, 62n.
WOOL, 27; from Australia, 56, 322; Wool Combing, 56, 66; Consumption of Raw, 160; Export of Prohibited, 39; Irish, 39; Imperial, 339; Imports of, 241; Merino, 41, 43; Shortage of, 56; Spanish, 39; Spinning of, '
WOOLLEN GOODS, International Comparisons, 160, 161.
WOOLLEN MANUFACTURE, Machinery in, 50ff; Combing Machine in, 56; Migration of, 51; Numbers employed in Factories, 59; Weaving, 53, 54, 55, 56, 65; in Yorkshire, 39, 61n, 65.
WORKMEN'S COMPENSATION, 150, 267.
WORKSHOPS, 52, 56, 58, 61.
WORSTED SPINNING, 51, 52; Weaving by Machinery, 56.

Y

YORKSHIRE, 39, 41, 61.

Z

ZINC, 340.